# Combine Harvesting

## JOHN DEERE

### FUNDAMENTALS OF MACHINE OPERATION

**Combine Harvesting**
**ISBN 0-86691-397-1**

FMO15106NC (2017)    (ENGLISH)

How to operate, maintain, and improve the efficiency of your combine.

## Deere & Company
PRINTED IN U.S.A.

## PUBLISHER

DEERE & COMPANY

JOHN DEERE PUBLISHING

One John Deere Place

Moline, IL 61265

http://www.JohnDeere.com/publications

### FUNDAMENTALS OF MACHINE OPERATION (FMO)

Fundamentals of Machine Operation (FMO) is a series of manuals created by Deere & Company. Each book and guide in the series was conceived, researched, outlined, edited, and published by Deere & Company.

### ACKNOWLEDGMENTS

Contributing Authors and Editors:
George A. Griffin
Frank Buckingham
Thomas J. Meyer
Thomas A. Hoerner, PhD.
Ralph J. Moens
Keith R. Carlson
John A. Conrads
John Deere gratefully acknowledges help from the following people and groups:
Mr. D. C. Bichel
Mr. R. M. Poterack
Mr. J. L. Crow

Mr. M. H. Krouse
Duetz-Allis Corporation, Batavia, IL.
American Association for Vocational Instructional Materials
American Society of Agricultural Engineers
Case-IH, Chicago, IL.
Iowa State University
Ohio State University
Ford-New Holland, New Holland, PA.
United States Department of Agriculture
University of Illinois
Almon, Inc. Waukesha, WI.

### FOR MORE INFORMATION

This text is part of a complete series of texts and visuals on agricultural machinery called Fundamentals of Machine Operation (FMO). For more information, of Manuals and Visuals. Send your request to John Deere Service Publications, Dept. FOS/FMO, John Deere Road, Moline, Illinois, 61265-8098.



ISBN 0-86691-397-1

FMO15106NC

OUO1023,000426B -19-01MAY17-1/1

## PREFACE

This newest version of *Combine Harvesting* covers the latest practices and machines used on farms across America. You will find a thorough discussion of the equipment and concepts behind combine harvesting. We hope you enjoy reading *Combine Harvesting*.

OUO1082,0000024 -19-24JUL13-1/1

# Contents



Continued on next page

*Original Instructions. All information, illustrations and specifications in this manual are based on the latest information available at the time of publication. The right is reserved to make changes at any time without notice.*





050217
PN=1

**Page**

## INTRODUCTION

DXP02701 —UN—23FEB11

DXP04046 —UN—08JAN13

Although many people may know what a combine is, few know the details of what it does or how it works. The combine and its operation are described in this book, but first let's look at the evolution of the combine from its humble beginning.

Continued on next page

OUO1023,000426A -19-01MAY17-1/6

## WHAT IS A COMBINE?

*Fig. 1 — The Harvester-Thresher of Great-Grandfather's Day (20 Mule Team)*

Early combines were known as harvester-threshers, machines that were pulled through the fields by teams of horses or mules as shown in Fig. 1. Some machines were known only as threshers (Fig. 2), because the grain was first cut and then brought to a thresher for threshing and separating the grain from the straw.

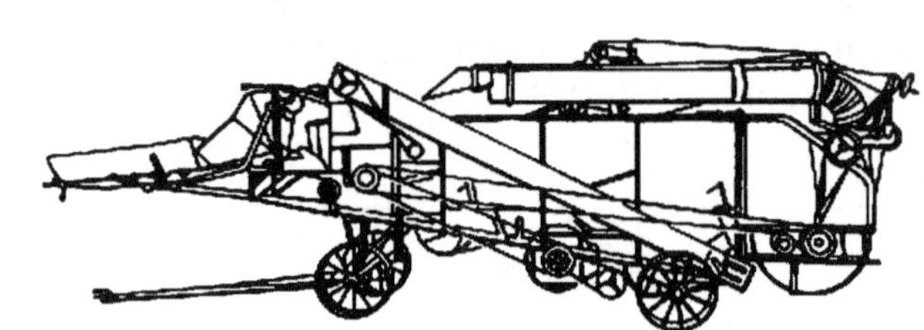

*Fig. 2 — The Thresher*

**Continued on next page**

OUO1023,000426A -19-01MAY17-2/6

*Fig. 3 — The Ultimate in Combining Wheat*

Later, the thresher was powered by a steam engine or tractor through a flat belt drive. Some harvester-threshers were propelled through fields by steam engines. These machines required a crew of several people to harvest and thresh what one person can handle with a modern combine.

The combine, as we know it today, is a machine used to harvest and thresh all kinds of grain in a variety of crop and field conditions. The name "combine" developed when the harvesting and threshing operations were "combined" into one complete machine.

Shown in Fig. 3 is the ultimate concept in combining wheat — from field to completed product that is ready for the table. While this seems farfetched to us today, in the late 1800s the concept of combining all harvesting and threshing operations into one machine seemed unbelievable.

The following chart depicts some of the major developments leading to the modern combine. Many machines were invented during the periods shown; those listed here are only typical developments. Each milestone is indicated by a period in time when important progress was made. The years shown do not always indicate when the machine was invented, but rather tell when the machine made its greatest impact in agriculture. The column on the right tells what happened in history during the same period.

**MAJOR DEVELOPMENTS LEADING TO MODERN COMBINES**

| Year | Mechanical Progress | What Else Was Happening Then |
| --- | --- | --- |
| To 1800 | Hand tools were the only means of harvesting grain for centuries. The sickle is the oldest tool and was used in our country during the colonial days. The scythe was developed later. It eliminated stooping and improved the speed of cutting. The cradle scythe was the first hand reaper that allowed the grain to be cut, collected, and deposited in untied bundles. | The ancient Egyptians used sickles more than 1400 years before Christ was born. The scythe was used during the Romans' rule of the world. |
| 1800 | Threshing was done by beating the grain from the straw with hand flails before the "ground hog" thresher was invented. This machine was a small stationary thresher that knocked the grain from the straw. The grain still had to be separated from the straw and chaff by hand winnowing. | Washington became the new capital of the United States. |
| 1830s | The McCormick Reaper mechanized the cutting and gathering of grain. It cut the grain and raked it from the platform into bunches. These bunches were collected onto a wagon by hand and taken to the "ground hog" for threshing. The Moore-Hascall Harvester was one of the first harvester-threshers built that performed the basic functions of cutting, threshing, and cleaning grain. | Battle of the Alamo occurred. First steel plow invented by John Deere at Grand Detour, Illinois. |
| 1864 | The Marvin Combined Harvester was patented. It was a four-wheeled carriage type. | Civil War was being fought. |

Continued on next page

OUO1023,000426A -19-01MAY17-3/6

**MAJOR DEVELOPMENTS LEADING TO MODERN COMBINES**

| Year | Mechanical Progress | What Else Was Happening Then |
|---|---|---|
| 1886 | The Hauser Harvester was patented. The separator was driven by a geared ground wheel. | Statue of Liberty was unveiled. |
| 1889 | The steam-powered combine was patented by Daniel Best. The steam engine was mounted on the combine, and a large steam tractor pulled the machine through the field. | The famous Oklahoma land rush occurred. |
| 1892 | Benjamin Holt patented a hillside combine that kept the separator on a level plane. | The first gasoline engine-powered automobile was made in the U.S. by Charles and Frank Duryea. |
| 1900–1935 | Binders were used extensively to cut and gather small grains. | Birth of mass automobile production. World War I fought. |
| 1920s | Tractor-drawn combines were becoming common in the wheat belt. | Women's Suffrage (19th) Amendment to the Constitution was declared constitutional. |
| 1930s | Tractor-drawn combines were becoming more popular across the continent because low-cost machines were developed. | The Great Depression occurred. |
| 1940s | Self-propelled combines came into popular use. | World War II ended. The United Nations was formed. |
| 1950–1970s | Sophisticated self-propelled combines were developed. Machines became gradually larger and more efficient. In 1955 self-propelled combines were adapted to harvesting corn. | America was experiencing the greatest technological advancements in history. |
| 1975 | First modern "rotary" combine marketed in the United States. John Deere introduced a sidehill combine. | Apollo/Soyuz rendezvous in space — a first for manned spacecraft built and launched by two different countries. |

**Continued on next page**

OUO1023,000426A -19-01MAY17-4/6

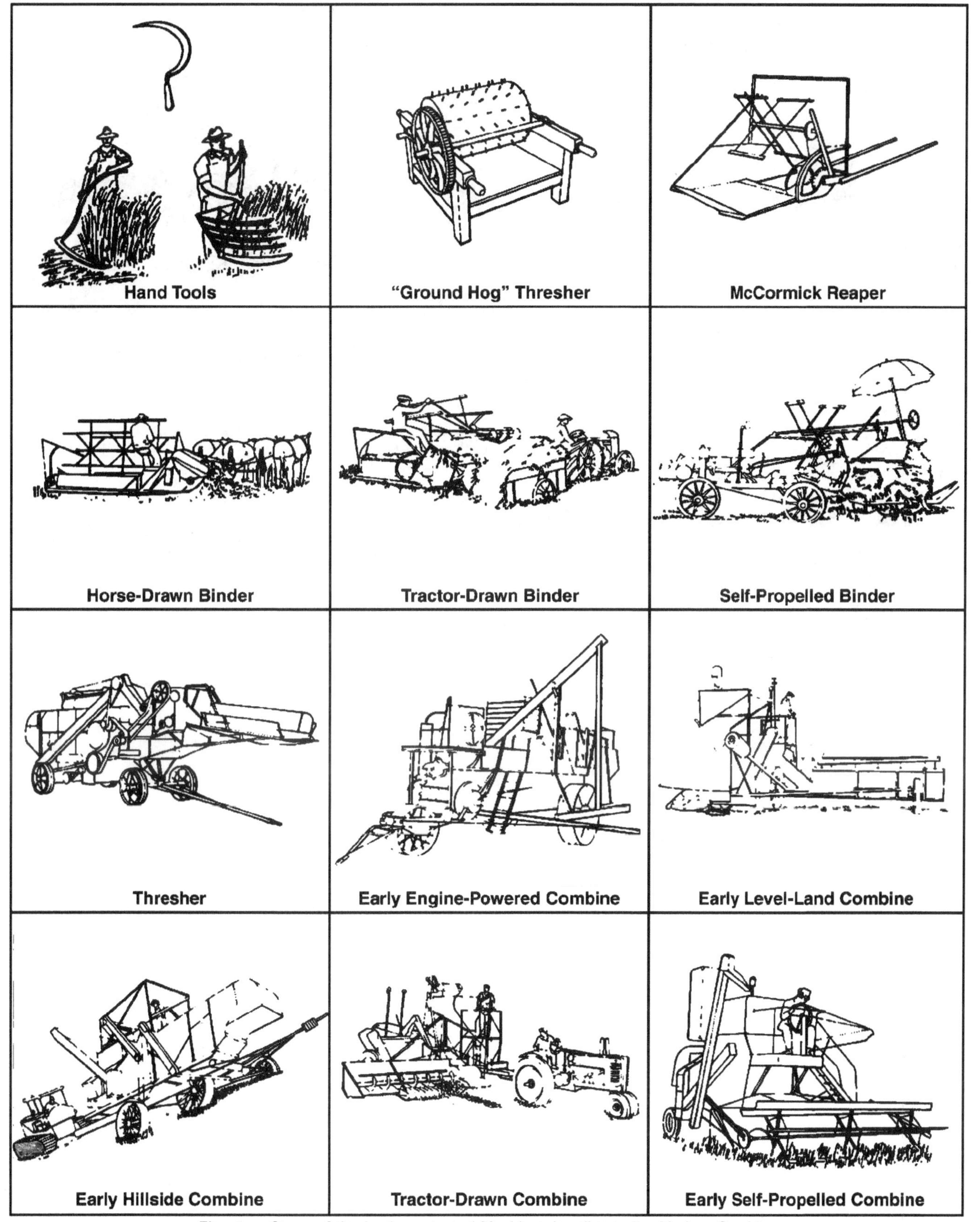



Fig. 4 — Some of the Implements and Machines Leading to the Modern Combine

Continued on next page

OUO1023,000426A -19-01MAY17-5/6

DXP04042 —UN—08JAN13



050217
PN=9

The harvester-threshers of yesterday were cumbersome machines that required very experienced operators. They lacked the capacity, efficiency, and crop flexibility that we have in our present machines.

The modern combine (Fig. 5) not only performs its basic functions better, but it also makes operation easier, safer, and more comfortable because:

- Convenient controls allow the operator to change speeds and settings from the operator seat.
- Monitoring systems permit the operator to check shaft speeds, concave clearance, cleaning shoe settings, and even crop moisture content, current yields, and separating losses.
- Operators cab provides temperature control to keep operators comfortable in any kind of weather and shield them from machine noise. Automated steering systems have been added to reduce operator fatigue and increase machine efficiency. Video monitors in the cab allow the operator to view items such as straw chopper spread and unloading auger operation.
- Refinements have been made to make the combine as trouble-free as possible. Not only has operation been made easier, but maintenance has been reduced. The

Fig. 5 — Modern Combine

development of the combine increased the harvesting of wheat, for example, from fractions of an acre per day with hand tools to 75 or more acres a day with modern combines. Let's take a look at some of the types and sizes.

OUO1023,000426A -19-01MAY17-6/6

## COMBINE TYPES AND SIZES

Modern combines are available in a wide range of types and sizes. Technology has given us a machine that is very flexible — a combine that can harvest many types of crops under various field and crop conditions.

Self-propelled combines can be described as either level-land combines or hillside combines. Self-propelled combines are available in a variety of models, depending on the crop to be harvested.

Combines may also be classified as conventional, with cylinder threshing and straw walker separation systems, or as rotary (sometimes called axial), in which a rotor or rotors replace the cylinder and straw walkers for threshing and separation of grain and straw. These differences will be explained in Harvesting Systems, chapter 2.

### *SELF-PROPELLED COMBINES*

The self-propelled combine didn't become widely available until the late 1940s. Many different types of self-propelled models were experimented with in those early years. Finding the right combination of engine horsepower and transmission of power was a difficult task.

Now this machine has been developed to where it has a husky engine that provides plenty of power to propel the machine through the toughest fields while combining the highest-yielding crops. It is also a machine operated by one person rather than a team.

Self-propelled units can harvest a swath up to 30 feet (9.1 m) wide in small grain or soybeans or up to 12 rows of corn.

Operators seated high on the machine, have a clear, direct view of their work with all controls conveniently located (Fig. 6), so that they can change the operation of the combine to adapt to the changing field and crop conditions. Most modern combines are equipped with cabs that protect the operator from cold, heat, dust, and rain.

*Fig. 6 — Controls Are Conveniently Located*

**Continued on next page**

KN52281,100453D -19-31JUL13-1/17

The fundamental design of the self-propelled combine is a straight-through type with the cut grain being delivered to the center of the platform and then up into the threshing unit (Fig. 7). This means that no standing grain is run down when opening a field. The operator may start combining anywhere, leaving green spots in the field to be harvested later, if desired.

The self-propelled combine is especially adapted for harvesting the large acreages of the Great Plains. Special equipment is available to meet conditions as they exist in most territories and crops. It is a machine that is used in all parts of the world wherever there is sufficient acreage to warrant its use.

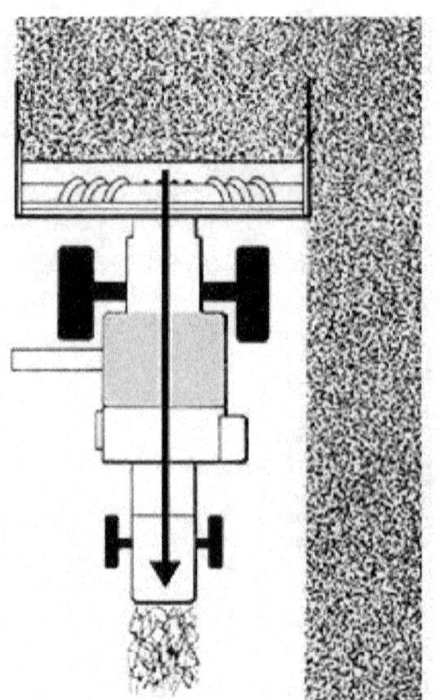

*Fig. 7 — Straight-Through Harvesting*

KN52281,100453D -19-31JUL13-2/17

Specific models (Figs. 8 through 10) are available for harvesting such crops as rice, edible beans, corn, soybeans, and grass seed. These models are determined by the special type of equipment used on the combine.

*Fig. 8 — Handling Grain*

**Continued on next page**

KN52281,100453D -19-31JUL13-3/17

For example, a rice combine will be equipped with different threshing parts than a corn combine. Also, the rice combine will have special rice tires or it may even be equipped with crawler tracks. An edible-bean combine has special features for eliminating dirt from the beans. A corn and soybean combine uses one header to gather the corn and another header to gather soybeans. Sacker attachments are used when it is desired to put grain or grass seed into sacks instead of hauling it in bulk to the market or storage.

As mentioned, the basic types of self-propelled combines are these:

- Level-land combines
- Hillside combines
- Sidehill combines

## LEVEL-LAND COMBINES

Level-land combines are used in flat areas or on slightly rolling hills.

*Fig. 9 — Combining Corn*

*Fig. 10 — Combining Rice*

**Continued on next page**

KN52281,100453D -19-31JUL13-4/17



PN=13

The level-land combine is supported by a fixed drive axle
(Fig. 11). When the combine operates over rolling ground,
the separator tilts with the contour of the ground.

A— Fixed Axle     B— Ground Line

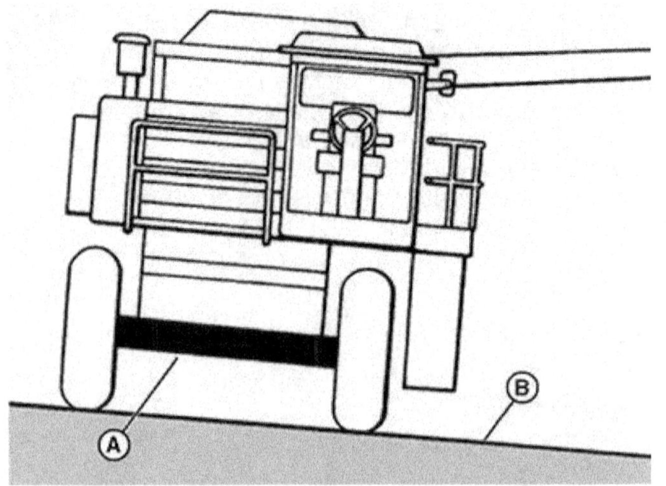

Fig. 11 — Level-Land Combines Have a Fixed Drive Axle

KN52281,100453D -19-31JUL13-5/17

The separating and cleaning units can handle some
degree of tilt, but if the combine becomes tilted too
much, the crop being harvested builds up on the low
side (downhill side) of the machine (Fig. 12). This will
result in poor separating and cleaning action; the material
may choke the machine or be pushed out the rear of
the combine with little or no separating action. This, of
course, causes crop losses.

A— Material Builds Up on Low B— Ground Line
  Side of Separator

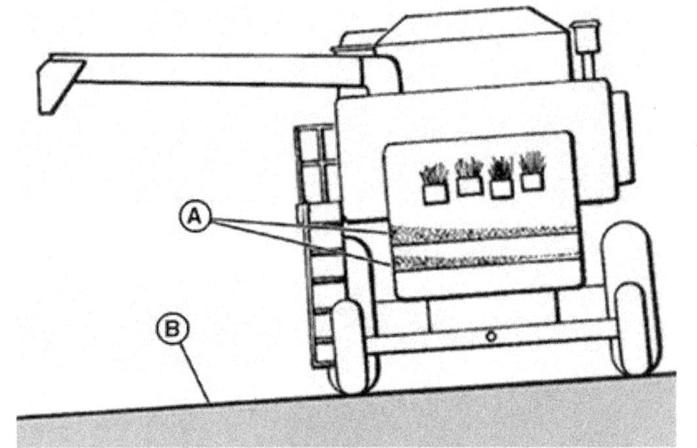

Fig. 12 — Level-Land Combine Separator Tilts with
the Contour of the Ground

**Continued on next page**    KN52281,100453D -19-31JUL13-6/17

Most manufacturers make available sidehill attachments for level-land combines that operate frequently on rolling ground. These attachments are usually shields or deflectors that prevent the material from falling to one side of the separator (Fig. 13). However, these devices can only reduce the material buildup; separating and cleaning action may still be affected when the combine operates on steep slopes.

## HILLSIDE AND SIDEHILL COMBINES

Hillside models are almost exclusively used on steep slopes such as those of the Pacific Northwest. Sizes vary considerably; this subject is discussed later in this chapter.

The hillside combine is supported by pivoting axles that adjust to the changing slopes of hillsides (Fig. 15). The separator levels automatically on grades up to 45 percent.

A— **Hillside Attachments Prevent Buildup of Material on Low Side of Separator**

B— **Ground Line**

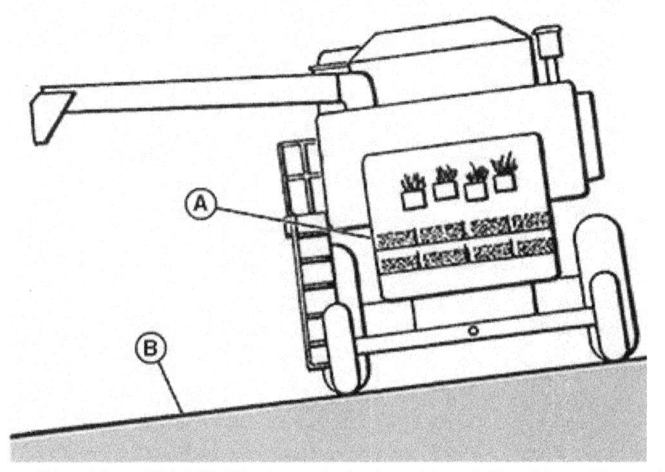



*Fig. 13 — Sidehill Attachments for Level-Land Combines Prevent Buildup of Material on Low Side of Separator*

KN52281,100453D -19-31JUL13-7/17

The sidehill combine (Fig. 14), introduced in 1975 by John Deere, is unique in that it levels to approximately 50 percent of what a hillside combine does. It was designed for use on slopes of up to 18 percent. This type of farm terrain is located primarily in the Midwest and a few other areas.

By keeping the separator level, the most efficient separating and cleaning action is maintained because material is distributed evenly throughout the separator — unlike the level-land combine separator, which tilts with the slope of the land and causes a buildup of material on one side.

*Fig. 14 — Sidehill Combine*

**Continued on next page**

KN52281,100453D -19-31JUL13-8/17

A combination of electrical and hydraulic power systems make up the self-leveling mechanism. Large hydraulic cylinders (Fig. 15), connected to the separator and axles, adjust the separator to keep it level. The rear axle is mounted on a pivot and adjusts freely to the slope of the ground. The cutting platform (header) also swivels on a pivot and conforms to the contour of the hill (Fig. 16). See chapter 3 for a full explanation of the operation of the hillside combine.

A— Separator Stays Level  
B— Pivoting Axle  
C— Ground Line  
D— Hydraulic Cylinder Extends

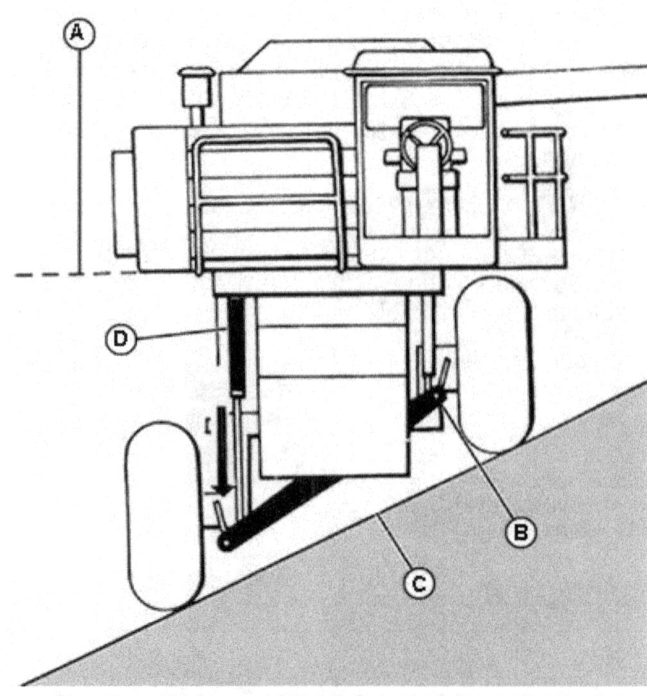

*Fig. 15 — Hillside and Sidehill Combines Have Pivoting Axles*

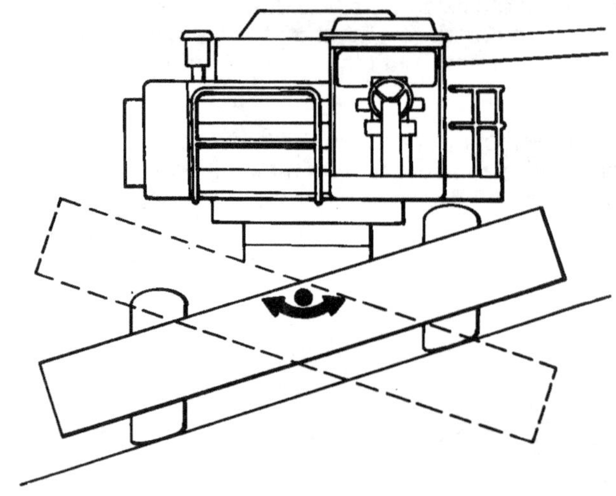

*Fig. 16 — The Platform Swivels Automatically to Conform to the Slope*

**Continued on next page**

KN52281,100453D -19-31JUL13-9/17

## COMBINE SIZES

How is the size of a self-propelled combine determined? The first factors that come to mind are power rating, header size, and separator size (Fig. 17). But these are only rough measures of combine size because manufacturers don't usually make combine components that measure exactly like those of other manufacturers. What the manufacturers are actually trying to describe is the capacity of the combine.

Capacity cannot be measured entirely in terms of dimensions of the combine or its horsepower. Yet all these elements contribute to the capacity.

Consider the following when trying to determine combine size and capacity:

- Engine power
- Separator width and length (total square inches or m$^2$ of area)

*Fig. 17 — How Is Size of Self-Propelled Combines Determined?*

- Type of threshing cylinder or rotor
- Header size
- Grain tank capacity

KN52281,100453D -19-31JUL13-10/17

## ENGINE POWER

New combine engines (Fig. 18) range in power from about 250 horsepower (185 kW) to over 450 horsepower (335 kW). When comparing combine sizes, consider engine power in relation to threshing method as well as separator width and length. For example, if one combine has a smaller separator but a larger engine than another combine of nearly the same size, the machine with the smaller separator may actually have more capacity when operating in crops with high yields.

*Fig. 18 — Typical Combine Engine*

Continued on next page

KN52281,100453D -19-31JUL13-11/17



050217
PN=17

## WIDTH AND LENGTH

Separator width and length in conventional combines
(Fig. 19) govern the maximum possible capacity of the
combine to separate and clean the crop. This, of course,
also depends on the efficiency of the design and the
available power. Separator width varies from less than
24 inches (0.61 m) to more than 60 inches (1.52 m);
separator length varies from about 105 inches (2.67 m) to
170 inches (4.32 m).

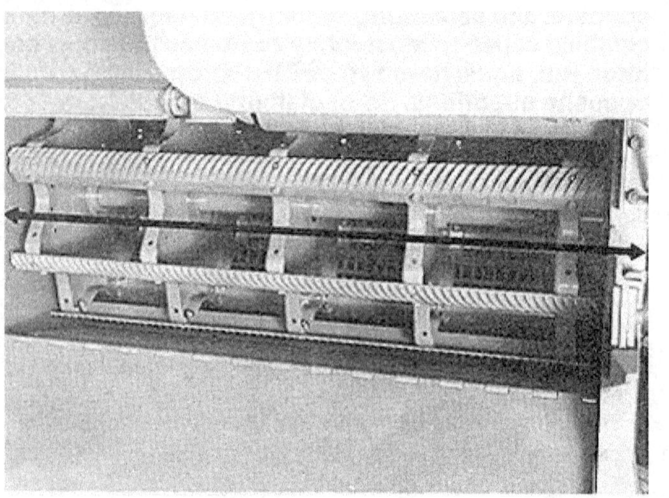

*Fig. 19 — Separator Width Affects Capacity of Conventional Combines*

KN52281,100453D -19-31JUL13-12/17

## TYPE OF THRESHING CYLINDER OR ROTOR

Capacity of rotary combines is affected by number,
length, and diameter of rotors as well as total concave
and separating grate areas. The design of the rotor
feeding system as well as that of the rotor, concave,
and separating system also helps determine combine
capacity. Most rotary combines have only one rotor.
But some have two parallel rotors that turn in opposite
directions. Rotor diameter ranges from 17 to 31.5 inches
(0.43 to 0.80 m) and length from 88 to 168 inches (2.24
to 4.27 m). Threshing cylinder or rotor diameter can also
affect capacity (Fig. 20). Small-diameter cylinders or
rotors must rotate much faster to provide equal threshing
action. They may lose momentum (ability to rotate at
proper speed) more easily than larger-diameter cylinders
or rotors when heavily loaded.

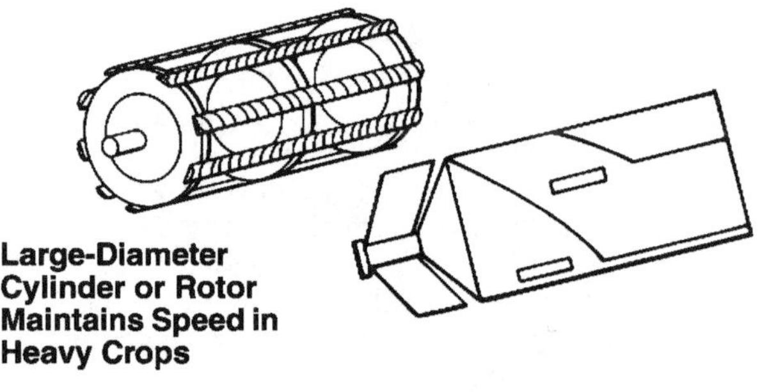


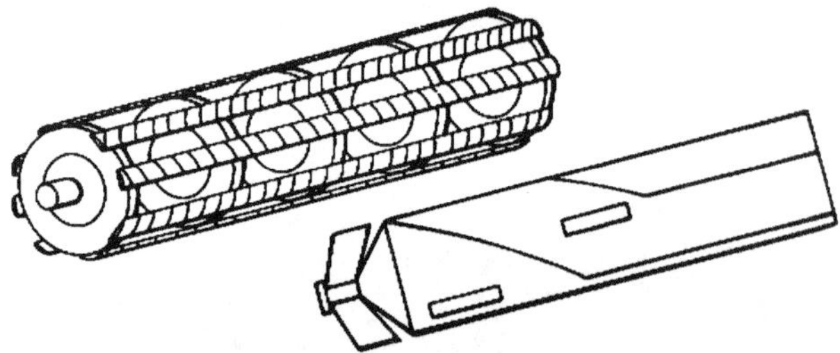


*Fig. 20 — Type of Threshing Cylinder or Rotor Affects Combine Capacity*

**Continued on next page**

KN52281,100453D -19-31JUL13-13/17

## HEADER SIZE

Header size (Fig. 21) is governed by the size of the separator and the power of the combine on which it is attached. For example, a 20-foot (6-m) header cannot operate properly on a combine with only 50 horsepower (37 kW). On the other hand, it would be inefficient to attach an 8-foot (2.4-m) header to a 150-horsepower (112-kW) combine. Headers are now available from 10 feet (3 m) wide to 40 feet (11 m) wide; corn heads vary from 2-row units up to 16-row units.

DXP04059 —UN—08JAN13

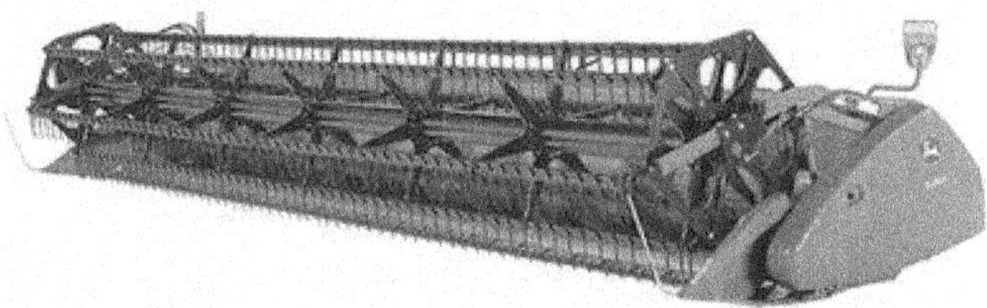

*Fig. 21 — Typical Header with Pickup Reel*

KN52281,100453D -19-31JUL13-14/17

## GRAIN TANK CAPACITY

Grain tank capacity usually corresponds with the overall size of the combine (Fig. 22). Grain tank capacities on current models are generally 300 to 350 bushels (2.4 m$^3$) on a 70-horsepower (52-kW) combine to more than 350 bushels (11 m$^3$) on a 270-horsepower (200-kW) combine. In summary, the capacity of each combine varies with the manufacturer, even though dimensions appear to be about the same. Determining the capacity of a combine is a matter of judgment, but the elements mentioned above must be considered individually and compared to those of other combines. Also, crop conditions and yields are factors that greatly affect the capacity of any combine.

DXP04060 —UN—16JAN13

*Fig. 22 — Grain Tank Capacity Is Important*

KN52281,100453D -19-31JUL13-15/17

## *WINDROWING METHOD*

When there are many weeds in the grain, when there is considerable moisture at harvest time, or when the crop ripens unevenly, it may be desirable to cut the grain with a windrower (Fig. 23) and thresh it later with the regular combine equipped with the windrow-pickup attachment. The windrower, performing the function of the reel, cutterbar, and platform of the combine, cuts the grain and lays it on the stubble. An opening at the end or center of the platform (depending on the type of windrower being used) permits the cut grain to be laid on the stubble in windrow form.

When the grain is properly cured or when the moisture content is low enough, a special windrow-pickup platform is attached to the combine, or a pickup unit is attached to the regular combine platform.

DXP04061 —UN—08JAN13

*Fig. 23 — Self-Propelled Windrower*

**Continued on next page**

KN52281,100453D -19-31JUL13-16/17



050217
PN=19

These attachments elevate the windrow onto the combine platform (Fig. 24). From this point on, the feeding, threshing, separating, cleaning, and handling processes are the same as described for the regular combine.

The windrow and pickup method are not used exclusively with the pull-type combine. Self-propelled combines also use this method of harvesting when conditions require it.

### SPECIAL COMBINES

Many special combines have been developed through the years to harvest special crops. Two excellent examples are the peanut combine and the castor bean combine.

### SPECIAL ATTACHMENTS FOR COMBINES

As harvesting grain with a combine has become more important, many special items of equipment have been developed to make the combine operator's task more productive. Special attachments make combines more adaptable and productive around the world.

Combines may now be equipped with straw choppers or spreaders and shaft speed sensing units to warn the

Fig. 24 — Windrow-Pickup Attachment in Operation

operator when the combine is malfunctioning. Signal devices that indicate when the combine is overloaded are also available. These are a few of the special items that have been developed. Many of these attachments are discussed later in this book.

KN52281,100453D -19-31JUL13-17/17

## SUMMARY: COMBINE TYPES AND SIZES

Modern combine types are:

- Level-land combines
- Hillside combines
- Sidehill combines
- Special combines

Combine sizes are determined by:

- Engine horsepower
- Separator width and length
- Type of threshing cylinder
- Header size
- Grain tank capacity

Combine sizes are a rough guide to combine capacity.

Windrow combining consists of harvesting a previously cut-and-windrowed crop.

Special attachments are available to:

- Chop or spread straw
- Monitor separator and shoe losses, moisture content, and current yields
- Warn the operator of combine malfunctions
- Adapt to different crops

KN52281,100453E -19-16JAN13-1/1

## TEST YOURSELF

### Questions

1. The name "combine" was developed for a machine that does what two jobs in one machine?
2. What are the two basic types of self-propelled combines?
3. Name the two ways combines may be classified according to the method of threshing.
4. Name three features of modern combines that make today's combine harvesting easier than it was 50 years ago.
5. List at least four factors that help determine the capacity and size of a combine.
6. Name three attachments available on today's combines.
7. Why are some crops cut and windrowed before combining?

KN52281,100453F -19-08MAY13-1/1

## INTRODUCTION

DXP02702 —UN—23FEB11

# 2

Today's combine is a complex machine. Not only are the harvesting and threshing units complicated, but add to that the engine, power train, electrical system, and hydraulic system, and it becomes one of the most complex machines in agriculture.

To understand the operation of a combine, look closely at each function of the machine. Once the operation of each of these components is understood, it becomes easier to understand how they relate to each other and the operation of the entire machine.

**Continued on next page**

OUO1082,000624A -19-22AUG13-1/3

Fig. 1 — Combine with Single Rotor and Concave

Fig. 2 — Double Rotor and Concave Arrangement

Continued on next page

OUO1082,000624A -19-22AUG13-2/3



PN=22

Fig. 3 — Transverse-Mounted Single Rotor and Concave Design

Figs. 1 through 3 are three variations of rotary combine designs. Cutting, cleaning, and handling functions are basically the same in all three models. The differences in threshing and separating functions will be described later in this chapter. This chapter describes each area of a typical combine.

All combines perform the following six basic crop harvesting functions:

1. Cutting or windrow pickup and feeding

2. Threshing

3. Separating

4. Cleaning

5. Handling

6. Residue disposal

OUO1082,000624A -19-22AUG13-3/3

## CUTTING AND FEEDING THE CROP

The mechanism that cuts or gathers the crop and feeds it to the combine separator is commonly known as a header (Fig. 4).

A— Gathering and Cutting Unit    B— Feeding Unit

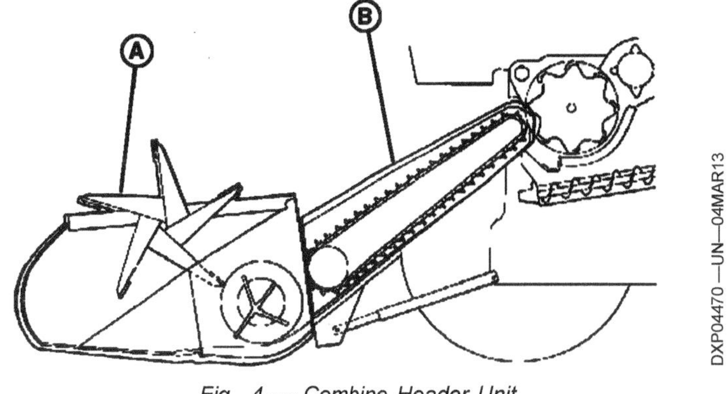

Fig. 4 — Combine Header Unit

Continued on next page

KN52281,1004542 -19-22AUG13-1/22

The header can be divided into distinct units: (1) the unit that cuts or gathers the crop may be a cutting platform, pickup platform, corn head, or row-crop head; (2) the unit that feeds the cut or gathered crop to the separator is a feeder conveyor. See Figs. 5 and 6 to identify these parts. Each of these units will be discussed in detail later.

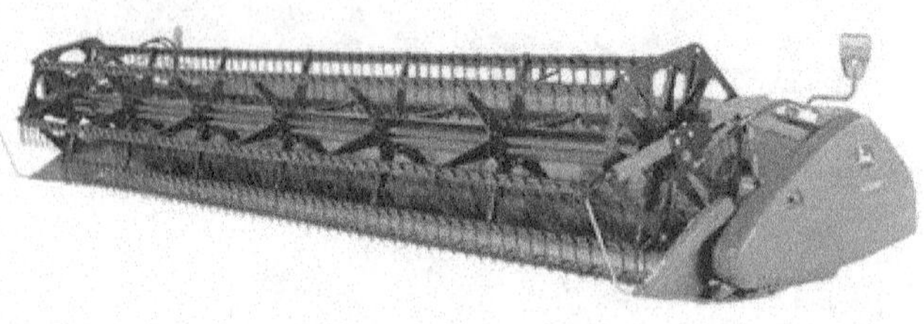

*Fig. 5 — Typical Cutting Platform*

*Fig. 6 — Typical Corn Head*

*Fig. 7 — Modern-Day Draper Platform*

KN52281,1004542 -19-22AUG13-2/22

The header is attached to the combine by a pivot device that allows the header to be raised or lowered (by hydraulic cylinders) to obtain the desired height of cut (Fig. 8).

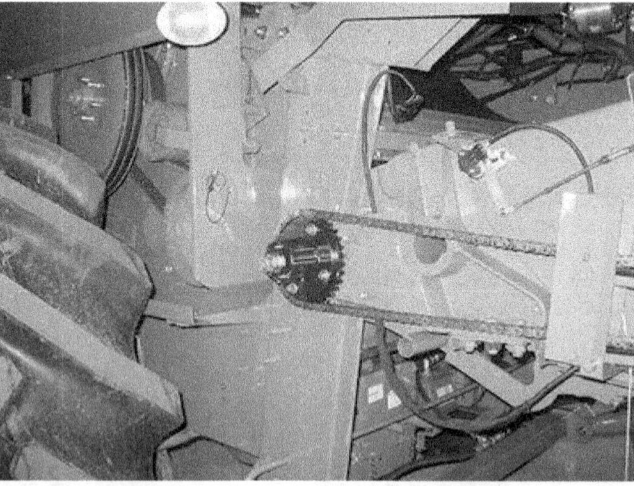

*Fig. 8 — Header Pivot Point*

**Continued on next page**

KN52281,1004542 -19-22AUG13-3/22

## CUTTING PLATFORM OPERATION

Depending on the crop, a combine may be equipped with either a regular cutting platform, which is used in most crops except corn and rice, or it may have a "draper" platform, which is used in rice (Fig. 9). The draper platform is similar to the regular cutting platform except it has a draper, or conveyor belt, between the cutterbar and the auger. The draper aids in picking up or getting more crop into the combine. Cutting platforms vary in width from 8 to 24 feet (2.40 to 7.20 m).

A— Auger  
B— End Sheet  
C— Reel  
D— Divider  
E— Cutterbar  
F— Draper

**Regular Cutting Platform**

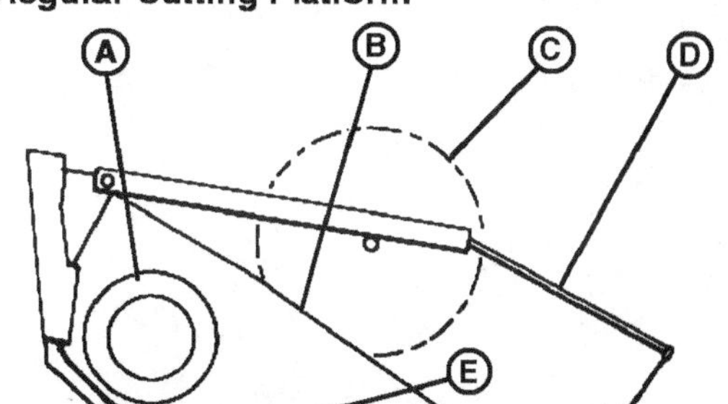

**Draper Platform**

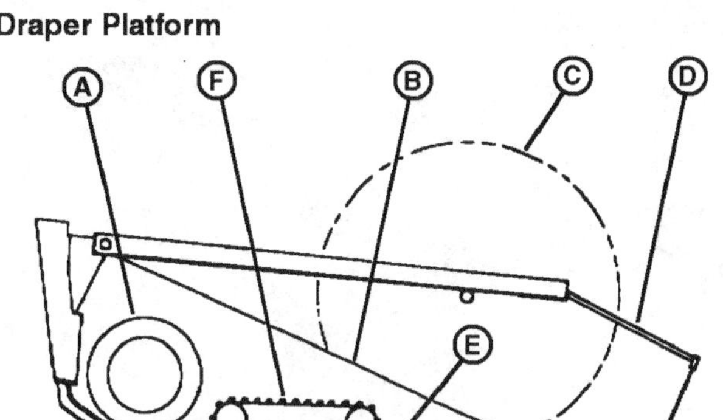

*Fig. 9 — Two Types of Platforms*

**Continued on next page**

KN52281,1004542 -19-22AUG13-4/22

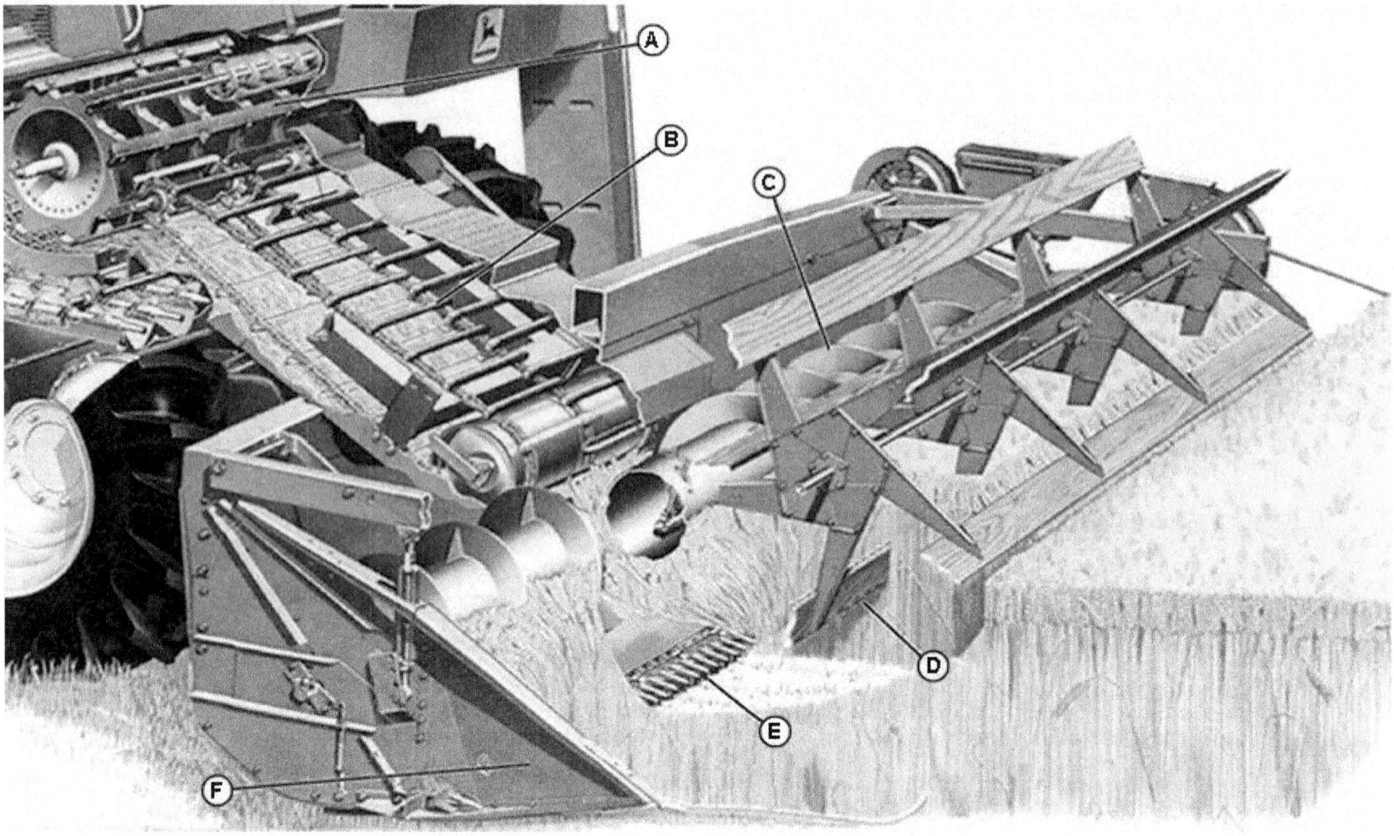

Fig. 10 — Cutting Platform Operation

| A— Cylinder | C— Spiraled Auger | E— Cutterbar |
|---|---|---|
| B— Feeder Conveyor | D— Reel | F— Dividers and End Sheets |

As the combine moves forward in the field (Fig. 10), the dividers and end sheets separate a swath from the rest of the crop. The reel parts a section of the crop and pushes it against the cutterbar. As the material is cut by the knife on the cutterbar, the reel continues to push the crop or lift it into the path of the spiralled auger. (On draper platforms, the reel lifts the crop onto the draper, or conveyor, which carries the material to the auger.) The auger moves the material to the center of the platform, where the feeder conveyor delivers it to the cylinder for threshing. The reel, cutterbar, auger, and feeder conveyor must work in proper relationship to cut and feed the crop evenly to the threshing cylinder without losing the kernels or seeds.

KN52281,1004542 -19-22AUG13-5/22

## REEL OPERATION

Two types of reels are available when combining various crops under different conditions:

- Bat-type or slat-type reel
- Pickup reel

The bat or slat reel consists of three to eight slats made of wood or steel (Fig. 11). The slats rotate against the standing crop to hold it until the crop is cut by the knife on the cutterbar. Then the slat lays the crop back into the path of the auger.

Fig. 11 — Slat-Type Reel

Continued on next page

KN52281,1004542 -19-22AUG13-6/22

The pickup reel has several steel tines or "fingers" attached to the slats (Fig. 12). The fingers pick up crops that have been blown down or have become badly tangled, such as rice or barley. A slat-type reel without fingers cannot pick up crops in these conditions.

The fingers on the pickup reel reach down into the crop and lift it so the cutterbar can get under it. The pickup reel is also used in very ripe crops, such as soybeans, because a slat-type reel would knock the beans out of the pods, causing heavy crop losses. The fingers of the pickup reel tend to gather the ripe crop gently rather than batting it into the auger.

Both types of reels are usually adjustable. The slats on the slat-type reels and the fingers on pickup reels may be adjusted to enter the crop at the proper angle. This adjustment helps prevent shattering of the crop and permits uniform delivery of the crop to the platform auger. This adjustment is very important on a pickup reel that is being used in a down or tangled crop.

REEL ADJUSTMENT

Either type of reel can be adjusted fore and aft as well as up and down. Both of these adjustments are important to ensure proper delivery of material to the cutterbar and auger.

In standing grain, the slat reel must be set so the slats, in their lowest position, strike just below the lowest grain heads and just slightly ahead of the cutterbar.

In crops that are down and tangled, the pickup reel must be set so it will pick up the crop and just clear the knife and conveying auger. This will ensure that the material is picked up, cut off, and swept back into the platform auger without losing grain.

The reel must also rotate at the proper speed to prevent shattering and loss of grain. Usually the speed for average conditions is 25 percent faster than ground travel

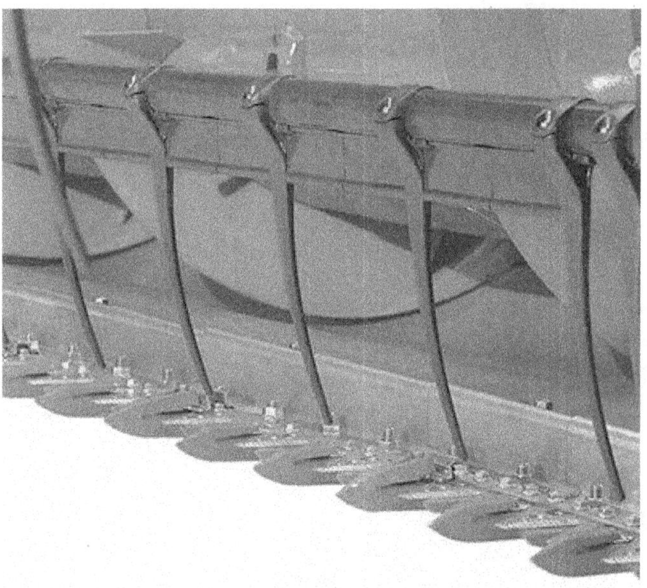

*Fig. 12 — Pickup Reel*

speed. Attachments are available for some combines that automatically adjust reel speed in relation to combine travel speed.

The procedures for making reel adjustment will be described in chapter 5, Field Operation and Adjustments.

REEL FORE/AFT OPERATION

Moving the reel forward or aft is controlled from the armrest control panel.

When the solenoid control valve is in the neutral position, oil in the system is trapped. The trapped oil maintains a constant reel fore/aft position.

When the control panel switch is positioned to move the reel forward or aft, the switch energizes solenoids to move valve spools, allowing oil to flow.

**Continued on next page** KN52281,1004542 -19-22AUG13-7/22

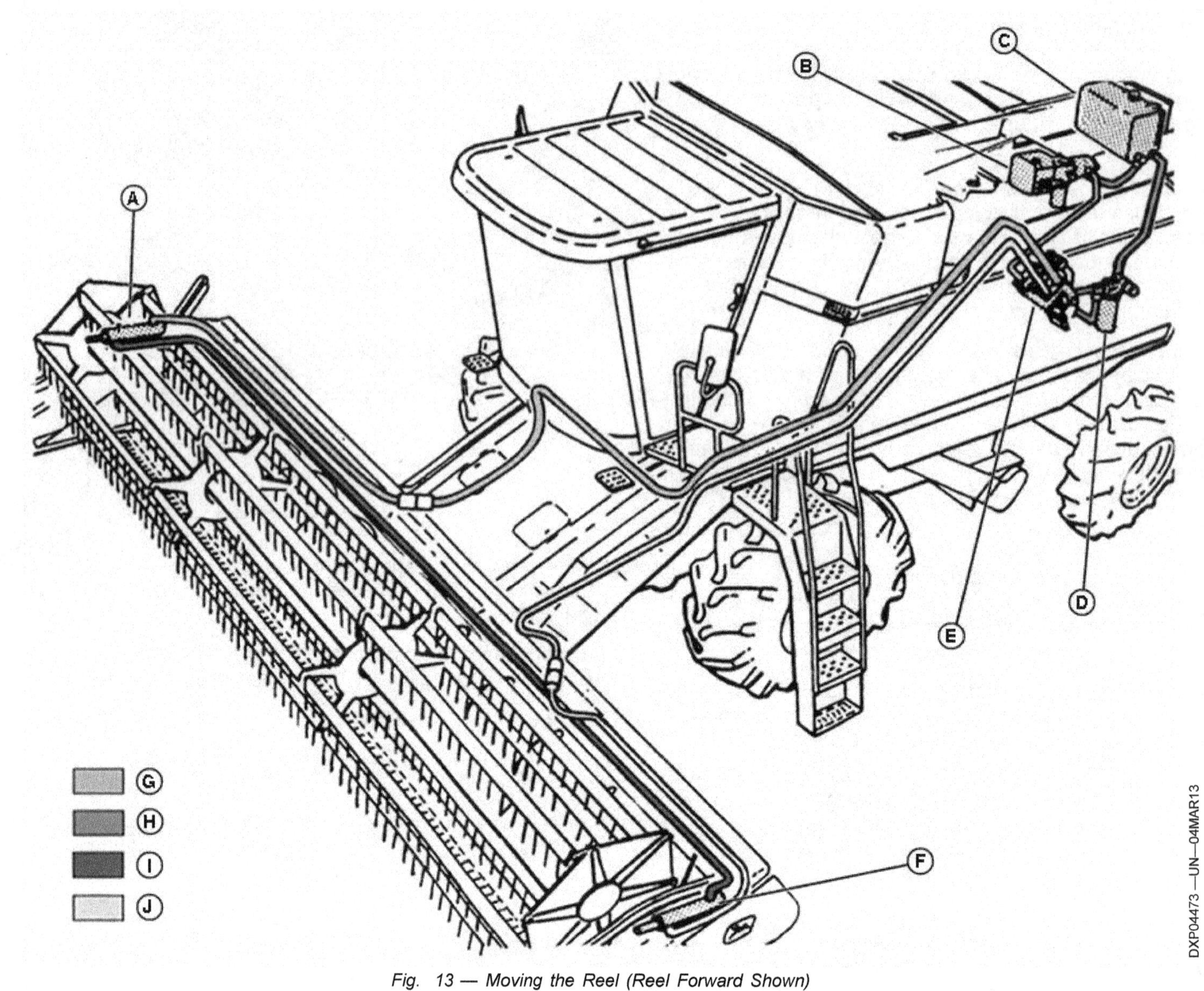

*Fig. 13 — Moving the Reel (Reel Forward Shown)*

A— **Right-Side Cylinder**
B— **Main Hydraulic Pump**
C— **Reservoir**
D— **Return Filter**
E— **Solenoid Control Valve**
F— **Left-Side Cylinder**
G— **Pressure Oil**
H— **Pressure-Free Oil**
I— **Trapped Oil**
J— **Medium-Pressure Oil**

In Fig. 13, pressure oil flows from main hydraulic pump, through the solenoid control valve, and to the right-side cylinder, extending the cylinder. Trapped oil is transferred from the right-side cylinder to the left-side cylinder. As oil flows, both cylinders are extending, moving the reel forward.

Medium-pressure oil flows from the cylinder to the solenoid control valve and then to the return filter and reservoir.

The direction of flow is reversed to retract both cylinders, moving the reel aft.

**Continued on next page**

KN52281,1004542 -19-22AUG13-8/22

## CUTTERBAR OPERATION

The reel holds the crop against the cutterbar as it is cut. The cutterbar consists of a steel angle bar on the front of the platform. Attached to the bar are stationary guards with slots through which the knife (sickle) moves back and forth, shearing or cutting the crop (Fig. 14). The knife is made up of several triangular blades (called knives or knife sections) that are riveted or bolted to a flat steel bar.

**A— Guard**  
**B— Knife Hold-Down Clip**  
**C— Knife Section**

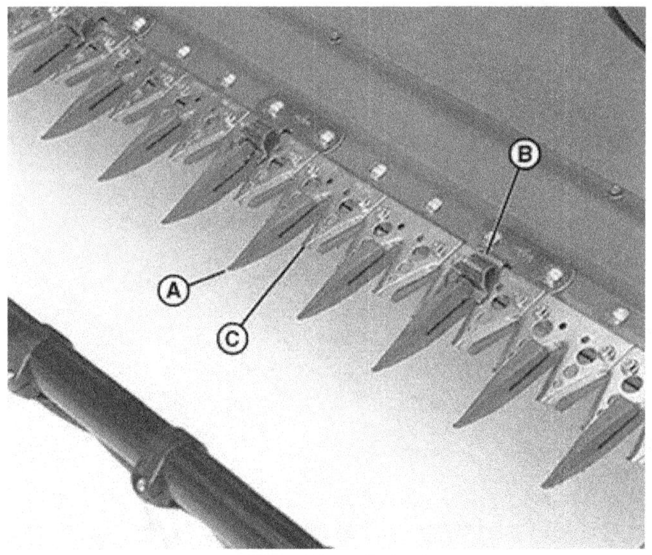
*Fig. 14 — Cutterbar*

KN52281,1004542 -19-22AUG13-9/22

One end of the knife is connected to a reciprocating drive mechanism (Fig. 15), which causes the knife to move back and forth at several hundred strokes per minute. Some wide headers have two knives, each one half the length of the header, with a timed knife drive at each end of the header. This permits faster knife operation and reduces knife weight per drive unit.

Wear plates and knife hold-down clips control the relationship of the knives to the guards and complete the cutterbar assembly.

To cut properly, the knife must be sharp and must run smoothly in the guards. Every knife section must rest on its guard to make a shear cut. This means that the guards, wear plates, and knife clips must be in good condition and set correctly. The knives must also register properly with the guards as they move back and forth.

Failure to achieve a clean shear cut results in a tearing or chewing action at the cutterbar. This needlessly agitates a ripe crop, shattering kernels or seeds, and possibly results in large crop losses. It also causes increased cutterbar plugging.

**A— Cutterbar**  
**B— Cutterbar Drive Case**

*Fig. 15 — Reciprocating Knife Drive Mechanism*

**Continued on next page**

KN52281,1004542 -19-22AUG13-10/22

A cross section of the cutterbar is shown in Fig. 16. Notice how closely the milled slot conforms to the section. The section is about 1/8 inch (3.2 mm) thick and the slot is about 3/16 inch (4.8 mm) wide. This helps prevent material from wedging between the cutting edges.

Occasionally, when the cutterbar fails to operate properly, it must be adjusted; this adjustment is covered in chapter 5, Field Operation and Adjustments.

A— Guard
B— Knife Clip
C— Wearing Plate
D— Knife Section

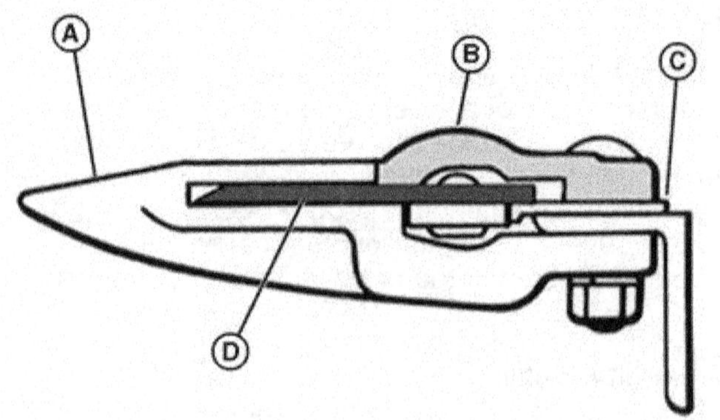


*Fig. 16 — Cutterbar in Cross-Section View*

KN52281,1004542 -19-22AUG13-11/22

## FLEXIBLE FLOATING CUTTERBAR OPERATION

The same basic header and cutting parts are used on headers with either rigid or flexible cutterbars. However, instead of being rigidly attached to the front edge of the header, skid shoes and a special linkage permit the flexible cutterbar to closely follow ground contours (Fig. 17). This permits cutting much closer to the surface across the full width of the cutterbar. More seeds are saved in crops, such as soybeans, that usually have pods growing very close to the ground. The bar may be locked up for rigid operation in crops such as sorghum or small grains. These flex cutter bars usually flex about 4 inches (102 mm).

A— Block

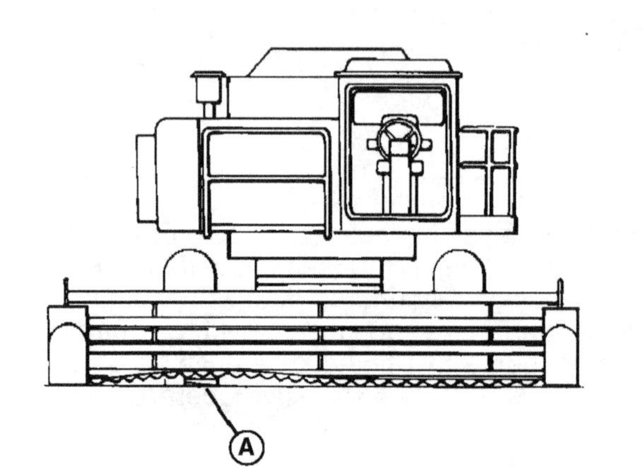


*Fig. 17 — Flexible Floating Cutterbar*

KN52281,1004542 -19-22AUG13-12/22

## AUTOMATIC HEADER HEIGHT CONTROL

Sensors under the combine header (Fig. 18) control height of cut when operating very close to the ground, for instance in soybeans. This relieves the operator of constantly adjusting header height and is particularly valuable when working over uneven ground or harvesting at night. The sensors may be adjusted to provide the desired height of cut under different harvesting conditions. Automatic header height control is standard on some flexible cutterbar headers and is optional on other models. Refer to the operator's manual for specific operating instructions.

A— Height Sensors

*Fig. 18 — Automatic Header Height Control*

**Continued on next page**

KN52281,1004542 -19-22AUG13-13/22



050217
PN=30

## PLATFORM AUGER OPERATION

After the crop is cut by the cutterbar, the reel lays the material back onto the floor of the platform. Here spiral flights of the platform auger sweep the crop toward the center of the platform, where the feeder conveyor is located (Fig. 19). Most augers have retracting fingers that move the material to the feeder conveyor for feeding the threshing cylinder. (Augers on draper platforms do not have these fingers.)

Smooth, even feeding of the material by the auger is essential for proper delivery to the feeder conveyor. The auger is adjustable to achieve correct feeding in most crops and conditions. This adjustment is covered in chapter 5, Field Operation and Adjustments.

This concludes the basic information on operation of the cutting platform. Next we will explain the operation of the pickup platform.

### *DRAPER PLATFORM OPERATION*

Draper platform designs have changed over the years, although they share some of the same components as a cutting platform, such as cutterbar and reel operation. Older styles of draper headers used drapers (conveyor belts) to move the crop away from the cutterbar and into the cross auger to aid in keeping the cutterbar clear in difficult harvesting conditions. Newer designs of draper heads replaced the cross auger with a fabric or rubber belts that move the crop to the center feed belt, which is then fed into the feed drum. This leads to faster feeding and higher throughputs while using less power. Draper heads have been found to work better than conventional heads in short or damp crops, such as rice and soybeans, as the draper belts pull material away, keeping the cutterbar clear. By evenly distributing the crop into the combine, farmers can harvest more acres in a day by starting earlier and finishing later when soybean stems begin to toughen due to moisture in the air, not to mention increased combine efficiency due to consistently feeding the material in head first. The operation of the reel retains

Fig. 19 — Platform Auger

**A— Retracting Fingers**          **B— Spiral Flights**

the same function, lifting the crop while dropping it onto the drapers in a consistent manner.

Two different types of draper headers exist, a flexible cutting platform and a rigid cutting platform. The rigid platform is intended for "near-ground" harvesting, while the flexible platform "flexes" to meet the contours of the ground and thus can cut crops much closer to the ground, reducing shattering losses on crops such as soybeans, rice, and other crops that may be lodged or hard to pick up.

Designs and layouts of draper platforms vary by brand (some draper headers have cross augers and some do not), although their operation remains the same.

Some draper platforms can be converted to do swathing operations with the aid of a kit from the manufacturer or by moving some components on the platform.

**Continued on next page**                    KN52281,1004542 -19-22AUG13-14/22

## PICKUP PLATFORM OPERATION

Pickup platforms are used to gather crops that have been cut and gathered into a windrow by a grain windrower. The pickup platform is essentially the same as a cutting platform, except it does not have a reel nor is a cutterbar required. It is equipped with a conveyor belt to which steel or plastic fingers are attached. These fingers pick up the windrowed crop and deliver it to the platform auger (Fig. 20).

A cutting platform may be converted to a pickup platform by installing a belt-pickup attachment.

Fig. 20 — Pickup Platform in Operation

KN52281,1004542 -19-22AUG13-15/22

## CORN HEAD OPERATION

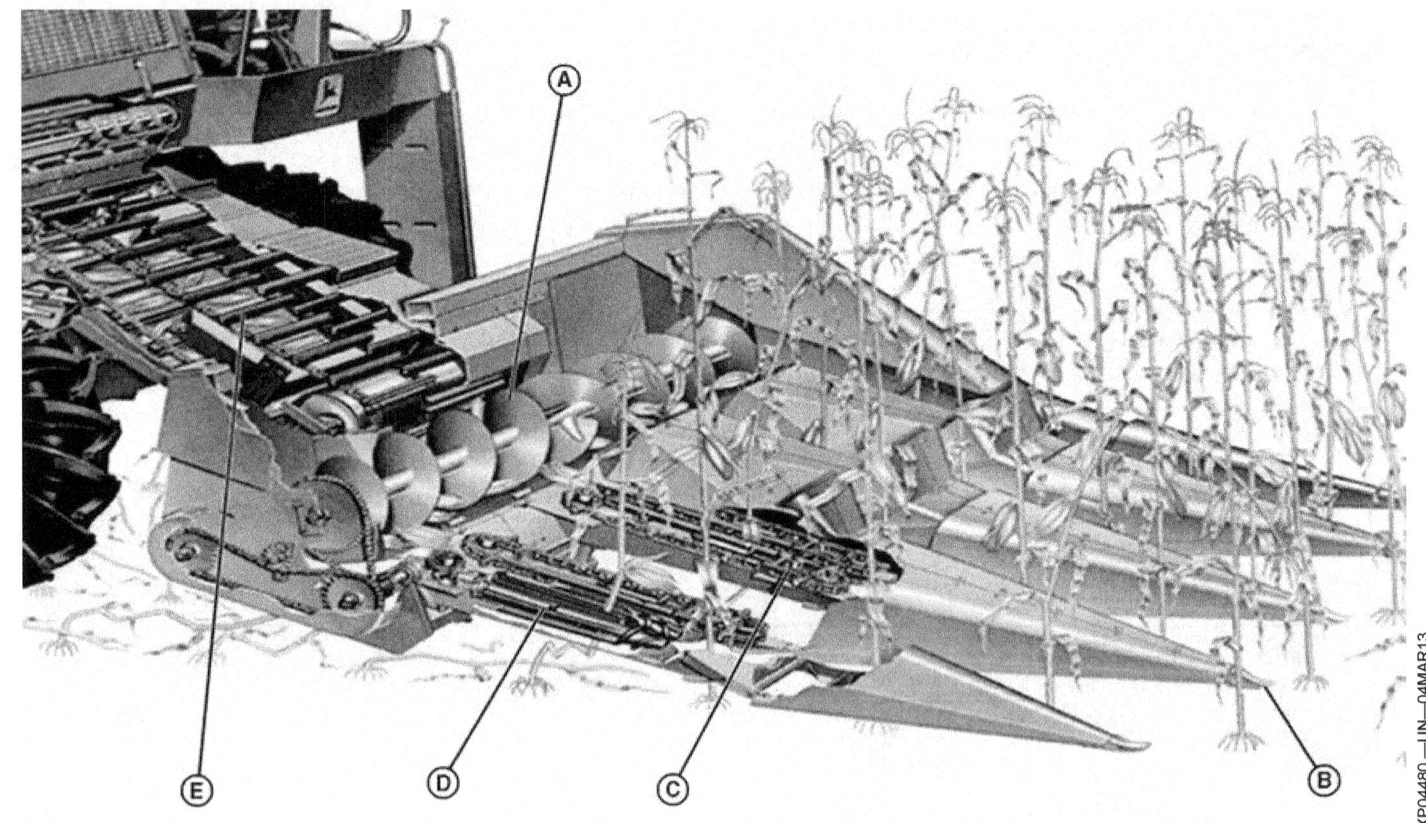

Fig. 21 — Corn Head in Operation

A— Cross Auger
B— Gatherer Points
C— Gatherer Chains
D— Snapping Rolls
E— Feeder Conveyor

The corn head is essentially a corn picker mounted to the feeder conveyor of the combine. Corn heads vary in size from two-row units up to twelve-row units.

As the combine moves through the field, the gatherer points are positioned between the rows of corn (Fig. 21).

Snapping rolls grab the corn stalks and pull them rapidly down between the rolls.

Continued on next page

KN52281,1004542 -19-22AUG13-16/22



PN=32

When an ear of corn reaches the snapping bar, the ear is prevented from going through because of the narrow opening (Fig. 22). The snapping rolls continue to pull on the stalk and snap the ear free of the stalk.

Gatherer chains catch the ears and carry them to a cross auger, which delivers the ears to the feeder conveyor. The feeder conveyor delivers the ears to the threshing cylinder.

The snapping rolls must operate at a speed in direct relationship to the forward speed of the combine in order to pull the stalks through the rolls before the combine runs over them. If the snapping rolls are operated too fast, the ears may bounce off the corn head and be lost on the ground. High speed may cause shelling at the snapping bars, which will also result in losses. It may also cause the entire stalk to be pulled into the combine, resulting in overloading the machine.

If the speed is too slow, the ears will be snapped at the back of the rolls, causing congestion and possible plugging of the head.

Fig. 22 — Gathering Mechanism

The corn head must operate close to the ground to pick up low ears. The gatherer points are adjustable for various crop or field conditions.

KN52281,1004542 -19-22AUG13-17/22

Some corn heads are equipped with additional knife blades below the stalk rolls for additional chopping of the corn stalks (Fig. 23).

In chapter 5, Field Operation and Adjustments, adjustments of the corn head are covered.

**A— Knife Blades**

Fig. 23 — Additional Knife Blades

**Continued on next page**

KN52281,1004542 -19-22AUG13-18/22

## ROW-CROP HEAD OPERATION

Row-crop heads (Fig. 24) are used in crops such as soybeans, grain sorghum, sunflowers, and similar crops to reduce cutting and gathering losses. The gathering points are very low and sloping so that they lift and guide lodged or tangled crops into the gathering area.

Rubber gathering belts on each row unit hold the plants as they are cut off by a knife. The knives are located just below the forward end of each set of gathering belts.

After the plants are cut, the gathering belts convey the crop to the cross auger at the rear of the header.

Each row unit is free to float up and down within a preset range to follow ground contour. This feature helps permit cutting of soybeans that grow close to the ground.

By eliminating a reciprocating knife and rotating reel, seed shattering is reduced. This permits faster operating speed than would be possible with a regular or flexible cutterbar header.

Fig. 24 — Row-Crop Head

KN52281,1004542 -19-22AUG13-19/22

## FEEDER CONVEYOR OPERATION

The movement of the material from the platform or corn head is done by the feeder conveyor chain or the rake in the feeder conveyor unit. The feeder conveyor takes the material from the platform or corn head and feeds it to the threshing cylinder (Fig. 25).

In some combines, a feeder beater is used to help move the material from the platform auger to the feeder conveyor. This beater consists of a round drum equipped with retracting fingers that are similar to the fingers used in the platform auger. In one type of feeder conveyor, the feeder beater feeds the material from the platform directly to the cylinder. Some combines may have a series of paddles that move the material to the cylinder. The most common type of feeder conveyor consists of the chain conveyor or a combination of a feeder beater and chain conveyor.

The chain-type feeder conveyor is allowed to "float" at the lower end to permit smooth feeding of both small and large masses of material. The conveyor chain is adjustable for various crop conditions. Adjustment is covered in chapter 5, Field Operation and Adjustments.

A variable-speed feeder control is available for some combines that permits the operator to adjust feeder speed

Fig. 25 — Feeder Conveyor

to crop conditions and travel speed. Some combines also have a reversible conveyor drive to aid in clearing the feeder conveyor in case it becomes plugged.

**Continued on next page**

KN52281,1004542 -19-22AUG13-20/22

## STONE PROTECTION

Most combines provide some form of protection from stones and other objects that might enter the machine and damage the cylinder, rotor, or concave. Some stone doors open automatically (Fig. 26) when a stone or some other solid object large enough to cause damage approaches the cylinder or rotor. When the door opens, a monitor light and alarm are activated and the operator must stop the machine to close the door. These stone doors rely on impact force of the object striking the door. This force is created by the object contacting the cylinder, rotor, or a special beater. Before closing the door, inspect these moving parts plus the feeder conveyor for possible damage. Make necessary repairs before resuming operation.

Successful operation of an automatic stone door requires proper adjustment and lubrication of the latch. This ensures that the door opens when needed and will not open unnecessarily. The monitor control must also be checked regularly so the operator is notified when the door opens to avoid heavy crop loss through the opening.

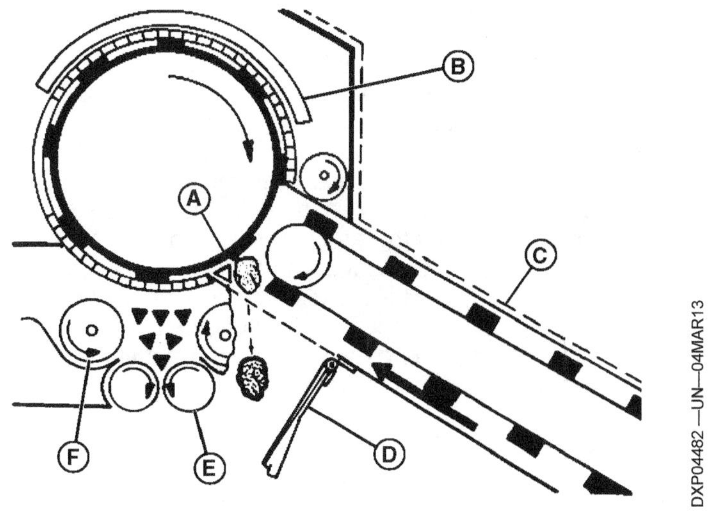

*Fig. 26 — Automatic Stone Door*

A— Cylinder  
B— Cage Sweep  
C— Rear Feed Conveyor  
D— Stone Door  
E— Accelerator Doors  
F— Distribution Augers

KN52281,1004542 -19-22AUG13-21/22

---

A cavity in front of the cylinder on some combines (Fig. 27) catches and holds stones and other hard objects. Depending on combine design, these traps may open and empty while the combine is operating, or they may require the machine to be stopped and the door opened manually from under the feeder house.

 **CAUTION: Be sure header lift cylinder safety stops are in place and the engine stopped before working under the header or feeder house.**

During normal combine operation these stone traps fill with straw, grain, and dirt. The impact of an object after striking the cylinder normally causes it to displace enough material from the trap to lodge there rather than pass through the cylinder. However, if the trap is not cleaned regularly, the buildup of material may become so compacted that stones cannot lodge in the material. Consequently the stones are carried into the cylinder and concave anyway. Therefore, to maintain protection, empty the stone trap regularly whether stones are present or not.

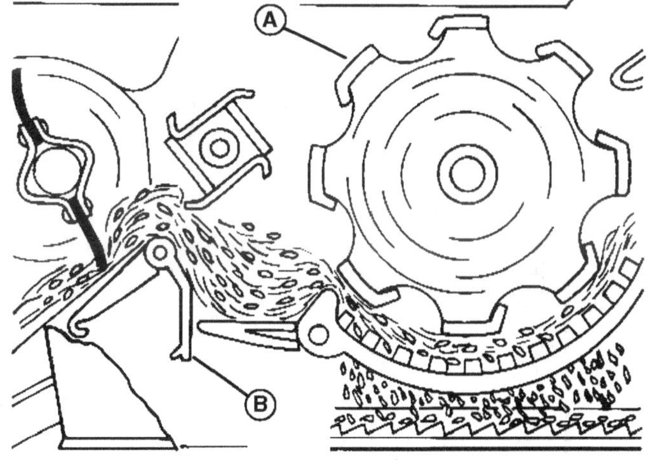

*Fig. 27 — Stone Trap Must Be Opened Manually*

A— Cylinder  
B— Stone Trap

KN52281,1004542 -19-22AUG13-22/22

# THRESHING THE CROP

The "heart" of any combine is the threshing section (Figs. 28 and 29). Webster's dictionary defines "threshing" as beating the grain from its husk or beating the grains from the heads, as in wheat. In the case of corn, this would be removing kernels from the cob, and in soybeans, it would be removing the beans from the pod. In the threshing section of the combine, all the grain is threshed from the stems or pods. Up to 90 percent of the grain is separated from its stem, cob, or pod as it passes between the cylinder or rotor and the concave. This vital area affects the entire operation of the combine, because if proper threshing isn't achieved here, the combine will fail to do its job.

A— Cylinder    B— Concave

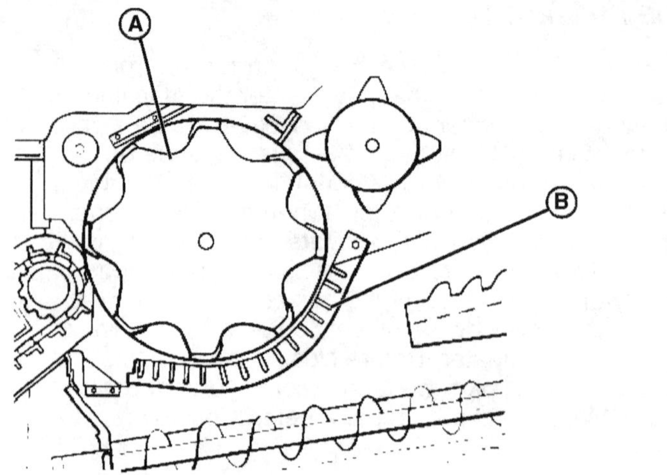

Fig. 28 — Threshing Section of a Conventional Combine

**Continued on next page**    KN52281,1004543 -19-23AUG13-1/15

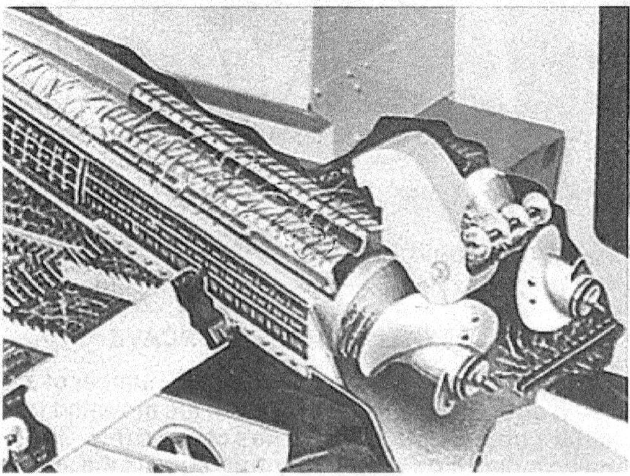

*Fig. 29 — Threshing Sections of Rotary Combines — Single Rotor (Top), Double Rotor (Left), Transverse-Mounted Rotor (Right)*

The threshing area of a combine is located in the body of the combine, which is known as the "separator." The components that make up the threshing mechanism (Fig. 28) consist of a cylinder and a concave, or rotor and concave (Fig. 29).

## TYPES OF THRESHING SYSTEMS

There are four basic types of threshing systems and matching concaves:

- Rasp-bar cylinder and concave
- Spike-tooth cylinder and concave
- Angle-bar cylinder and concave
- Single or dual rotor and concave

The most common design now is the rasp-bar cylinder and concave, because almost all crops can be threshed with it. The spike-tooth design is used almost exclusively in rice or edible beans. The angle-bar design is found in a few machines and is used mostly for small-seed crops such as clover and alfalfa. Cylinder sizes may vary from 15 to 22 inches (38 to 56 cm) in diameter and from 24 to 60 inches (61 to 152 cm) in length.

Rotary combines with rasp-bar rotors and matching concaves are now available. They can thresh essentially all the same crops harvested by combines with rasp-bar cylinders. Rotors range from 17 to 31.5 inches (43 to 80 cm) in diameter and from 88 to 168 inches (2.24 to 4.27 m) in length. Both single and dual rotor models are available.

Here are more detailed discussions of each type.

**Continued on next page**

KN52281,1004543 -19-23AUG13-2/15



PN=37

## RASP-BAR CYLINDER AND CONCAVE

The rasp-bar cylinder consists of a number of corrugated steel bars attached to the outer circumference of a series of hubs (Fig. 30). These hubs are mounted on a cross shaft at the front of the separator. The shaft is carried on ball bearings.

The cylinder is driven at speeds ranging from 150 to 1500 rpm. This variety of speeds is necessary to meet the different threshing actions required for various crops and crop conditions. Most crops can be threshed at speeds between 400 and 1200 rpm.

The concave consists of a series of parallel steel bars held together by curved side bars and rods. The concave is mounted under and slightly to the rear of the cylinder. The curvature of the concave generally conforms to the outer circumference of the cylinder.

The rasp-bars have corrugations running in opposite directions on adjacent bars (diagonally). Usually, the direction of the corrugations is referred to as right or left to identify a particular bar. These corrugations provide rubbing or rasping action on the crop as it passes through the threshing area.

*Fig. 30 — Typical Rasp-Bar Cylinder*

The rasp-bar cylinder and concave will handle damp weeds with less tearing of the stems than a spike-tooth cylinder. This means drier and cleaner grain in the tank and less overloading of the cleaning shoe.

KN52281,1004543 -19-23AUG13-3/15

## SPIKE-TOOTH CYLINDER AND CONCAVE

The spike-tooth cylinder consists of a number of steel teeth attached to metal bars that are mounted to the outer circumference of a series of hubs (Fig. 31). The basic mounting and drive configuration is the same as described for the rasp-bar cylinder.

*Fig. 31 — Typical Spike-Tooth Cylinder*

**Continued on next page**

KN52281,1004543 -19-23AUG13-4/15

The concave also has teeth (Fig. 32). The teeth are attached to bars that are held in place by curved side bars. As with the rasp-bar concave, this concave is mounted under and slightly to the rear of the cylinder.

A— Cylinder
B— Cylinder Tooth
C— Open Gate Concave
D— Spike-Tooth Concave
E— Concave Tooth

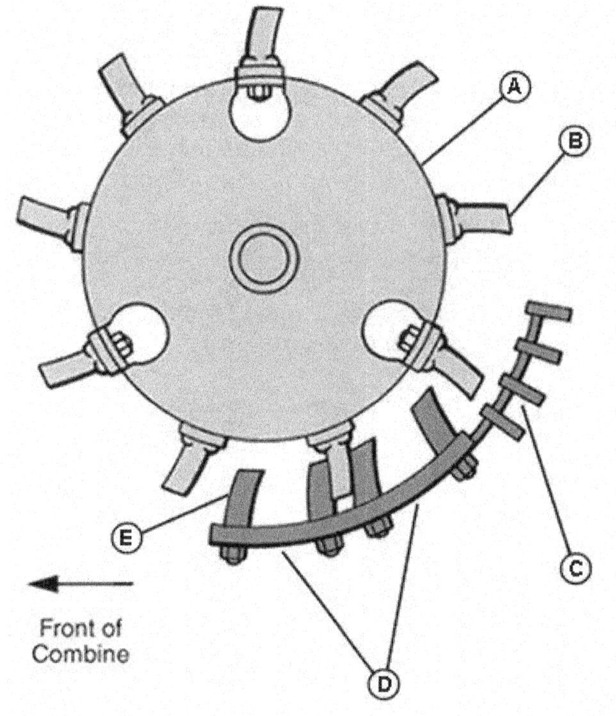


*Fig. 32 — Spike-Tooth Concave*

KN52281,1004543 -19-23AUG13-5/15

The teeth in this design tear and shred the material rather than rub and flail it as does a rasp-bar design. As the cylinder rotates, its teeth pass between the stationary teeth of the concave, which causes the threshing action (Fig. 33).

A— Cylinder
B— Concave Tooth (First Row)
C— Equal Clearance
D— Concave
E— Concave Tooth (Second Row)
F— Cylinder Tooth

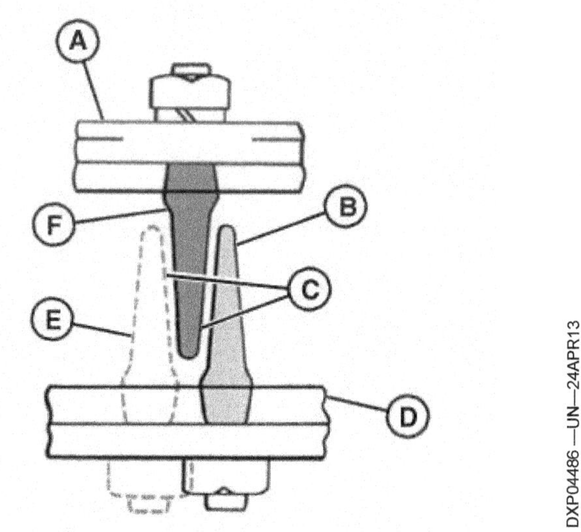

*Fig. 33 — Spike Cylinder Teeth Pass Between Concave Teeth*

**Continued on next page**

KN52281,1004543 -19-23AUG13-6/15

## ANGLE-BAR CYLINDER AND CONCAVE

The angle-bar cylinder consists of helically mounted angle-iron bars attached to hubs (Fig. 34). Both the bars and the concave have rubber faces. The basic mounting and drive configuration of this design is the same as the others described above.

This design flails the grain rather than rubbing and flailing as does the rasp-bar type. The threshing action is considerably more gentle than either the rasp-bar or spike-tooth cylinder. It is commonly used for small-seed crops, such as clover or alfalfa.

A— **Rubber-Faced Cylinder Bar**    B— **Rubber Concave Bars**

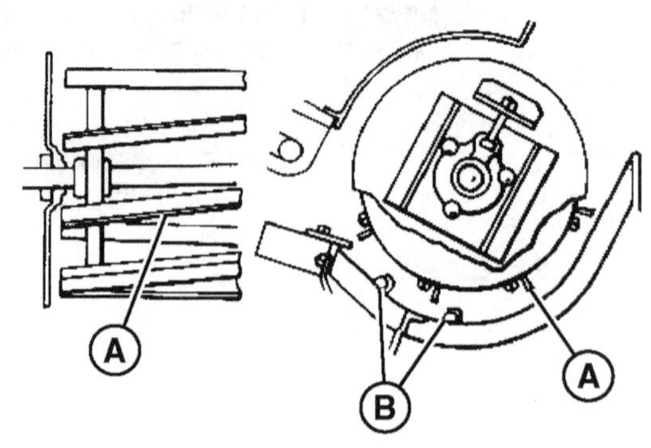

Fig. 34 — Typical Angle-Bar Cylinder and Concave

KN52281,1004543 -19-23AUG13-7/15

## ROTOR AND CONCAVE

Rasp-bars, similar to those used on rasp-bar cylinders, are attached to the input end of the cylindrical rotor in rotary combines. A matching concave under the rasp-bar portion of the rotor permits separation of up to 90 percent of the grain from the straw at that point. Rotary combines may have a single rotor or two parallel rotors. Most rotors are mounted longitudinally in the combine and are fed from the front end of the rotor. Some models have a rotor mounted transversely (crosswise) that is fed from the side at one end of the rotor, much like a conventional cylinder. Typical rotor configurations are shown in Fig. 35.

Rotor speeds range from about 200 to almost 1600 rpm, depending on machine design and the threshing requirements of a variety of crops. Variable speed controls permit adjustment of rotor speed to meet changing crop conditions as the machine moves through the field.

Unlike the cylinder-concave design of conventional combines, the crop is exposed to more than one pass between the rotor and concave. Therefore, the clearance between rotor and concave is usually somewhat wider than conventional cylinder-to-concave spacing for the same crop and conditions. This wider spacing may result in less grain damage than is caused by conventional cylinders, particularly in high-moisture crops.

A— **Rotors**          C— **Separating Grates**
B— **Concaves**        D— **Cage**

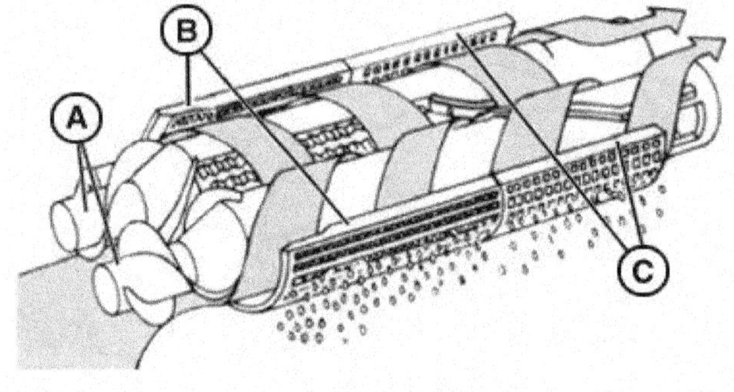

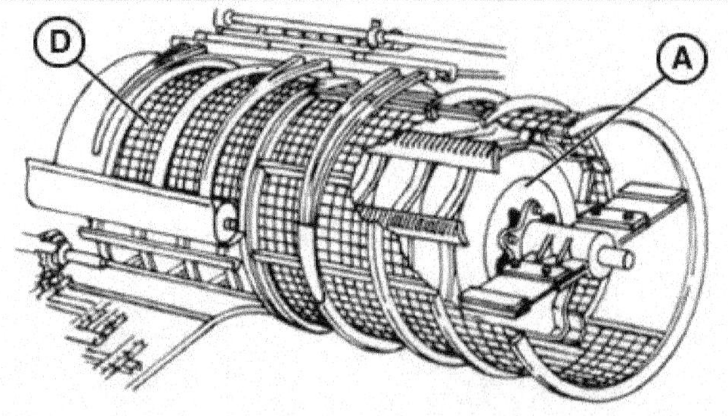

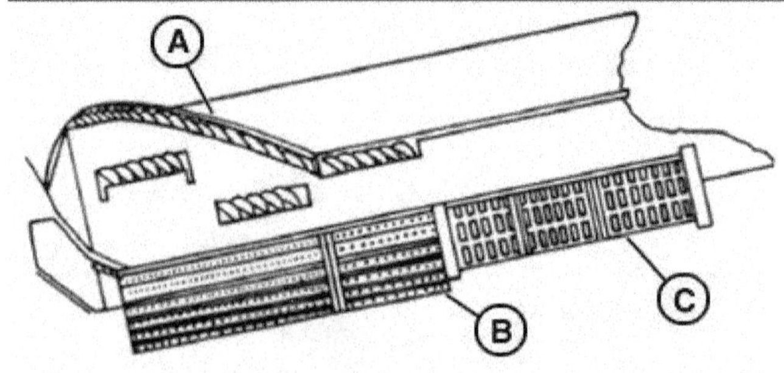

Fig. 35 — Typical Rotor Configurations

**Continued on next page**

KN52281,1004543 -19-23AUG13-8/15

## THRESHING CYLINDER OR ROTOR AND CONCAVE FUNCTION

The feeder conveyor delivers the crop to the threshing area of the combine. The crop is fed into the opening between the cylinder or rotor and concave. As the cylinder rotates, the material comes into contact with the rapidly moving rasp-bars and this impact shatters the grain or seed from the stem, cob, or pod. Additional threshing occurs by a rubbing action as the material is accelerated and passes through the restriction between the cylinder or rotor and the concave (Fig. 36).

The impact and rubbing action of the rotor and concave on the crop are essentially the same in both rotary and conventional combines. And, regardless of the machine type, the speed at which rasp-bars travel must be about the same for the same crop and conditions. For instance, corn usually requires an impact speed of about 3000 feet (914 m) per minute. This means that a 17-inch (43-cm) diameter rotor would need to operate at approximately 675 rpm, compared with 520 rpm for a 22-inch (56-cm) conventional cylinder to provide equal rasp-bar impact.

Up to 90 percent of the seeds that have been broken loose by impact or rubbing are separated from the straw through the concave openings or grates. The amount of separation that takes place here directly affects the overall capacity of the combine. For instance, if very little separation occurs through the concave, more of the grain is thrown onto the straw walkers or straw racks, which causes the limit of the separating capacity of these units to be reached much faster. This means that a higher percentage of grain will be lost at the rear of the combine at a given ground speed.

*Fig. 36 — Threshing Action*



KN52281,1004543 -19-23AUG13-9/15



PN=41

For hard-to-thresh crops, concave cover plates can be used to increase the threshing action (Fig. 37).

On some rotary combines, concave filler wires must be installed when threshing small grain and other small-seed crops, and removed again when threshing large-seed crops such as corn and soybeans.

Combines with no provision for separation at the concave have other devices such as chain separators ahead of the straw walkers.

### THRESHING CYLINDER OR ROTOR AND CONCAVE ADJUSTMENTS

Two basic adjustments are provided for the threshing cylinder or rotor and concave. These are:

- Cylinder or rotor speed
- Concave spacing

Making these adjustments properly is vital for harvesting high-quality seed with a minimum of losses.

### CYLINDER OR ROTOR SPEED

Cylinder or rotor speed basically affects two things:

- How many seeds are threshed from the straw, cob, or pods
- How many seeds are broken or damaged in threshing

The so-called easy-to-thresh crops are those seeds that require a low impact to thresh them but can tolerate a high impact before they are damaged. When threshing this type of crop, speed adjustment is not very critical.

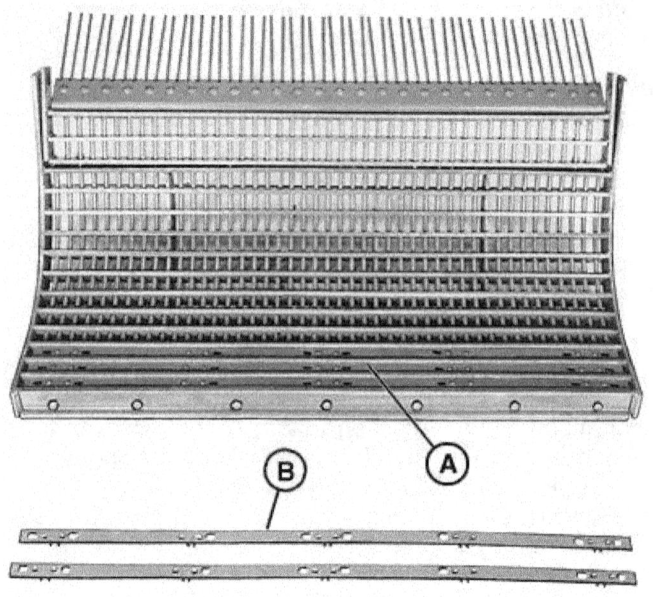

Fig. 37 — Concave Cover Plates

**A— Cover Plate Installed**

But when the impact required to thresh the crop is almost the same as the impact that will break the seed, then the speed must be carefully adjusted to get the most complete threshing with the least damage.

Cylinder or rotor speed is usually controlled from the operator's platform.

**Continued on next page**                    KN52281,1004543 -19-23AUG13-10/15

## CONCAVE SPACING

Concave-to-cylinder or concave-to-rotor spacing is controlled from the operator's platform (Fig. 38). On most combines, including most rotary models, the concave is moved up or down as desired to get the proper spacing (Fig. 39). On other combines, the cylinder is moved up or down for this spacing; on these models the operator must make this adjustment with a wrench. The concave on some rotary combines is adjusted with a ratchet on the side of the combine. Most combines have adjustments for the front and rear of the concave.

**A— Concave Spacing**

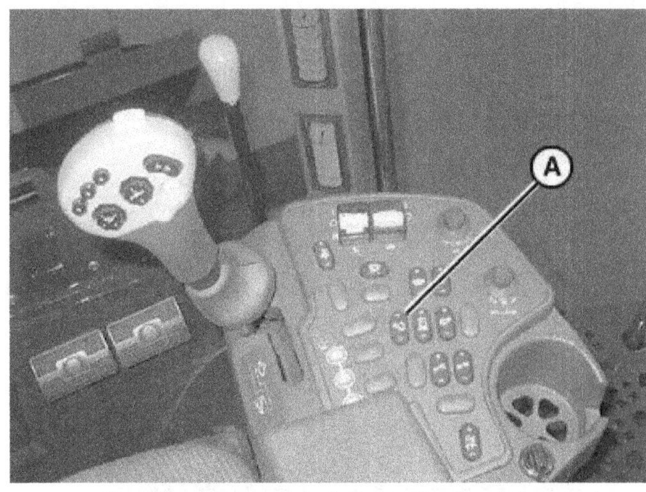

Fig. 38 — Concave-to-Cylinder Spacing Adjustment

**Concave Type Adjustment**
(Rotary Combine)

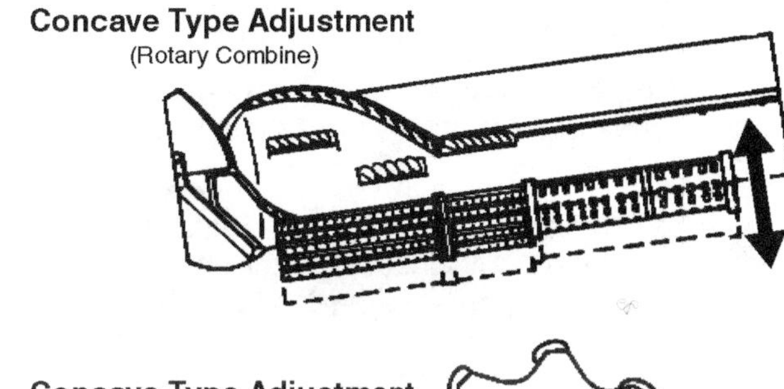

**Concave Type Adjustment**
(Only Concave Moves Up and Down)

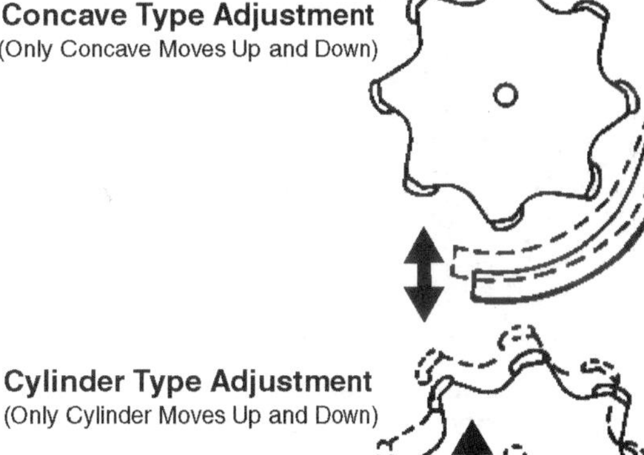

**Cylinder Type Adjustment**
(Only Cylinder Moves Up and Down)

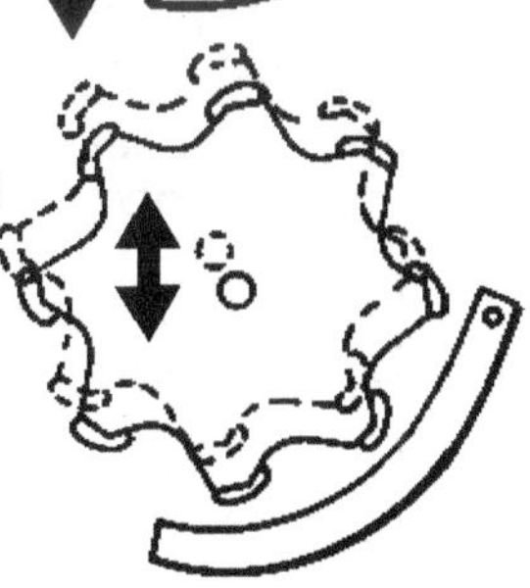

Fig. 39 — Two Methods of Adjustment

**Continued on next page**

KN52281,1004543 -19-23AUG13-11/15

Always refer to the operator's manual to determine the proper front and rear spacing. Spacing at the front of the concave may vary from a fraction of an inch (or centimeter) to 1-1/2 inches (3.8 cm). The spacing at the rear of the concave is affected by the spacing at the front. Normally, the rear spacing is about one-half the dimension selected at the front (Fig. 40). On rotary combines, the rotor-to-concave spacing may be the same along the full length of the concave. Most on-the-go adjustments change front and rear spacing at the same time and maintain the same clearance ratio through the full range of adjustment.

**A— 1-Inch Opening**     **B— 1/2-Inch Opening**

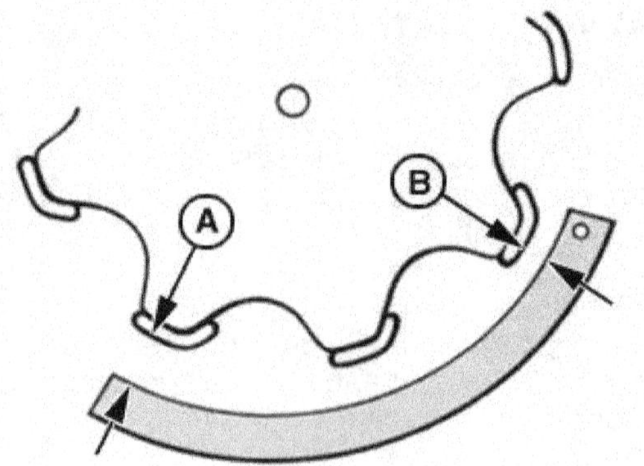

Fig. 40 — Typical Front and Rear Spacing Proportions

KN52281,1004543 -19-23AUG13-12/15

---

Concave spacing affects:

- How well seeds are threshed
- How many seeds are separated from the straw through the concave grate

When threshing is difficult, spacing is reduced to make the band of straw between the cylinder or rotor and concave thinner, causing more of the seed heads to come in contact with the rasp-bars. A narrow spacing between the concave and cylinder or rotor can also result in more grain being separated from the straw (Fig. 41). Seeds will move through the thinner mat of straw with the narrow spacing much more readily than through the thicker mats at the wide spacings. With proper clearance, more threshing occurs at the front of the concave, and more of the seed falls through the grate before it is discharged from the concave area. With wide concave spacing, threshing occurs farther toward the rear of the concave, and the grain does not have time to be separated through the concave.

As we mentioned earlier, the concave is usually adjusted so that the straw entry end is spaced farther from the cylinder or rotor than the discharge end. This gives a wedge or funnel effect and helps material to feed uniformly and without hesitation into and through the threshing area.

To achieve the optimum operation of the cylinder and concave, the components must be adjusted together. When harvesting small-grain or small-seed crops such as wheat, barley, rye, maize, alfalfa, or clover, the concave spacing is adjusted so that the crop is completely removed from the straw or stem without excessively tearing up the straw or stem. This is judged by the condition of the straw discharged by the combine.

If the concave is set too close for the crop being harvested, the straw will be excessively torn up and more horsepower will be required to thresh the crop.

If the concave is set too wide, the crop will not be completely threshed.

After the concave is properly adjusted, the cylinder or rotor speed is then adjusted to achieve maximum threshing

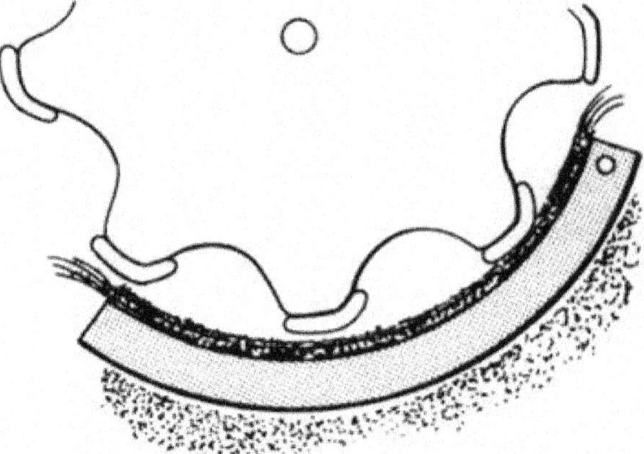


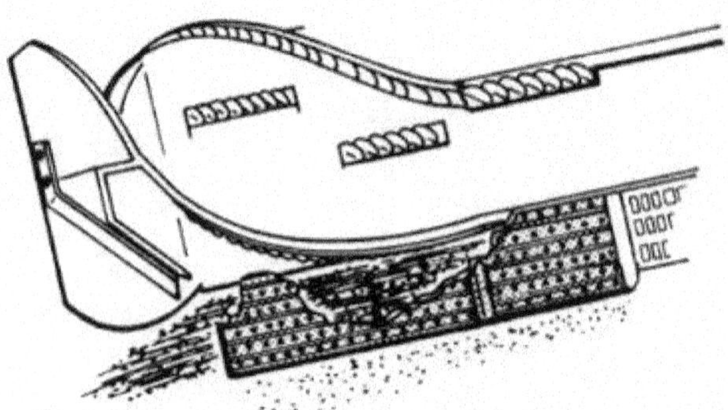


Fig. 41 — Cylinder-Concave Spacing for Small Grains

with the least crop damage. If crop damage does occur, do not widen the concave clearance. Instead, reduce the cylinder speed. Concave spacing in these crops has very little effect on seed damage.

Continued on next page

KN52281,1004543 -19-23AUG13-13/15

In the large-seed crops such as soybeans and corn, the same procedure holds true except that the concave spacing is somewhat wider (Fig. 42). In corn, this spacing is dictated by the condition of the cobs and ease of shelling. Usually this space is adjusted open far enough to keep the cobs whole or as large as possible during the threshing process.

If some varieties of corn are harvested early when the cob is soft, concave spacing must be closer for satisfactory shelling. Seeds of the large-seed crops usually are damaged very easily.

Because concave spacing is often critical, be sure that the concave is kept parallel with the cylinder or rotor (Fig. 43). This will help ensure uniform threshing across the complete width of the cylinder or the full rotor length.

Often where there is a problem of grain loss due to "carrying over" at the shoe, it can be corrected by leaving the shoe adjustments alone and going back to the cylinder or rotor to make needed corrections.

EFFECTS OF SEPARATOR SPEED AND CYLINDER (ROTOR) — CONCAVE SETTINGS

Regardless of the crop harvested, the combine must be run up to speed. Many operators will reduce or increase engine speed in an attempt to reduce cracking at the cylinder or rotor or to get better threshing of the crop. This completely upsets the balance of other units in the combine.

Reducing overall speed also reduces speed of platform, straw walkers, shoe, and elevators. This sluggishness can result in clogging of the entire combine and increased grain losses.

Increasing overall speed will cause material to pass through the machine too rapidly, resulting in grain losses and, naturally, more strain and wear on all moving parts.

Never alter the basic recommended speed of the combine. Use the correct speed and concave setting to help ensure thorough threshing. Always refer to the combine operator's manual for the basic settings recommended. From there, slight adjustments both in speed and spacing may be necessary to correspond with the crop and crop conditions.

**IMPORTANT: All of the crop is threshed at the cylinder or rotor and 90 percent can be separated there.**

Ideally, the cylinder or rotor and concave should be adjusted as follows:

1. In small grain and beans, adjust to remove all grain from the stalk with no damage to the grain, and to produce no broken straw or chaff so separation and cleaning are more efficient.
2. In corn, adjust to remove all kernels from the cob, with no damage to the kernels, and to pass as many whole cobs through as possible.

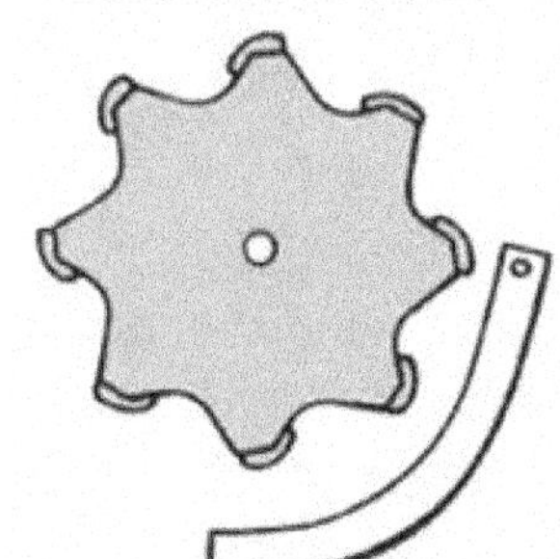
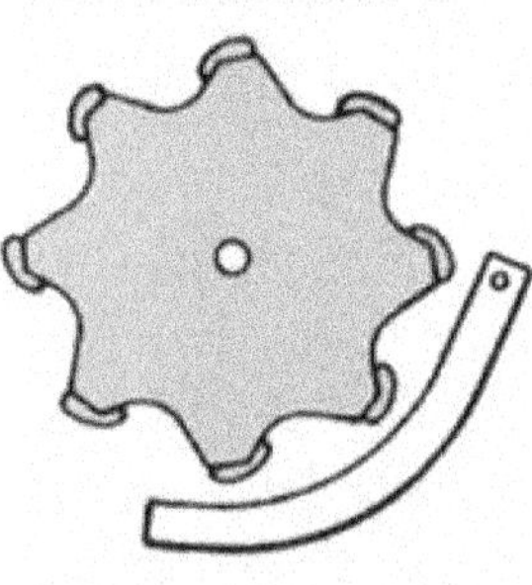


Fig. 42 — Concave-Cylinder Positions for Small- and Large-Seed Crops

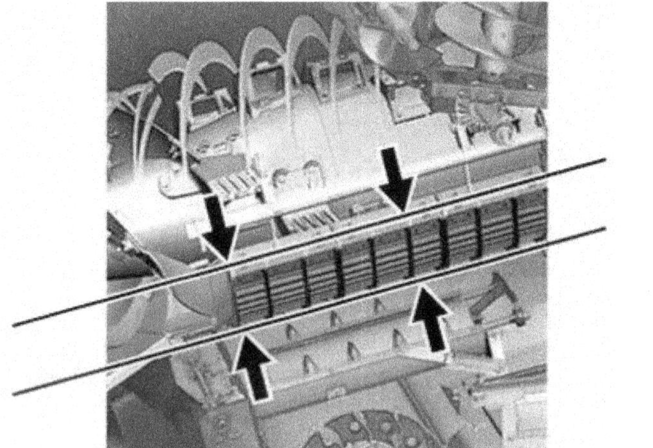

Fig. 43 — Concave Must Be Parallel with the Cylinder

The grain in the tank will show the amount of damaged grain and foreign material. However, if kernel damage is excessive, it may be caused by too many threshed kernels being rethreshed from the tailing elevator rather than improper adjustment of the cylinder and concave.

It is nearly impossible to achieve perfect performance. However, by adjusting to near the point of underthreshing, very satisfactory performance may be obtained. Adjustments to the cylinder or rotor and concave are discussed in more detail in chapter 5.

Continued on next page  KN52281,1004543 -19-23AUG13-14/15

## DETERMINING THRESHING ACTION

The only method of determining threshing action is by examining the material in the grain tank and tailings elevator, and the straw discharged from the rear of the combine (Fig. 44).

The straw discharged from the rear of the combine should contain only an occasional low-quality kernel in the heads, and the straw should be as whole as possible.

These problems and how to correct them are discussed in chapter 5 of this book.

A— Grain Tank     C— Rethreshed Kernels
B— Discharged Straw

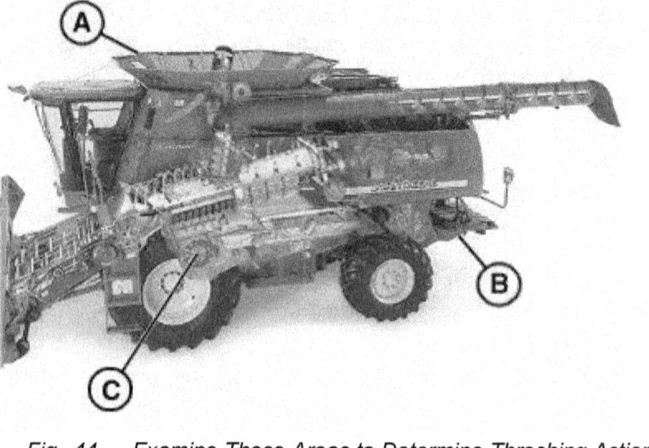

*Fig. 44 — Examine These Areas to Determine Threshing Action*

KN52281,1004543 -19-23AUG13-15/15

## SEPARATING THE CROP

The beater or rotary deflector is located directly behind and usually slightly above the threshing cylinder or rotor (Figs. 45 and 46). It has a small diameter and about the same width as the threshing cylinder or rotor.

Up to 90 percent of the grain may be separated from the straw, cob, or husks as the crop is threshed by the cylinder or rotor. In conventional combines (Fig. 45), the remaining loose grain is separated by the beater, finger grate, and straw walkers. The finger grate and straw walkers are eliminated from rotary combines. Final separation of grain and straw takes place as material flows through the rotor chamber and a beater (Fig. 46).

We'll look first at the separating process in conventional combines, then study how it is done in rotary combines.

A— Cylinder     D— Straw Walker
B— Beater     E— Finger Grate
C— Finger Bar or Grate

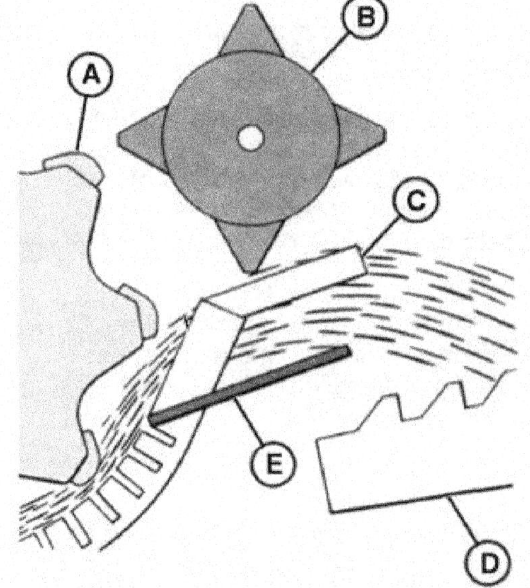

*Fig. 45 — Separating Section of a Cylinder-Concave Combine*

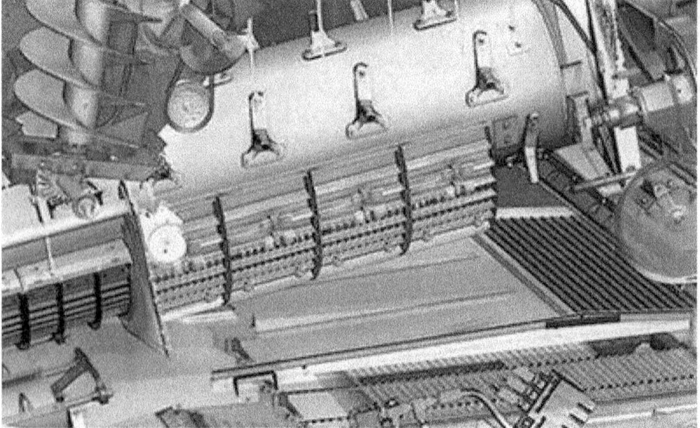

*Fig. 46 — Separating Section of a Rotary Combine*

**Continued on next page**     KN52281,1004544 -19-27AUG13-1/13



050217
PN=46

## CONVENTIONAL SEPARATION

Only loose kernels can be separated from the straw by the beater, finger grate, and straw walkers — no threshing occurs in this area. Thus, unthreshed kernels will remain attached to the straw.

### BEATER AND FINGER BAR OR GRATE

Four types of beaters or deflectors (Fig. 47) are:

- Wing-type
- Drum-type with teeth (or covers removed)
- Drum-type with removable wings (cover teeth)
- Drum-type with non-removable wings

The beater has two jobs:

- It slows material as it leaves the cylinder and concave.
- It deflects this material down onto the front of the straw walkers.

If the material to be separated is not deflected down to the extreme front of the straw walker surfaces, valuable separating area is lost.

The finger grate at the rear of the concave (Fig. 45) holds the material up so the beater can deflect it onto the straw walkers. Without the fingers, much of the material could fall down into the cleaning section, and overload it. The fingers also allow loose grain to fall through to the cleaning area.

### STRAW WALKERS OR RACKS

Effective separation in a combine is determined by how the crop is shaken as it travels through the separating area. The straw walkers (or straw rack) not only provide the agitation to separate the remaining grain, but they also remove the straw and trash by "walking" it out the rear of the combine.

The two most common types of straw carriers are:

- One-piece oscillating straw rack
- Multiple straw walkers

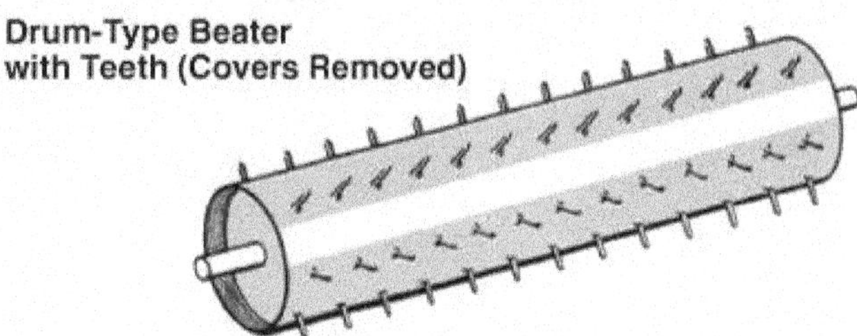


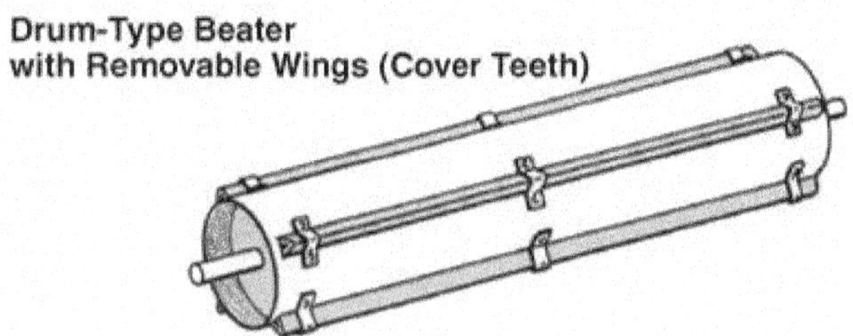


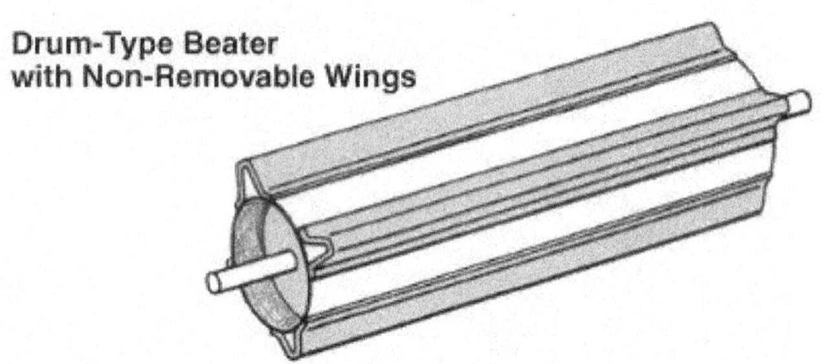


*Fig. 47 — Four Types of Beaters or Rotary Deflectors*

KN52281,1004544 -19-27AUG13-2/13

The straw rack fits across the width of the separator and is mounted so that it oscillates back and forth as a unit (Fig. 48).

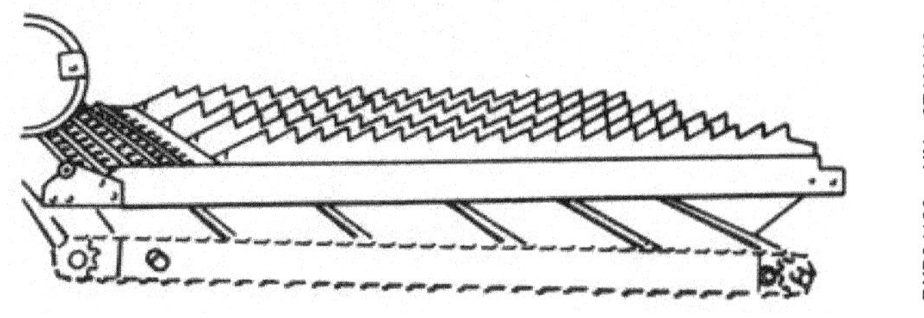

*Fig. 48 — Typical Straw Rack*

**Continued on next page**  KN52281,1004544 -19-27AUG13-3/13



050217
PN=47

The most popular design of straw carrier today is the straw walker. Straw walkers are attached to multiple-throw crankshafts at the front and rear (Fig. 49). From three to five straw walkers are mounted in the separator, depending on the width of the combine. Each walker is positioned by the crankshaft at 90 or 120 degrees around the circle of crankshaft rotation (Fig. 50).

A— Rear Crankshaft      C— Front Crankshaft
B— Return Pan

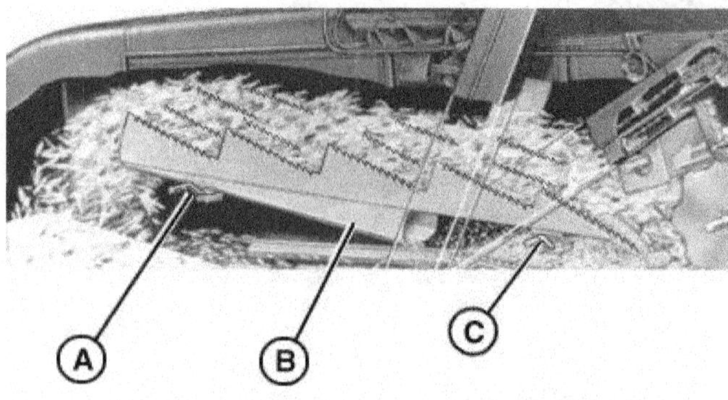

*Fig. 49 — Straw Walker with Return Pans*

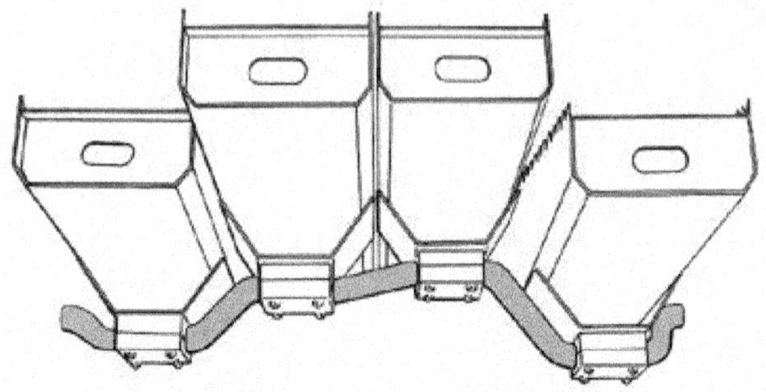

*Fig. 50 — Straw Walker Crankshaft*

KN52281,1004544 -19-27AUG13-4/13

---

Some straw walkers or racks have return pans under them, which allow the grain to travel or roll forward down the pan to an opening located just over the front of the cleaning section (Fig. 49). Other designs are open along the entire bottom, and the free grain can fall through onto a series of augers or a grain conveyor, which moves the grain to the cleaning area (Fig. 51).

Different types of straw walkers are used in different crop conditions. The most common straw walker is a step-type walker, which provides excellent tumbling and walking action; two types are shown in Figs. 49 and 51. Sawtooth edges along each side of the walker help in agitating the straw as it moves through the combine.

A— Straw Walker with Open      B— Cleaning Area
     Bottom

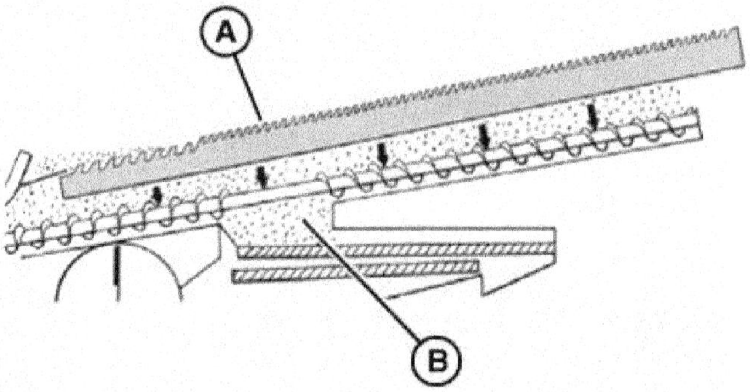

*Fig. 51 — Straw Walker with Open Bottom*

**Continued on next page**

KN52281,1004544 -19-27AUG13-5/13

Another walker, which is sometimes used in corn and eliminates plugging of heavy green stalk material, is called a one-step walker (Fig. 52).

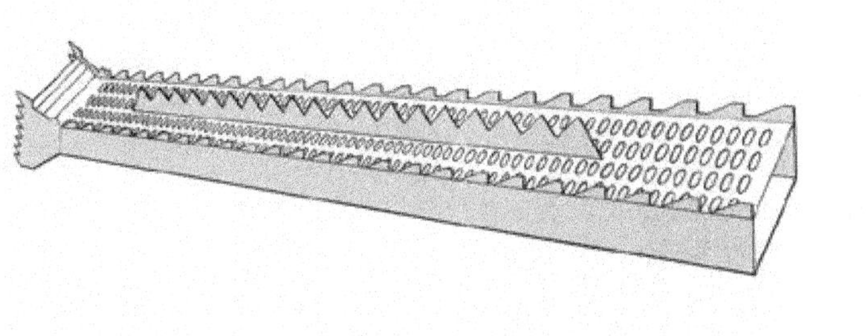

*Fig. 52 — One-Step Straw Walker*

KN52281,1004544 -19-27AUG13-6/13

Straw walkers have holes of different shapes and sizes to allow grain to fall through, yet prevent straw and trash from going through (Fig. 53). For many years square or rectangular openings had been used. But because much corn is now harvested with combines, a hooded or lip-type opening has been developed to reduce plugging by stalks and cobs.

**A— Rectangular Openings**      **B— Lip-Type Openings**

*Fig. 53 — Hole Shapes in Straw Walkers*

Continued on next page      KN52281,1004544 -19-27AUG13-7/13

For some crop conditions, "fishbacks" (Fig. 54) are attached to the walkers to provide additional agitation and reduce the speed of the material traveling over the walker surfaces. However, if the straw is extremely fluffy or damp, the fishbacks may prevent material from moving away from the beater. If material is held too long at this point, the buildup will be picked up by the beater and cylinder and will cause backfeeding. If this happens, remove the fishbacks.

STRAW WALKER ACTION

After straw is deposited onto the straw walkers, it is tumbled and tossed as it is propelled to the rear (Fig. 55). The loose grain falls down through the openings in the walkers or rack and is carried to the cleaning shoe. The straw continues to be agitated along the length of the straw walkers until it reaches the rear of the combine and falls to the ground.

The straw walkers throw the straw in an upward and rearward direction during a portion of the agitation cycle. This leaves the straw momentarily in midair. The material then falls onto a section of the walker nearer to the discharge end. Each cycle "walks" the straw a little farther toward the rear.

Each agitating cycle occurs 150 to 250 times per minute depending on the combine. If the speed is either too fast or too slow, grain losses may increase. Always refer to the operator's manual to determine the proper speed. The speed of the straw walkers or racks is usually not variable but is determined by the basic speed of the separator.

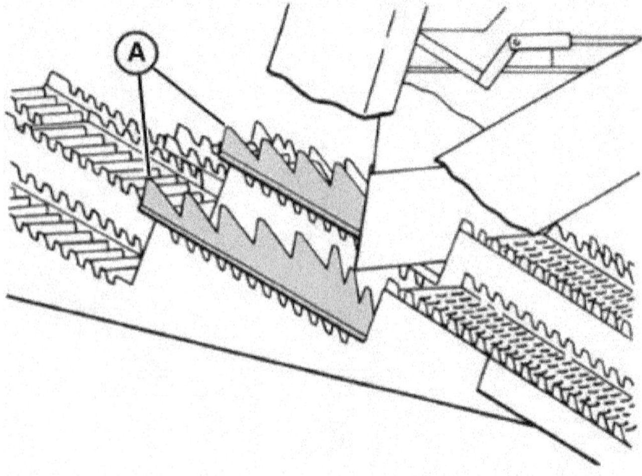

*Fig. 54 — Straw Walker "Fishbacks"*

**A— Fishbacks**

Some operators have the misconception that better separation can be achieved by speeding up the walkers, but actually this can result in poor separation. With increased speed, straw moves through the machine too quickly for all the grain to fall through the walker openings. This, of course, results in greater grain losses.

**Continued on next page**                KN52281,1004544 -19-27AUG13-8/13

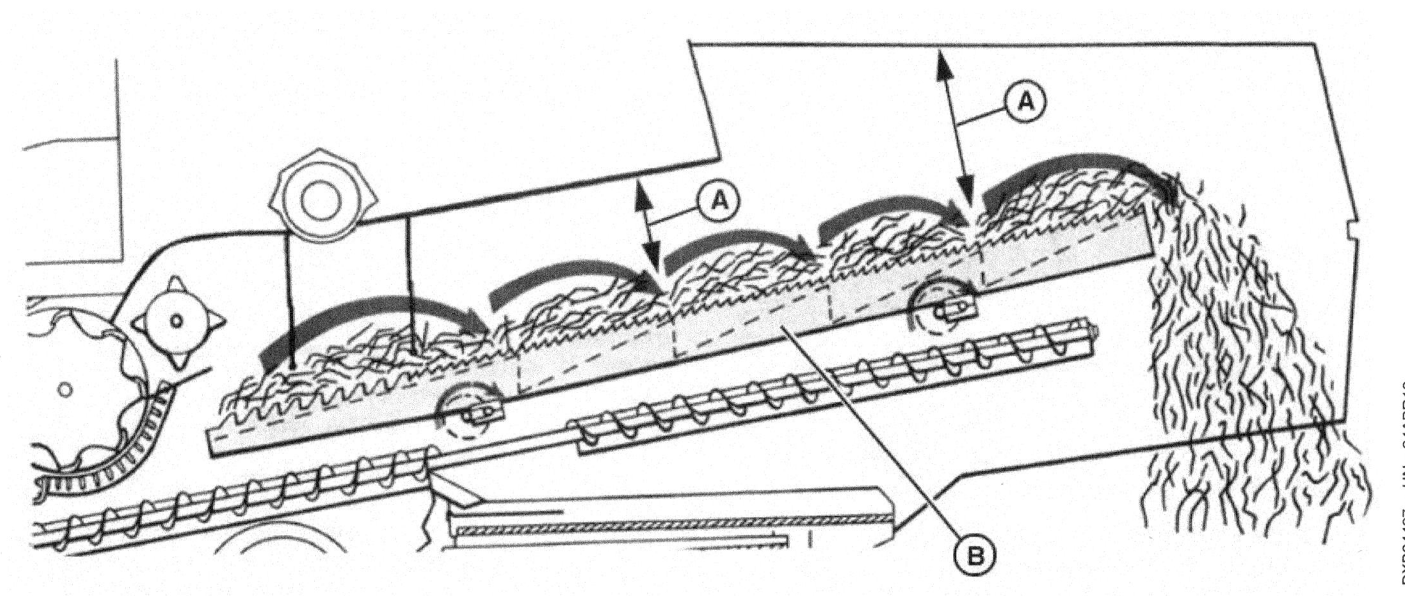

Fig. 55 — Straw Walker Action

**A— High Clearance**     **B— Straw Walker**

Most combines have high clearance or headroom above the straw walkers to ensure that the tossing action in heavy crops is not hindered (Fig. 55).

Curtains (or retarders) over the straw walkers or rack help to retard or slow down the flow of material, giving more time to agitate the straw and release grain or kernels. They also help prevent the grain from being thrown by the cylinder over the walkers and out of the combine. These curtains are made of rubber or canvas material and go the full width of the combine separator. Usually one to three curtains are used. Curtains may be added or removed depending on the condition of the crop. If the crop is damp and there is difficulty in getting the grain to fall freely from the straw, additional curtains may be required to slow down the flow of material for better separation.

IMPORTANCE OF GOOD SEPARATING ACTION

The function of transporting the straw is a very important one. The straw must be moved through the machine quickly enough for good material-handling capacity, but not so fast that the grain cannot be separated from it.

KN52281,1004544 -19-27AUG13-9/13

As we mentioned before, if the straw is transported too slowly, the beater may catch the pile of straw and feed it back to the cylinder, which will then carry it completely around and back through the concave (Fig. 56). This backfeeding is undesirable because the backfed material added to the new material entering the threshing cylinder causes overloading. Such an overload results in wasted power and plugged cylinders. Straw moving too slowly also results in a very deep mat of straw on the straw rack or straw walkers and reduces the amount of grain falling to the cleaning shoe.

In nearly all crops, the effective capacity of a combine is limited by the loss of free grain not separated from the straw before it is discharged from the combine. In other words, separation capacity of a combine is usually reached before its straw-handling capacity is reached. Combines are purposely designed this way to reduce the chances of plugging the combine when it is running at capacity.

To obtain the maximum grain separation in a combine, the concave clearance and cylinder speed must be properly

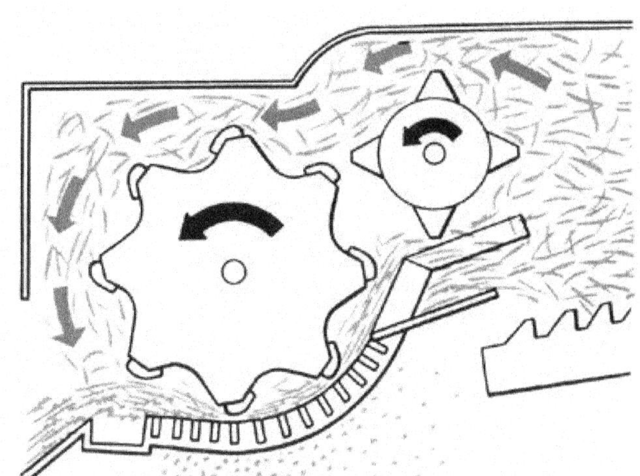

Fig. 56 — Backfeeding Caused by Straw Buildup

adjusted because most of the grain is separated by these components. Poor separation in this area cannot be completely overcome by the straw walkers.

Continued on next page

KN52281,1004544 -19-27AUG13-10/13



050217
PN=51

The chart in Fig. 57 shows the typical separating loss versus feed rate characteristics of one combine model operating in wheat, rye, or other similar crops.

For example, if this combine were to travel through the field at a speed that feeds straw and chaff in at the rate of 500 pounds (227 kg) per minute, 3 percent of the grain would not be separated from the straw. The capacity of this combine, in the particular crop that produced this curve, is from 400 to 500 pounds (181 to 227 kg) of straw per minute. As the straw input increases over this rate, the separating loss increases quite rapidly. This is typical of all combines, because as the input of the straw is increased, a point is eventually reached at which the mat of straw flowing between the concave and cylinder becomes so compressed and dense that much less grain can be separated from the straw.

Also, the mat of straw on the walkers reaches such a depth that very little grain can be shaken out.

The "separation curve" for a given combine varies for the different crops as well as for conditions of the crop. However, all combines experience the characteristic rapid increase of losses beyond a certain feed rate. The amount

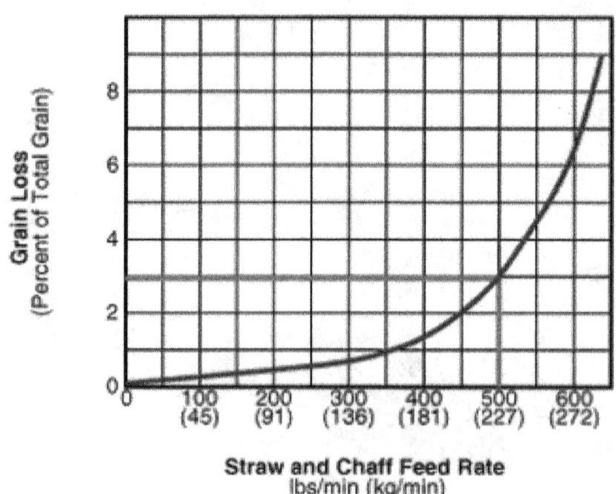


*Fig. 57 — Typical Separating Loss Versus Feed Rate*

of grain lost by the separating system is determined by the amount of straw fed through it. If a combine is operated at high travel speeds, it may handle the straw without plugging, but the separating loss will be too high. How to correct grain losses is covered in more detail in chapter 5.

KN52281,1004544 -19-27AUG13-11/13

## ROTARY SEPARATION

As a crop spirals through the rotor chamber, past the rotor rasp-bars and concave, centrifugal force moves grain outward through the layer of straw and chaff to the outside of the chamber (Fig. 58). When grain reaches the separation grates around, or partially around, the rotor, it passes through the grate, and straw continues to move toward the rotor discharge. Curved fins or vanes attached inside the chamber act like rifling in a gun barrel to spiral straw and chaff around the chamber. Vanes on the rotor keep material agitated and moving toward the discharge. This aids separation by preventing straw or husks from matting and carrying grain through the combine.

A— Rotors
B— Concaves
C— Separating Grates



*Fig. 58 — Rotary Separation System*

**Continued on next page**

KN52281,1004544 -19-27AUG13-12/13



050217
PN=52

As straw leaves the rotor chamber of most rotary combines, a beater or impeller directs material toward the rear of the machine (Fig. 59). The action of the beater or impeller also helps remove any remaining grain from the straw and deflects it through a grate, where it is carried to the chaffer.

As with conventional combines, overloading the machine makes it more difficult to obtain complete threshing and separation of grain and straw. Therefore, the feed rate or operating speed of rotary combines must be matched to crop density and condition to ensure satisfactory performance.

Overthreshing results in more damage to grain. Chewed-up straw passes through the concave and grates to overload the chaffer. Underthreshing can cause grain to pass through the combine, still attached to the head or cob. Therefore, proper concave and separation-grate clearance and rotor speed are major factors in determining the separating efficiency of rotary combines.

Separated grain leaving the rotor chamber is collected in an outer shell and falls directly onto the shoe or into a reciprocating grain pan under the separation grates. This grain is carried to the shoe for cleaning. On some machines, a series of parallel auger conveyors catch

Fig. 59 — Beater Removes Some Grain from Straw

grain falling through the concave and from part of the separation grate, and deliver it to the shoe.

Because rotary combines rely on centrifugal separation, operation on sloping land doesn't cause material to pile up on the downhill side of the combine.

KN52281,1004544 -19-27AUG13-13/13

## CLEANING THE CROP

After the crop is threshed and separated, the grain and chaff must be delivered to the cleaning area of the combine by gravity or a conveying system. Four basic methods are used to deliver grain to the cleaning area:

- Gravity feed
- Conveyor belts or chains
- Multiple augers
- Reciprocating grain pan

### GRAVITY FEED

A basic gravity feed design is shown in Fig. 60. Grain threshed by the cylinder and concave falls directly into the cleaning unit. Grain and chaff, separated from the straw, are returned to the cleaning area by the straw walker return pan.

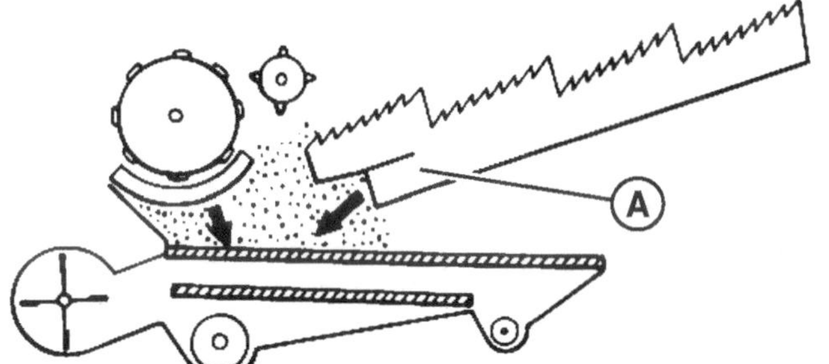

Fig. 60 — Gravity Feed to Cleaning Unit

A— Return Pan

Continued on next page

KN52281,1004545 -19-24JUL13-1/6

## CONVEYOR BELTS OR CHAINS

Two basic conveyor belt or chain designs are shown in Figs. 61 and 62. One design utilizes a grain conveyor under the straw walkers (Fig. 61). The grain threshed by the cylinder and concave falls directly into the cleaning unit. The straw walkers are open on the bottom, and separated grain and chaff fall onto the conveyor (sometimes called raddle), which delivers it to the cleaning unit. Another design utilizes a grain conveyor under the concave (Fig. 62). The grain conveyor is located under the cylinder and concave. The conveyor delivers threshed grain to the cleaning unit. The straw walkers have return pans, which return the separated grain and chaff to the cleaning area.

A— Grain Conveyor

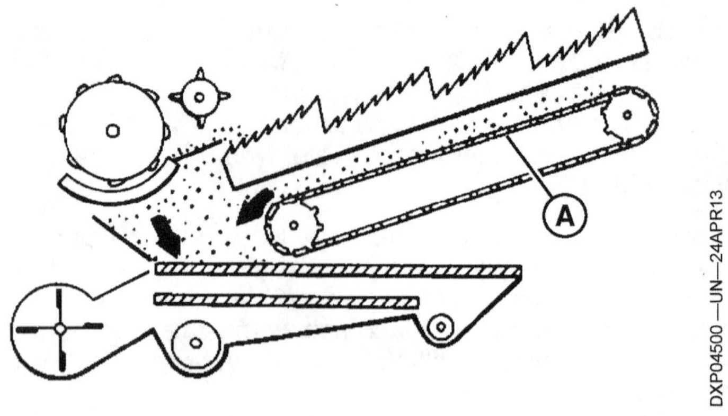

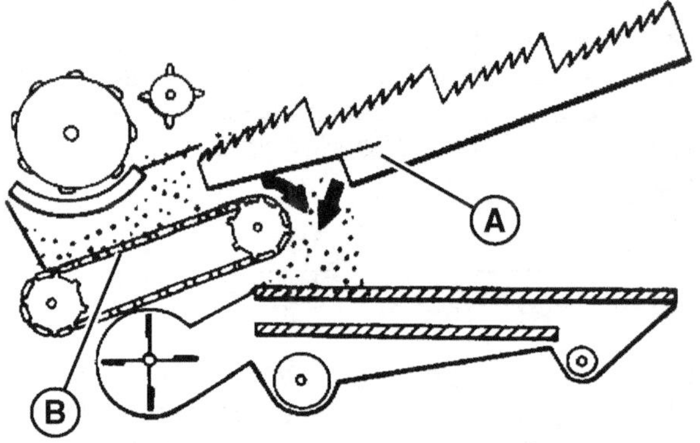

*Fig. 61 — Conveyor Belts or Chains Deliver Grain to Cleaning Unit*

KN52281,1004545 -19-24JUL13-2/6

## GRAIN CONVEYOR UNDER WALKERS

A— Return Pan          B— Grain Conveyor

*Fig. 62 — Conveyor Belts or Chains Deliver Grain to Cleaning Unit*

KN52281,1004545 -19-24JUL13-3/6

## MULTIPLE AUGERS

Multiple augers (Fig. 63) running the length of the threshing and separating areas move the grain to the cleaning unit. Grain separated by the cylinder and concave is moved back to the cleaning area by augers at the front. Augers at the rear of the separator have reverse flighting that pushes the grain separated by the straw walkers forward to the cleaning unit.

A— Augers

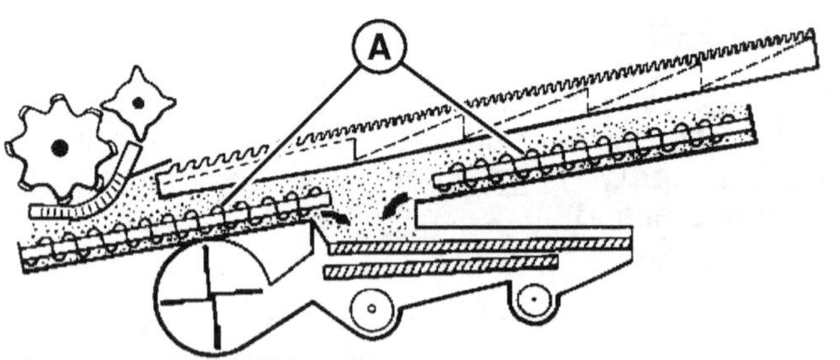

*Fig. 63 — Multiple Augers Deliver Grain to Cleaning Unit*

Continued on next page

KN52281,1004545 -19-24JUL13-4/6

## RECIPROCATING GRAIN PAN

A stepped reciprocating grain pan (Fig. 64) catches grain from the rotors of some combines. Shaking action of the pan separates material into layers of grain on the bottom, then chaff and small pieces of straw on top. Pan motion carries the material rearward, and a finger grate on the end of the pan permits grain to drop directly onto the chaffer, while air from the fan carries chaff and straw out the rear of the combine.

A— Reciprocating Grain Pan

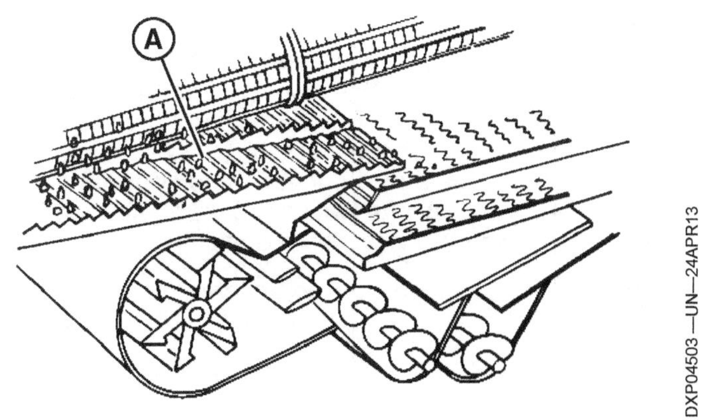

*Fig. 64 — Reciprocating Grain Pan Delivers Grain to Cleaning Unit*

KN52281,1004545 -19-24JUL13-5/6

Some machines use a combination of reciprocating grain pans and augers (Fig. 65). For example, augers might carry grain to the cleaning shoe from the threshing area under the cylinder or front of the rotor, while a reciprocating pan may carry grain to the cleaning shoe from the walker or separating area of the rotor.

A— Auger        B— Reciprocating Grain Pan

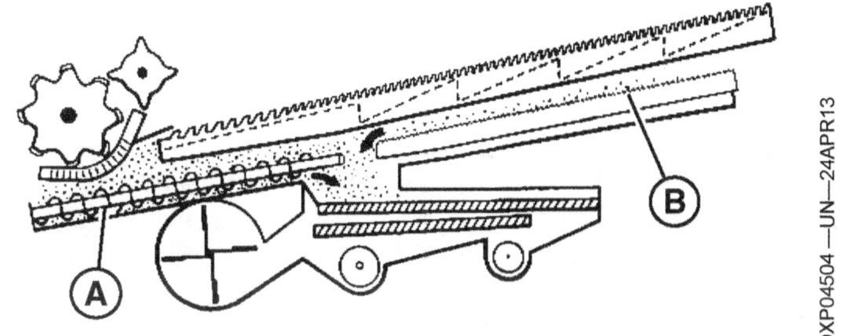

*Fig. 65 — Combination of Reciprocating Grain Pans and Augers Deliver Grain to Cleaning Unit*

KN52281,1004545 -19-24JUL13-6/6

## CLEANING UNIT

After threshing and separation, some chaff and straw are still mixed with the grain. The cleaning unit removes this trash from the grain. To do this, most combines have three basic components (Fig. 66) that make up the cleaning unit:

- Fan
- Chaffer
- Sieve

The fan has its own housing. The chaffer and sieve are in the unit usually known as the cleaning shoe.

These components will now be discussed in more detail.

A— Chaffer        C— Fan
B— Sieve

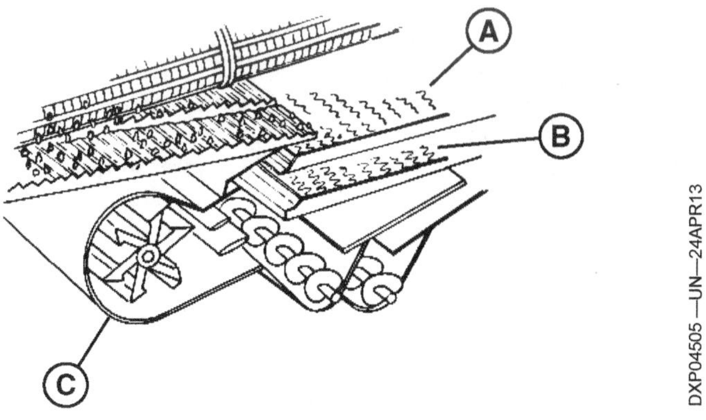

*Fig. 66 — Cleaning Unit Components*

Continued on next page

KN52281,1004546 -19-24JUL13-1/8



PN=55

### Cleaning Fan

The cleaning fan is a multiple-bladed fan mounted in front of the cleaning shoe (Fig. 67). The air blast from the fan removes most of the chaff and straw from the grain. Fan speeds may be adjusted from about 250 to 1500 rpm on most combines, depending on crops and conditions. However, on some combines, fan speed is constant and air flow is controlled by changing fan openings.

The amount of air can be controlled by three methods:

• Shutters
• Windboards
• Fan speed

Some cleaning fans are equipped with fan shutters and windboards.

A— Cleaning Fan

Fig. 67 — Cleaning Fan Location

KN52281,1004546 -19-24JUL13-2/8

Shutters (Fig. 68) are used to control the amount of air taken in and delivered by the fan. These shutters are partially closed when cleaning light seed and are wide open when cleaning heavy seed.

In many combines today, the fan housing and fan throat are designed to direct the air to the section of the shoe where it will do the most effective cleaning job. The need of windboards or side shutters has been eliminated by effective design and variable fan speeds.

Windboards (Fig. 68), located in the fan throat, control the direction of the fan blast to the chaffer and sieve. Usually the fan blast is directed well to the front of the cleaning shoe in heavy crops and directed more to the rear of the shoe in light crops. Adjustment of the windboards is critical because if the air blast is too far to the front of the shoe, material will accumulate at the rear of the chaffer and grain will be carried out of the combine. If the air blast is too far to the rear of the shoe, material will accumulate at the front of the chaffer and will result in poor cleaning action.

Fan speed is adjusted along with the chaffer and sieve openings. In dry crops, the chaffer and sieve are opened more than normal for the crop being combined. This helps prevent the chaffer and sieve from riding grain out the rear of the combine or into the tailings. The fan is

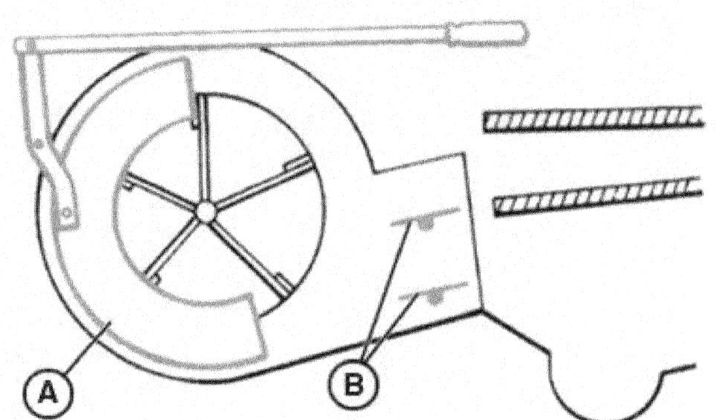

Fig. 68 — Shutters and Windboards

A— Shutter            B— Windboards

then adjusted for maximum usable air volume without blowing the grain into the tailings or out the rear of the combine. Slight adjustments are then made to get the most effective cleaning action. Problems of poor cleaning and grain losses caused by fan adjustments are covered in chapter 5.

Continued on next page

KN52281,1004546 -19-24JUL13-3/8

## CLEANING SHOE

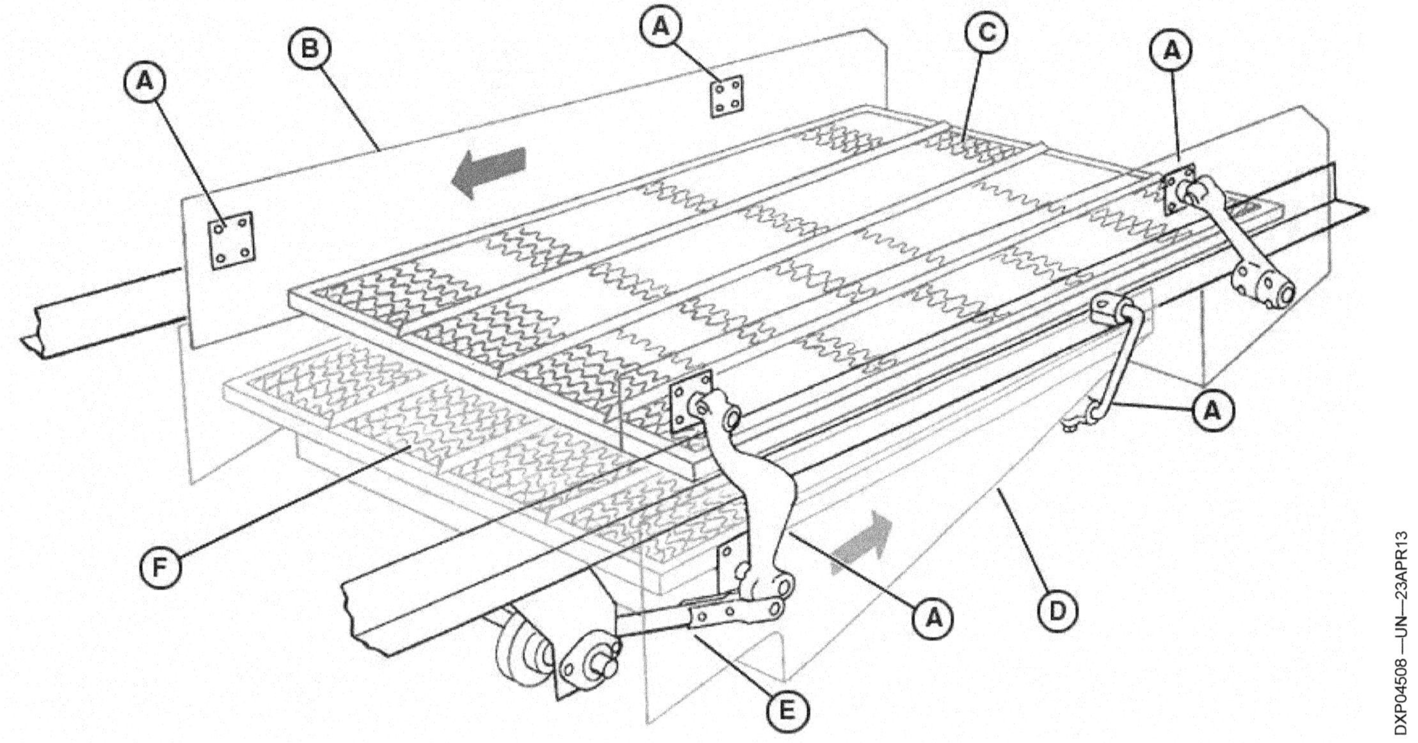

*Fig. 69 — Reciprocating Shoes Move Chaffer and Sieve*

A— Hanger  
B— Upper Shoe  
C— Chaffer  
D— Lower Shoe  
E— Pitman  
F— Sieve

The cleaning shoe, which contains the chaffer and sieve, is a housing mounted under the main frame of the combine separator. The bottom of the shoe usually contains the lower tailings auger and the lower clean grain auger, which will be discussed later in this chapter.

The chaffer and sieve are suspended on hangers, mounted on rubber bushings attached to the sides of the cleaning shoe (Fig. 69). The chaffer is mounted in the upper shoe; the sieve is mounted in the lower shoe. The chaffer and sieve are moved back and forth by a pitman-type drive attached to the hangers.

**Continued on next page**

KN52281,1004546 -19-24JUL13-4/8

The three types of typical shoe action are called reciprocating, shaker, and cascading. In the reciprocating shoe, the chaffer and sieve move in opposite directions from each other — opposed motion (Fig. 69). In the shaker shoe design, the chaffer and sieve move in the same direction at the same time (Fig. 70). The third type is a cascading shoe, which also uses chaffers and a sieve positioned so that the material drops from one unit to the other in a cascading or rolling motion as it is cleaned (Fig. 71).

A— Chaffers    B— Sieve

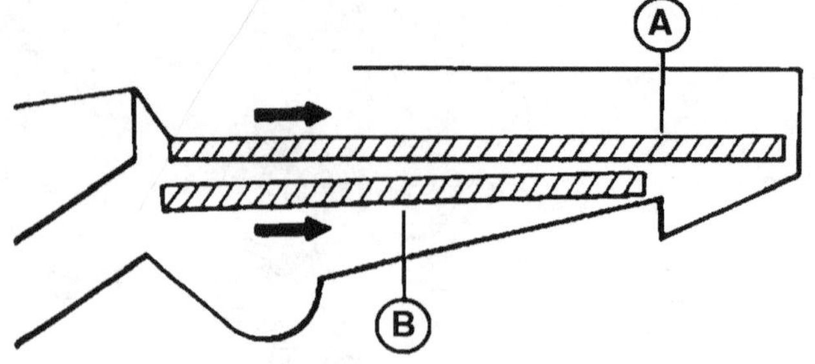

Fig. 70 — Shaker Shoe

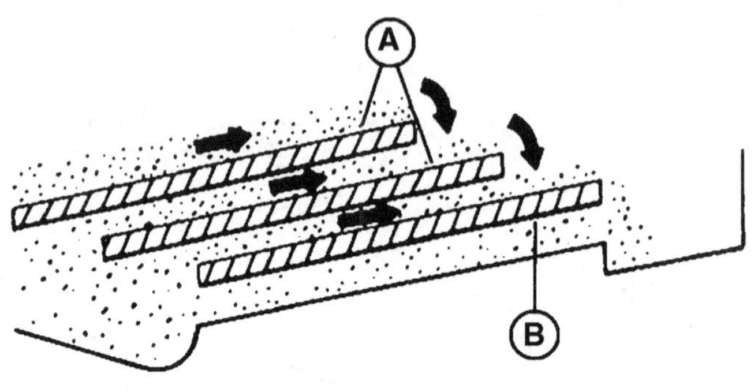

Fig. 71 — Cascading Shoe

KN52281,1004546 -19-24JUL13-5/8

## CHAFFER

Chaffers are available as either adjustable type or nonadjustable type. The adjustable chaffer is made up of a series of cross pieces of overlapping metal louvers with lips or teeth (Fig. 72). These louvers are mounted on rods and fastened together so that they may be adjusted simultaneously to the desired openings. They are usually available in various shapes of louvers and different spacings to accommodate various crops and conditions.

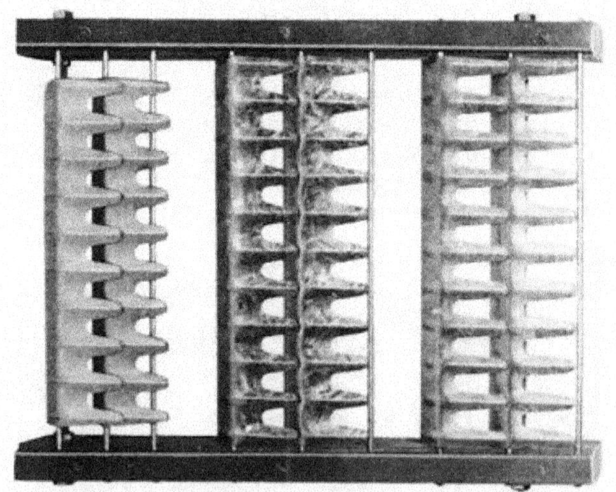

Fig. 72 — Three Types of Adjustable Chaffers

**Continued on next page**

KN52281,1004546 -19-24JUL13-6/8

Nonadjustable chaffers are available in a number of designs (Fig. 73). The openings and louvers have different shapes and sizes to meet the problems encountered in different crop conditions. In normal conditions, however, the adjustable chaffer will provide the most effective operation.

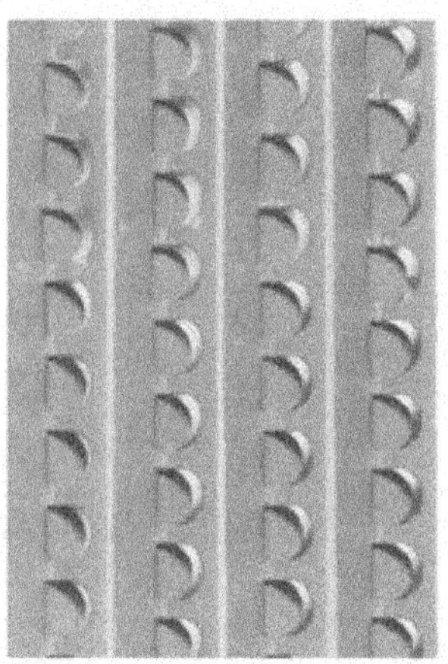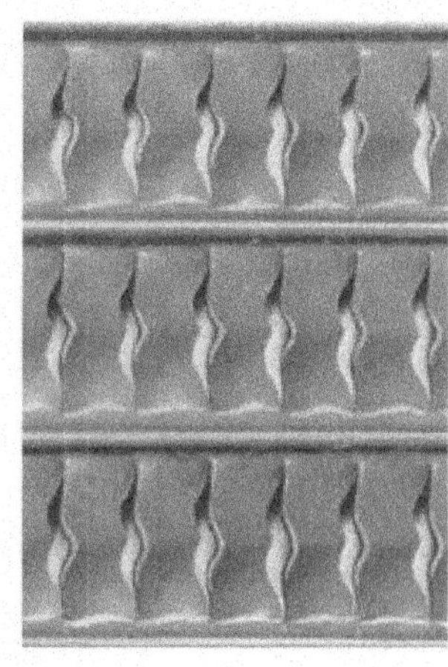

*Fig. 73 — Two Types of Nonadjustable Chaffers*

**Continued on next page**

KN52281,1004546 -19-24JUL13-7/8

## CHAFFER OPERATION

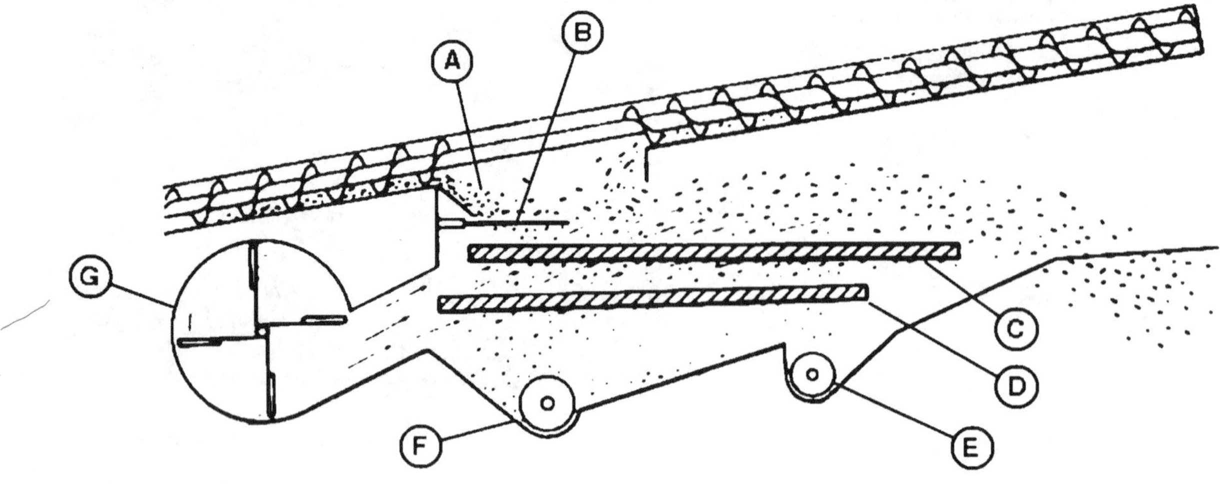



*Fig. 74 — Chaffer and Sieve Operation*

A— Grain and Chaff Mixture  
B— Finger Bar  
C— Chaffer  
D— Sieve  
E— Tailings Auger  
F— Clean Grain Auger  
G— Fan  

The grain and chaff mixture is delivered to the front of the chaffer (Fig. 74) over the finger bar or "scalper." The finger bar holds the incoming layer of grain and chaff mixture over the front part of the chaffer and allows the relatively high velocity of air from the fan to break up the layer. The lighter chaff is suspended in air and carried out of the combine. The grain and heavier particles rain down onto the chaffer. The oscillating motion of the chaffer carries those particles and grain toward the rear of the chaffer. The grain and smaller heavy particles fall through the chaffer louvers onto the sieve, and the lighter particles are carried rearward until they either fall through the chaffer extension into the tailings auger or to the ground off the end of the chaffer extension.

### SIEVE

The sieve is similar to the chaffer except that the louvers and openings are smaller. The final job of cleaning is done here.

There are several types of sieves, but the two most common types are the adjustable louver type (similar to the chaffer) and the nonadjustable round-hole type. The round-hole type sieve is available in hole sizes from 1/10 inch up to 9/16 inch (2.5 to 14.3 mm) according to the crop being harvested.

The sieve is located below the chaffer, and material that falls through the chaffer falls directly onto the sieve (Fig. 74). The sieve oscillates to move this material rearward. In some combines, it oscillates with the chaffer, and in others, it oscillates in a direction counter to the chaffer to help reduce the buildup of straws that poke down through the chaffer. The fan forces air through the sieve to help separate the tailings from the grain. The grain falls through the sieve to the clean grain auger and is carried to the grain tank. The unthreshed grain heads or tailings are carried to the tailings auger by the action of the sieve. The tailings are then carried back to the cylinder for rethreshing.

KN52281,1004546 -19-24JUL13-8/8

# HANDLING THE CROP

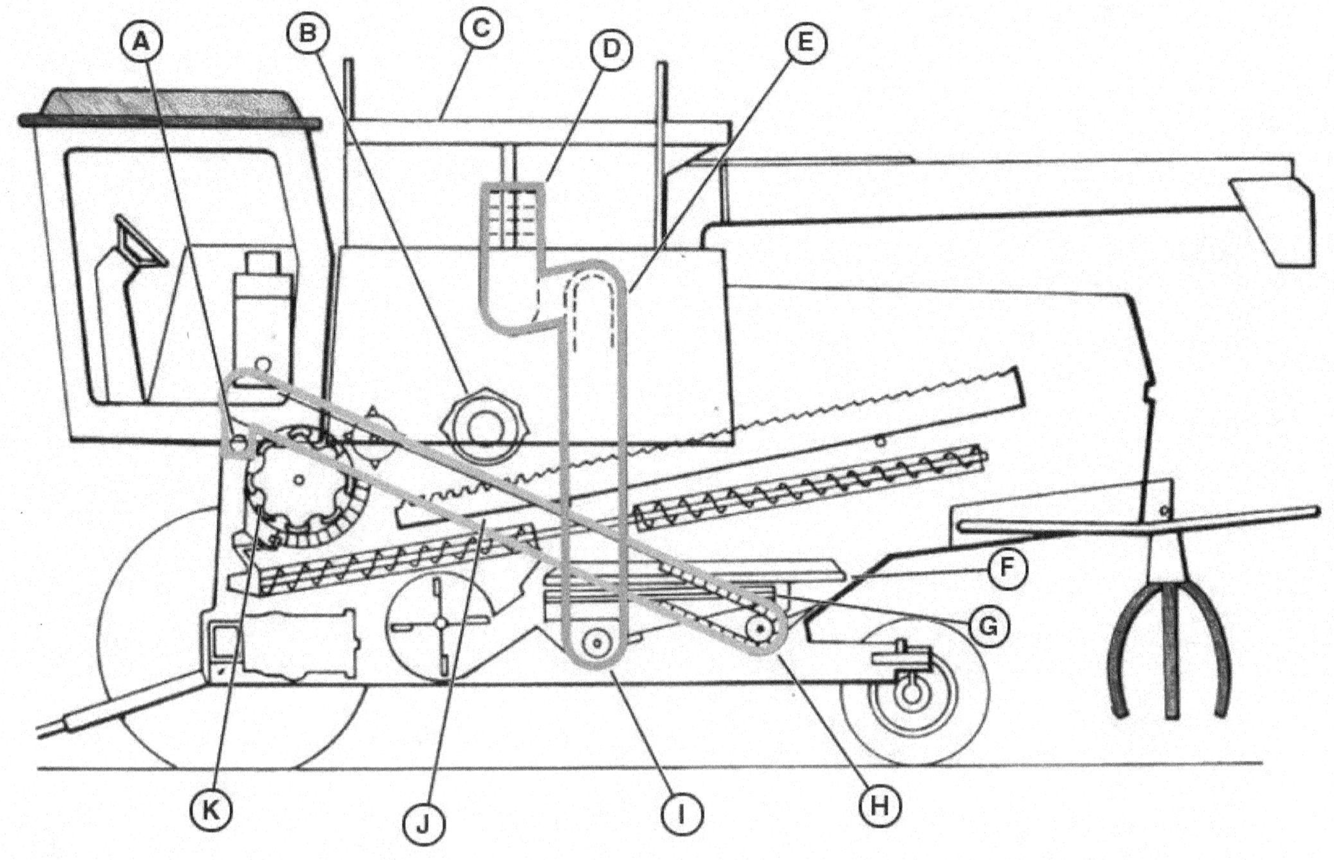

*Fig. 75 — Typical Clean Grain and Tailings Handling System*

A— Upper Tailings Auger
B— Grain Tank Unloading Auger
C— Grain Tank
D— Grain Tank Loading Auger
E— Clean Grain Elevator
F— Chaffer
G— Sieve
H— Lower Tailings Auger
I— Lower Clean Grain Auger
J— Tailings Elevator
K— Cylinder

Handling the crop means moving the threshed, separated, and cleaned crop from the cleaning shoe to the grain tank, and then from the grain tank to a wagon or truck for transporting. However, rethreshing of the tailings is another phase of grain handling that must be included.

The crop handling components (Fig. 75) include:

- Lower clean grain auger
- Clean grain elevator
- Grain tank loading auger
- Lower tailings auger
- Tailings elevator
- Upper tailings auger
- Grain tank
- Grain tank unloading auger

## CLEAN GRAIN ELEVATOR AND AUGERS

After the grain has been cleaned by the shoe, the lower clean grain auger delivers it to the clean grain elevator (Fig. 75). The elevator carries the grain to the upper clean grain auger or grain tank loading auger, which deposits the grain in the center of the grain tank or grain bin.

The augers are usually 4 to 6 inches (10.2 to 15.2 cm) in diameter with the flights 3 to 5 inches (7.6 to 12.7 cm) apart. The elevator has a series of rubber or steel paddles attached to a drive chain that moves at about 350 feet (107 m) per minute. This slow speed helps prevent damage to the grain. In some crops, such as edible beans, metal buckets are used to ensure careful handling of the beans.

## TAILINGS ELEVATOR AND AUGERS

Tailings are unthreshed or unseparated material that falls through the chaffer extension at the end of the chaffer and off the rear of the sieve (Fig. 75). Here the lower tailings auger moves the tailings to the tailings elevator. The elevator then carries the material to the upper tailings auger, which drops the tailings in the center of the separator just above the threshing cylinder. Here the material is rethreshed and later separated and cleaned.

The tailings elevator and augers are similar to the clean grain components, except that the tailings components are smaller because they do not carry as much material.

Continued on next page

KN52281,1004547 -19-27AUG13-1/5

Some combines return the tailings to a small tailings threshing cylinder located near the shoe for rethreshing, and then the rethreshed material is delivered to the chaffer and goes through the cleaning cycle again.

An inspection door is provided so that the operator can examine the tailings to determine whether or not the combine is adjusted correctly. Ideally, there should be very little material returned in the tailings; this indicates a properly adjusted cleaning shoe and that proper threshing is taking place.

KN52281,1004547 -19-27AUG13-2/5

## GRAIN TANK

The grain tank (Fig. 76) or bin is the storage compartment for the clean grain. Grain tanks come in many shapes and sizes and may be located on top, on one side, or on both sides of the combine. Sizes vary from about 70 to 315 bushels (3 to 11 $m^3$), depending on the size of the combine.

One problem in combine design has been how to increase the capacity of the tank without increasing the height of the combine. Height is a problem because of low storage-building doors and limits on bridges and overpasses when trucks transport combines.

To solve the height problem, engineers have (1) flattened the top of the tank and expanded the sides out; (2) used small tanks (one on each side as a saddlebag effect); (3) lowered the entire tank; or (4) located the threshing cylinder forward and lower to the ground, allowing more room for the grain tank. All of these variations have greatly reduced the height of the combine and have produced a neat, low-profile machine.

Not only has lowering the grain tank provided a lower profile, but the center of gravity has also been lowered. This has reduced the possibility of tip-over that can occur in taller combines.

To increase capacity of the grain tank, side extensions may be installed. Care must be taken when side extensions are used so that the power train, wheels, and

Fig. 76 — Grain Tank

separator are not overloaded. Undue strain may damage these components.

## GRAIN TANK UNLOADING AUGER

The auger for unloading should be positioned before beginning to harvest. The auger swing path must be free of obstruction. Keep combine on level ground so auger weight can be handled with minimum effort.

⚠ **CAUTION: No one should be in the grain tank when the combine engine is running.**

**Continued on next page**

KN52281,1004547 -19-27AUG13-3/5



050217<br>PN=62

When the grain tank becomes full, it is necessary to unload the grain into a wagon or truck (Fig. 77). The harvested grain can then be transported to a storage area or to the elevator for marketing.

An unloading auger system is normally used to unload the grain from the tank. This system usually consists of a large auger across the bottom of the tank. Connected to it is an outer auger (Fig. 77) that unloads the grain from the tank to the truck. The outer auger is usually tilted at an angle to extend up over the sides of the truck.

*Fig. 77 — Grain Tank Unloading Auger*

KN52281,1004547 -19-27AUG13-4/5

On some combines, a sump-type system is used (Fig. 78). Here the grain is augered from the bottom of the tank vertically up the side of the combine. When the grain reaches the top of the vertical auger, it is then carried by a horizontal auger to the truck or wagon.

Because higher-yielding crops are being produced today, combine grain tanks now fill much more rapidly and must be unloaded more frequently. Therefore, unloading time must be kept to a minimum. For this reason, large unloading augers (up to 15.5 inches [39 cm] in diameter) are used and are run at a high speed to unload the grain fast. On some combines a 300-bushel (10.6-m$^3$) tank can be emptied in slightly over a minute.

To save additional time, some combine owners unload on-the-go. With this method, the truck or wagon is driven alongside the combine, and the grain tank is unloaded as the combine moves through the field. This allows the combine to keep operating while unloading the tank. On some models, an automatic hydraulic latch allows the operator to swing and latch the unloading auger without leaving the cab to save more time.

Cooperation between combine operator and hauler is essential. The hauler must position the truck or wagon without getting too close to the combine. The hauler must be prepared for unexpected stops and leave plenty of

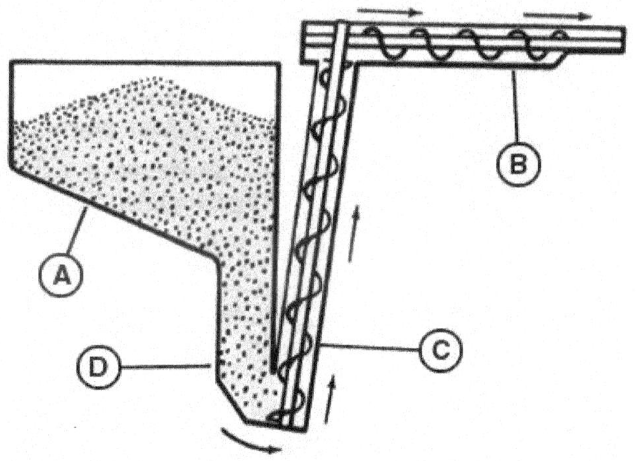

*Fig. 78 — Sump-Type Unloading Auger*

A— Grain Tank  
B— Horizontal Unloading Auger  
C— Vertical Unloading Auger  
D— Sump

room for the combine to turn at the ends of the field. The combine operator must stop unloading in time for the hauler to turn corners and drive around obstacles.

KN52281,1004547 -19-27AUG13-5/5

# RESIDUE DISPOSAL

The final function of the combine is residue disposal. Depending on the type of crop grown and the tillage practices used by the farmer, straw and chaff exiting the separating and cleaning areas of the machine can be chopped or spread evenly across the field or even windrowed for baling.

A— Tailboard
B— Chopper
C— Vanes

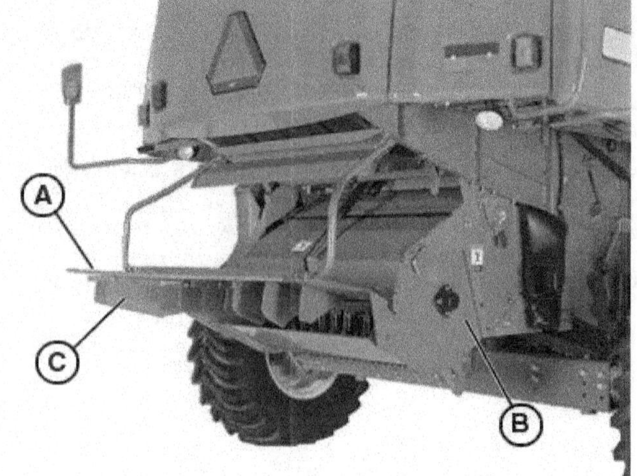

*Fig. 79 — Chopper and Tailboard Assembly*

KN52281,1004548 -19-27AUG13-1/4

Choppers and spreaders are two attachments used for residue disposal. Depending on machine configuration, a combine may be equipped with only one of these attachments that can handle both chaff from the cleaning shoe and straw from the separating area (Fig. 79), or may be equipped with a combination of these (Fig. 80), such as a chopper for straw residue from the separating area and a separate spreader attachment to spread chaff that leaves the cleaning shoe.

A— Spreader
B— Chopper

*Fig. 80 — Spreader and Chopper Attachments*

**Continued on next page**

KN52281,1004548 -19-27AUG13-2/4

## CHOPPERS

Many machines are equipped with a chopper attachment that chops the straw and chaff residue into finer pieces as it exits the combine. The chopper consists of a rotating drum surrounded by several replaceable knives (Fig. 81). The chopper is driven from the engine gear case by a series of belts and sheaves, and operates whenever the separator is engaged. The chopper can also be placed in a raised position so that residue may be windrowed for baling instead of being chopped. Most choppers are equipped with a tailboard assembly (Fig. 79). The tailboard mounts to the rear of the chopper and has directional vanes to aid in spreading the residue that leaves the chopper. Often these vanes are adjustable to help control the spread of material.

*Fig. 81 — Chopper Knives*

KN52281,1004548 -19-27AUG13-3/4

## SPREADERS

Spreaders are used to spread residue evenly when fine cutting is not required. The spreader in Fig. 82 consists of two large disks driven by hydraulic motors. The speed of the spreader can be adjusted from the cab to control how wide the straw or chaff is spread. The spreader can also be placed in a raised position so that residue may be windrowed for baling.

A— Hydraulic Motors  
B— Spreader Disks

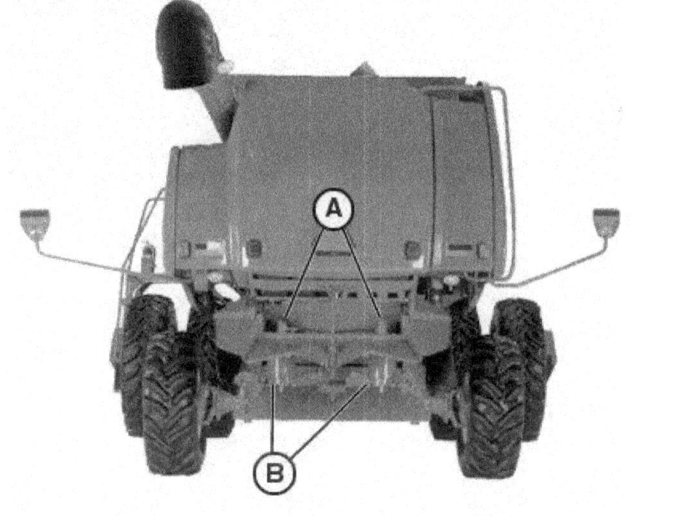

*Fig. 82 — Combine Equipped with Spreader Attachment*

KN52281,1004548 -19-27AUG13-4/4

## SUMMARY: OPERATION OF COMPONENTS

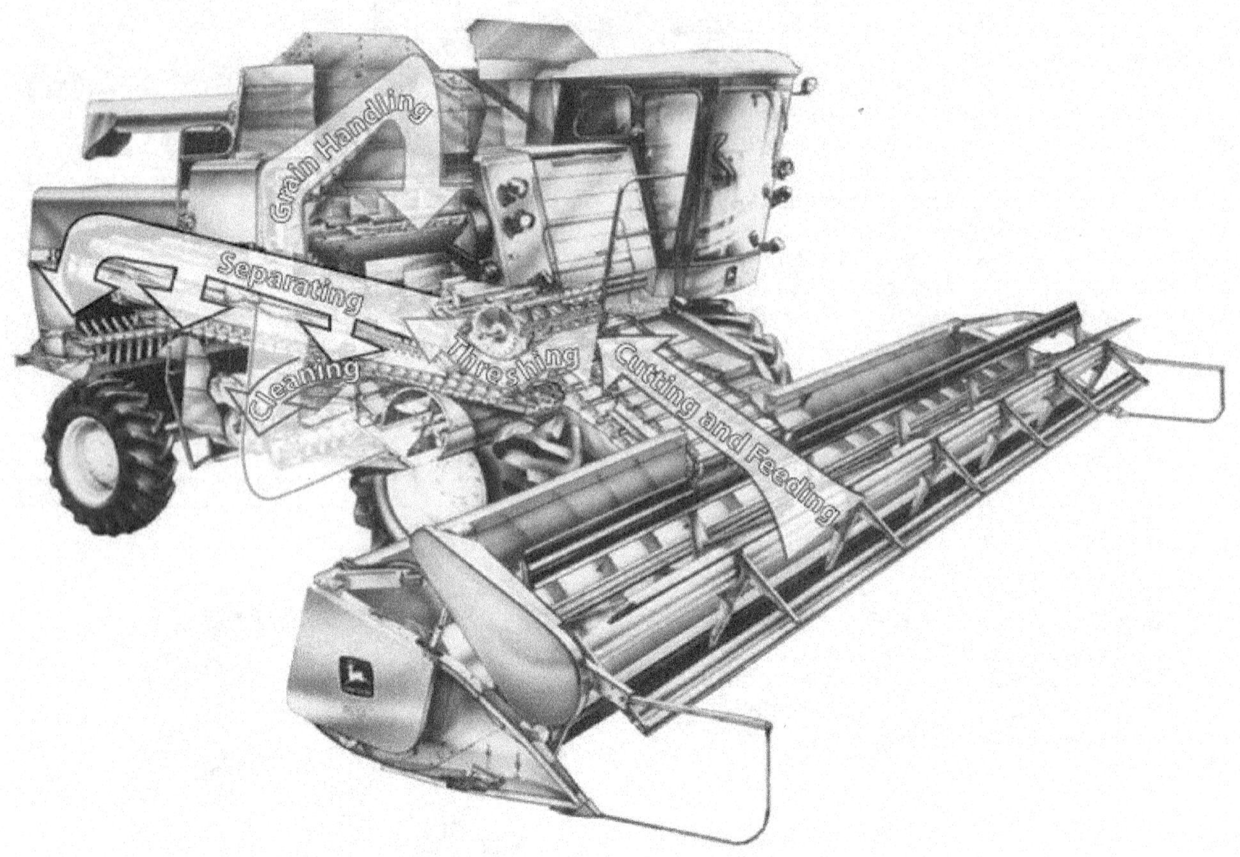


*Fig. 83 — Flow of Material Through a Combine*

The operator must know the fundamentals of how a combine works in order to operate it properly. The flow of material through a combine is shown in Fig. 83. In summary, these are the six basic crop harvesting functions of a combine:

- Cutting and feeding
- Threshing
- Separating
- Cleaning
- Handling
- Residue disposal

In other chapters, these functions are given more meaning when we discuss field operation, adjustments, and troubleshooting.

An animation of the material flow in a combine is also available with the companion Instructor Art Catalog (FMO15106TD) to this textbook.

KN52281,1004549 -19-27AUG13-1/1

## TEST YOURSELF

### Questions

1. What are the six basic functions of a combine?

2. (Fill in the blank.) The mechanism which cuts or gathers the crop and feeds it to the combine separator is known as a _______________.

3. What platform is used to gather a crop that has been cut and laid into a windrow?

4. (Fill in the blank.) A reel and cutterbar are necessary to harvest _______________ grains.

5. (Fill in the blank.) A series of parallel steel bars held together by curved side bars and mounted under the threshing cylinder or rotor is called a _______________.

6. What are the two adjustments that affect threshing action?

7. (True or False) To reduce the ground speed on a combine, the engine should be slowed down.

8. (True or False) The straw walkers move the grain to the rear of the machine by oscillation.

9. What are the three parts of a cleaning unit on a conventional combine?

10. (Fill in the blank.) Grain heads that are not completely threshed at the threshing cylinder are called

_______________.

11. List two attachments used for residue disposal.

KN52281,100454A -19-06SEP13-1/1

## INTRODUCTION

Combines of the past depended on simple forms of power, such as horses, mechanical devices, and manpower. Today's combines have enormous power requirements. Horses cannot pull a modern combine through the fields efficiently, because the equivalent of 150 work horses would be needed to pull some of the largest machines.

OUO1082,000624B -19-27AUG13-1/2

In fact, it takes three power systems to operate a modern self-propelled combine:

- Engine
- Power train
- Hydraulic system

The engine is the power source, the power train transmits power to the drive wheels or tracks to propel the combine through the field, and the hydraulic system converts the engine power to hydraulic force to operate the hydraulic components.

*Fig. 1 — Modern Diesel Engine for Self-Propelled Combine*

OUO1082,000624B -19-27AUG13-2/2

# ENGINES

Small self-propelled combines may have engines as small as 70 horsepower (52 kW), while larger machines may require 250 horsepower (185 kW) or more. The engines provide power to propel the machine through the field, to drive the harvesting components, and to power the hydraulic systems.

The engine may have four, six, or eight cylinders. Engines are mounted in different positions on different combines — on the top, on the side, or underneath the combine. Some manufacturers position the engine low to obtain a low center of gravity and provide ground level accessibility. Others prefer to mount it high, out of the dust and dirt. This can also provide a better distribution of weight, which is important for good traction.

With the modern diesel engines (Fig. 1), there is no premixing of air outside the cylinder. Air only is taken into the cylinder through the intake manifold and compressed (Fig. 3). Fuel is then sprayed into the cylinder and mixed with air as the piston nears the top of its compression stroke. The heat produced by the compression of air results in ignition.

## *COMPRESSION RATIOS*

Compression ratio compares the volume of air in the cylinder before compression with the volume after compression.

A 16-to-1 ratio is common for diesels (Fig. 2).

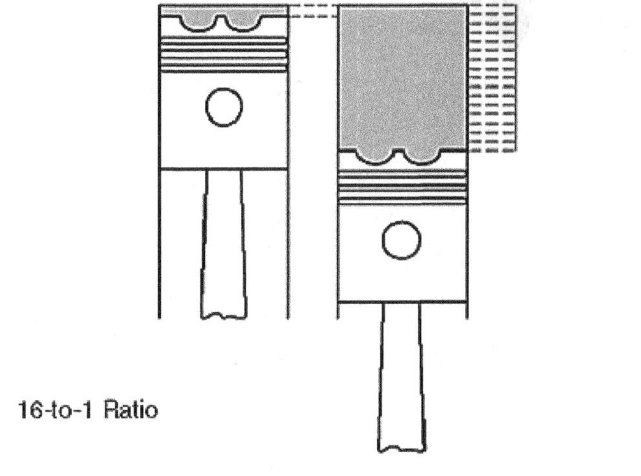


*Fig. 2 — Engine Compression Ratios*

The higher compression ratio of the diesel raises the temperature of the air high enough to ignite the fuel without a spark.

This gives the diesel better efficiency because the higher compression results in greater expansion of gases in the cylinder following combustion. Result: a more powerful stroke.

**Continued on next page**

KN52281,100454B -19-24JUL13-1/2

The higher efficiency that results from diesel combustion must be offset by the need for sturdier, more expensive parts to withstand the greater forces of combustion.

## GRADES AND TYPES OF FUEL

Fuel energy is measured in standard heat units or British Thermal Units (Btu) or kilojoule (kJ) and gives a comparison of the power possible from each fuel.

Diesel fuel has more heat units (Btu) per gallon (kJ/L) than gasoline, and so gives more work per gallon (liter) of fuel. In addition, diesel fuel normally costs less than gasoline.

However, diesel fuel-injection equipment is more expensive than gasoline equipment.

A deciding factor relating to choosing an engine often is how much fuel is consumed per year in the engine operation. The more fuel that is burned, the greater the potential savings with the greater efficiency of the diesel engine.

## OPERATION OF ENGINES

Here is a brief description of how an engine operates. For more information, refer to John Deere's Fundamentals of Service Manual — Engines.

For an engine to operate, a definite series of events must occur in the correct sequence. They are:

1. Fill the cylinder with a combustible mixture.

2. Compress this mixture into a smaller space.

3. Ignite the mixture and cause it to expand, producing power.

4. Remove the burned gases from the cylinder.

The sequence generally is:

- Intake
- Compression
- Power
- Exhaust

To produce sustained power, the engine must repeat this sequence over and over again. One complete series of these events in an engine is called a cycle.

Most engines on modern combines are four-stroke-cycle engines. Four strokes of the piston — two up and two down — are needed to complete the cycle. As a result, the crankshaft will rotate two complete turns while one cycle is being completed.

### INTAKE STROKE

The intake stroke starts with the piston near the top and ends as it reaches the bottom of the cylinder. The intake

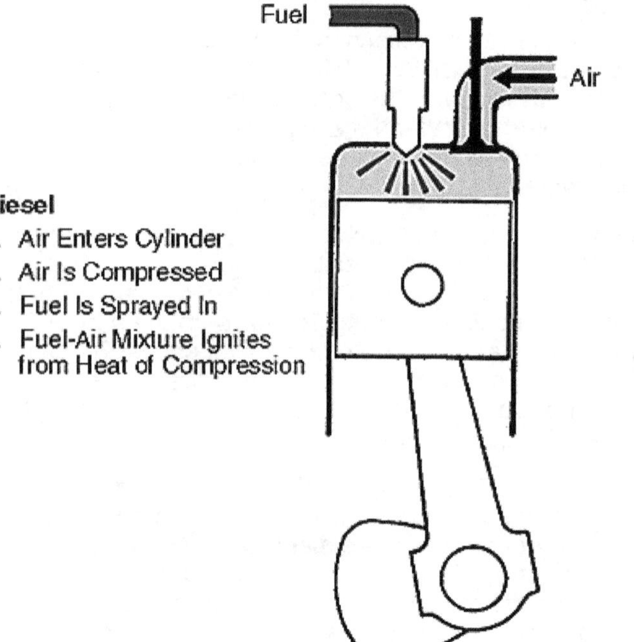


*Fig. 3 — Methods of Supplying and Igniting Fuel*

valve is opened, allowing the cylinder, as the piston moves down, to receive the fuel-air mixture. The valve is then closed, sealing the cylinder.

### COMPRESSION STROKE

The compression stroke begins with the piston at the bottom and rising up to compress the fuel-air mixture. Because the intake and exhaust valves are closed, there is no escape for the fuel-air and it is compressed to a fraction of its original volume.

### POWER STROKE

The power stroke begins as the piston is almost at the top of the cylinder in the compression stroke and the fuel-air mixture is ignited. As the mixture burns and expands, it forces the piston down on its power stroke. The valves remain closed so that all the force is exerted on the piston.

### EXHAUST STROKE

The exhaust stroke begins when the piston nears the end of its power stroke. The exhaust valve is opened and the piston rises, pushing out the burned gases. When the piston reaches the top, the exhaust valve closes and the piston begins a new four-stroke cycle of intake, compression, power, and exhaust. As it completes the cycle, the crankshaft has gone all the way around twice.

KN52281,100454B -19-24JUL13-2/2

# ENGINE SYSTEMS

Here are other systems required for engine operation:

- Fuel system
- Intake and exhaust systems
- Lubricating system
- Cooling system
- Electrical systems
- Governing system

We will now discuss each of these systems.

## ENGINE FUEL SYSTEM

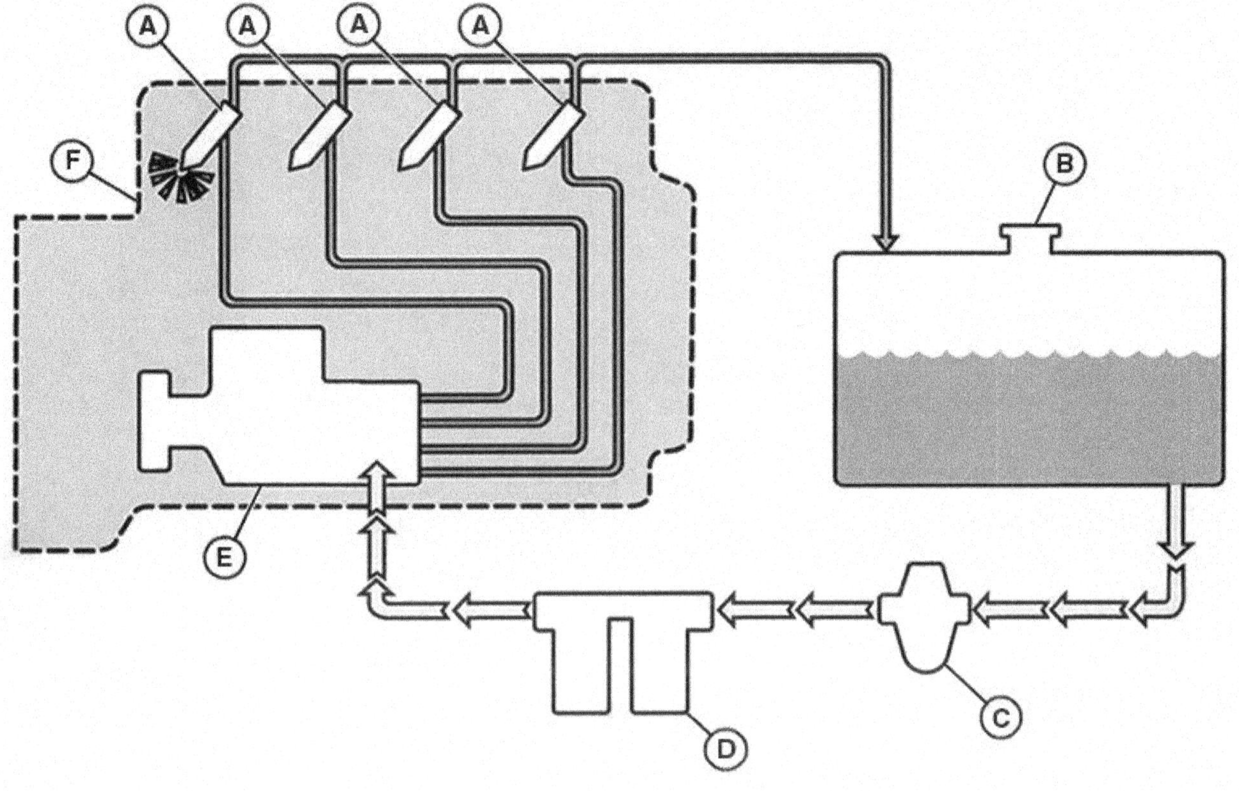


*Fig. 4 — Diesel Fuel System*

A— Injection Nozzle
B— Fuel Tank
C— Fuel Transfer Pump
D— Fuel Filters
E— Injection Pump
F— Combustion Chamber

A fuel system must deliver clean fuel, in the quantity required, to the fuel intake of an engine. It must also provide for safe fuel storage and transfer.

In the diesel fuel system (Fig. 4), fuel is sprayed directly into the engine combustion chamber, where it mixes with hot compressed air and ignites. No electrical spark is used to ignite the mixtures (as in gasoline and LP-gas engines).

Instead of a carburetor, a fuel injection pump and injection nozzles are used.

The major parts of the diesel fuel system are:

- Fuel tank — stores fuel
- Fuel transfer pump — moves fuel to Injection pump
- Fuel filters — help clean the fuel
- Injection pump — times, measures, and delivers fuel under pressure
- Injection nozzles — atomize and spray fuel into cylinders

In operation, the fuel transfer pump moves fuel from the tank and pushes it through the filters. Clean fuel free of water is very vital to the precision parts of the diesel injection system. Extra filters are often used to ensure clean fuel, but buying clean fuel and storing it properly are also prime needs for efficient operation.

The fuel is then pushed on to the injection pump where it is metered, put under high pressure, and delivered to each injection nozzle in turn.

The nozzles each serve one cylinder. They atomize the fuel and spray it under controlled high pressure into the combustion chamber at the proper moment.

High-pressure fuel is needed at each nozzle to get a fine spray of fuel. This atomization of fuel ensures good mixing of fuel with the hot compressed air for full combustion.

**Continued on next page**

KN52281,100454C -19-24JUL13-1/7



PN=71

### *Engine Intake and Exhaust Systems*

The intake system carries the fuel-air mixture into the engine, and the exhaust system removes the exhaust gases after combustion (Fig. 5).

INTAKE SYSTEM

The intake system supplies the engine with clean air of the proper quantity, temperature, and mix for good combustion.

The intake system has four to five parts:

- Air cleaners
- Turbocharger (if used)
- Intake manifold
- Intake valves
- Intercooler (if used)

### Air Cleaners

Air cleaners filter dust and dirt from the air passing through them en route to the intake manifold. Precleaners prevent larger particles from reaching the air cleaner and plugging it.

### Turbocharger

The turbocharger increases power by forcing more air or fuel-air mixture into the engine cylinders.

### Intake Manifold

The intake manifold transports the cleaned air to the engine cylinders.

### Intake Valves

Intake valves admit air to an engine cylinder. They are normally opened and closed by mechanical linkage from the camshaft.

### Intercooler

The intercooler helps increase power, improve fuel efficiency, and reduce engine noise by using engine coolant to reduce the temperature and increase the density of air compressed by the turbocharger.

EXHAUST SYSTEM

The exhaust system collects the exhaust gases after combustion and carries them away.

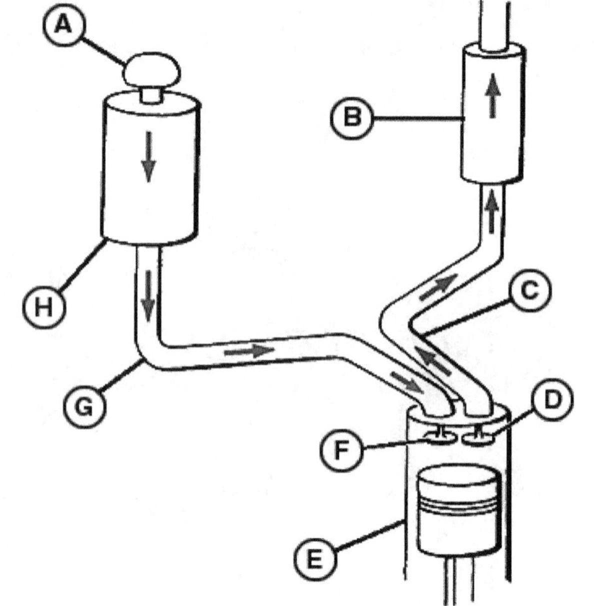

Fig. 5 — Intake and Exhaust Systems

A— Precleaner
B— Muffler
C— Exhaust Manifold
D— Exhaust Valve
E— Cylinder
F— Intake Valve
G— Intake Manifold
H— Air Cleaner

An exhaust system has three basic parts:

- Exhaust valves
- Exhaust manifold
- Muffler

### Exhaust Valves

Exhaust valves open to release the burned gases on four-cycle engines. The valves are normally driven by the camshaft.

### Exhaust Manifold

The exhaust manifold collects the exhaust gases and conducts them away from the cylinder.

### Muffler

The muffler reduces the emission sounds of the engine during the exhaust cycle.

**Continued on next page**                KN52281,100454C -19-24JUL13-2/7

## ENGINE LUBRICATING SYSTEM

The engine lubricating system (Fig. 6) reduces friction, dissipates engine heat, and helps keep the engine parts clean.

The lubricating system contains the following parts:

- Crankcase oil reservoir
- Oil pump
- Oil filter
- Oil passages
- Pressure regulating valve

The crankcase oil reservoir holds the oil that circulates in the engine system. Oil is pushed through the system by the oil pump, and oil is cleaned by the oil filter, which must be either cleaned or replaced periodically. Oil reaches engine parts through oil passages as shown in Fig. 6. Pressure of oil is controlled by the pressure regulating valve.

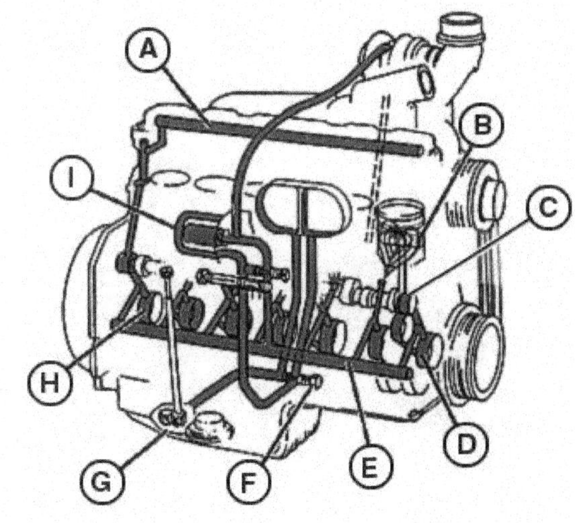

Fig. 6 — Engine Lubricating System

A— Tappet Lever Shaft
B— Piston Pin Bushings
C— Camshaft Bushings
D— Connecting Rod Bearings
E— Main Oil Gallery
F— Pressure Regulating Valve
G— Oil Pump
H— Crankshaft Main Bearings
I— Oil Filter

**Continued on next page**

KN52281,100454C -19-24JUL13-3/7

## ENGINE COOLING SYSTEM

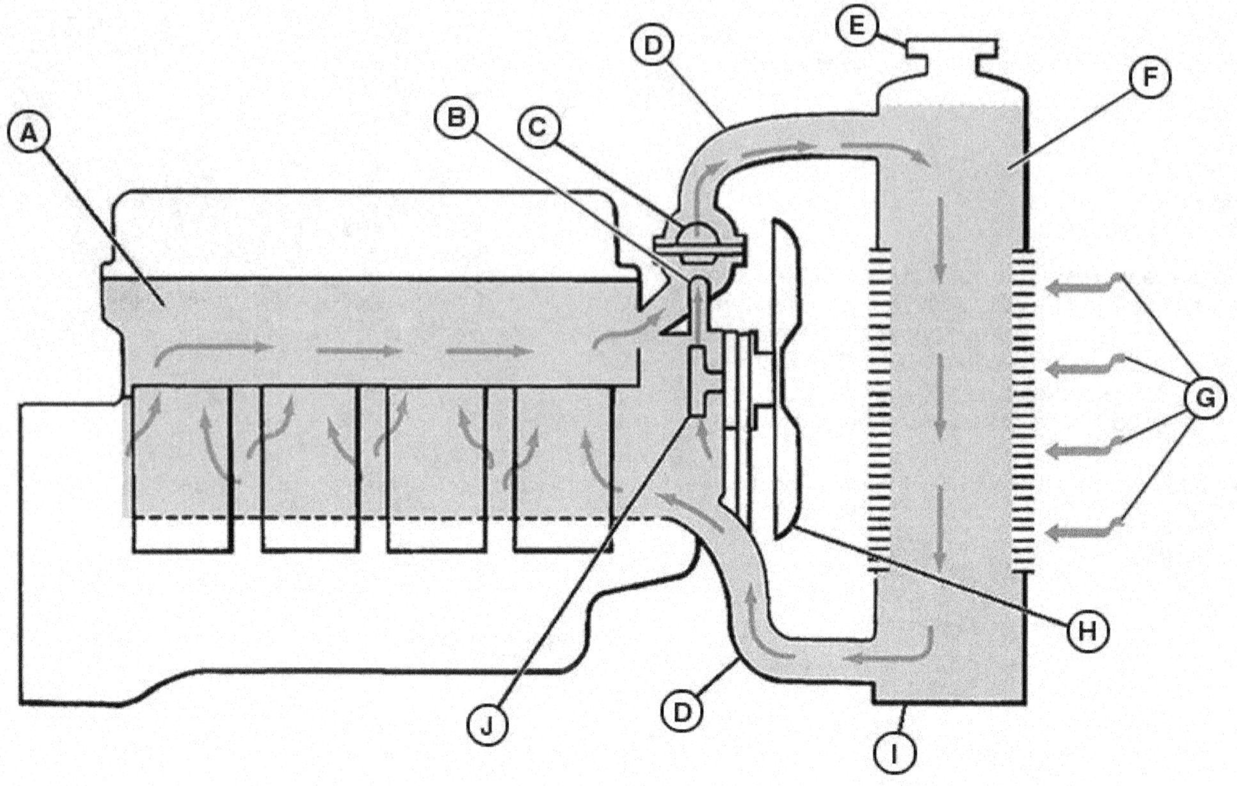


*Fig. 7 — Engine Cooling System*

A— Engine Coolant Jacket
B— Bypass
C— Thermostat
D— Hose
E— Pressure Cap
F— Coolant
G— Air Flow
H— Fan
I— Radiator
J— Water Pump

### Engine Coolant Jacket

Allows coolant to circulate around cylinders and head.

### Thermostat

If coolant is cold - closes to circulate in engine only.

If Coolant hot - opens to circulate coolant to radiator for cooling.

### Bypass

Allows coolant to circulate throughout engine when thermostat is closed.

### Pressure Cap

Holds pressure in system and releases excessive pressure.

### Air Flow

Removes heat from coolant as it passes through radiator.

### Fan

Device that generates air flow for engine cooling.

### Radiator

Reservoir and heat exchanger for coolant.

### Hose

Connects radiator to engine.

### Water Pump

Circulates the coolant throughout engine and radiator.

### Coolant

Contains a specified ratio of water, antifreeze, and inhibitors.

### EXHAUST SYSTEM

The exhaust system collects the exhaust gases after combustion and carries them away.

The cooling system prevents overheating of the engine and regulates its temperature at best levels.

Continued on next page

KN52281,100454C -19-24JUL13-4/7

The cooling system (Fig. 7) normally uses water as a coolant. In cold weather, antifreeze solutions are added to the water to prevent freezing. The coolant circulates in a jacket around the cylinders and through the cylinder head.

As heat radiates, it is absorbed by the coolant, which then flows to the radiator. Air flow through the radiator cools the coolant and dissipates heat into the air. The coolant then recirculates into the engine to pick up more heat.

**Continued on next page**

KN52281,100454C -19-24JUL13-5/7

## ENGINE ELECTRICAL SYSTEMS

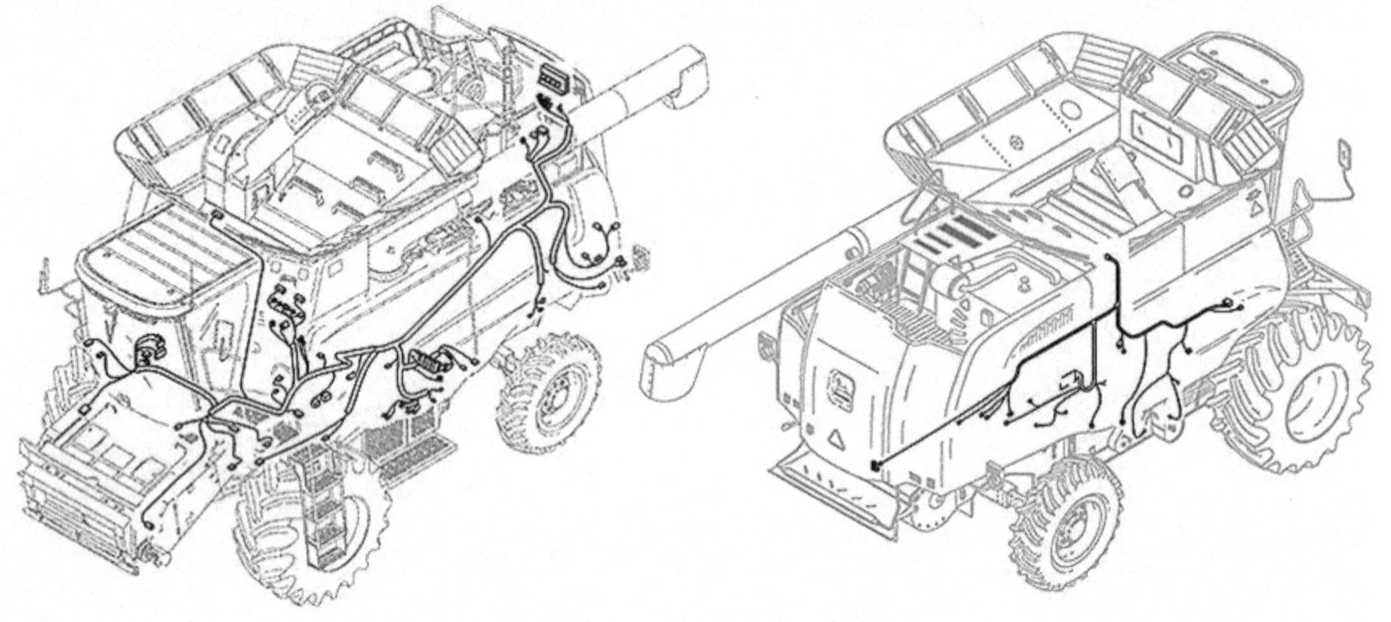

Fig. 8 — Typical Electrical System on a Modern Combine

The design of the electrical systems for a combine is similar to that for other types of farm equipment.

The complete engine electrical system (Fig. 8) consists of the following circuits:

- Starting circuit
- Charging circuit

Here is a brief look at these circuits.

### CHARGING CIRCUIT

The charging circuit recharges the battery and generates current during engine operation. It has these parts:

- Battery
- Voltage regulator
- Alternator

The alternator is really an AC generator. Like the generator, it produces alternating current, but rectifies it electronically using diodes. Alternators are generally more compact than generators of equal output, and supply a higher current output at low engine speeds. They have replaced generators on most modern engines.

The voltage regulator in AC charging circuits limits the alternator voltage to a safe, preset level. Transistorized models are used in many of the modern circuits.

### STARTING CIRCUIT

The starting circuit converts electrical energy from the battery into mechanical energy at the starting motor to crank the engine.

A basic starting circuit has four parts:

- The battery supplies energy for the circuit.
- The starter switch actuates the circuit.

- The motor switch (solenoid) engages the motor drive with the engine flywheel.
- The starting motor drives the flywheel to crank the engine.

### ELECTRICAL POWER FOR ACCESSORIES

The battery that supplies power for the engine also supplies power for the accessories of a modern combine. Many years ago, on engine-driven combines, the only electricity required was to start and run the engine. Now modern combines require much more electrical power to operate lights, air conditioning, radios, horns, instruments, and electrical clutches.

Complex electrical systems are required to operate today's combines. A typical electrical system for a self-propelled combine is shown in Fig. 8. Electrical power for pull-type combines is supplied by the tractor's electrical system.

## ENGINE GOVERNING SYSTEM

The governing system keeps the engine speed at a constant level. It does this by varying the amount of fuel or fuel-air mixture supplied to the engine, according to the demands of the load.

For earlier diesel engines, the level of engine speed was controlled by the position of the speed control lever, connected by linkage to the governor.

The function of the governor is to get the engine's power to match the load at all times to keep the speed at a steady level.

On modern diesel engines, the engine speed is controlled electronically by a microprocessor-based computer. The computer senses inputs from various points on the engine and the combine to best match speed to load.

Continued on next page

KN52281,100454C -19-24JUL13-6/7



PN=76

Governors can be either mechanical, hydraulic, electrical, or electronic.

KN52281,100454C -19-24JUL13-7/7

# ENGINE GEAR CASE OPERATION

The engine gear case/steering pump (Fig. 9) draws pressure-free oil from the gear case sump. Regulated pressure is maintained by a regulator valve. Regulated pressure oil flows from the pump to the steering control valve. Regulated pressure oil flows from the steering system, through the engine gear case filter, and into the control valve. Oil from the regulator valve flows through the oil cooler.

Lubrication pressure is maintained by a lube pressure valve. The filter bypass valve allows continued lube oil flow if the filter is plugged. An oil cooler bypass valve allows oil flow to continue to the lube circuit if oil flow is restricted through the cooler.

The unloading auger engage cylinder has pressure-free oil when disengaged. When engaged, regulated pressure oil is sent to the unloading auger engage cylinder.

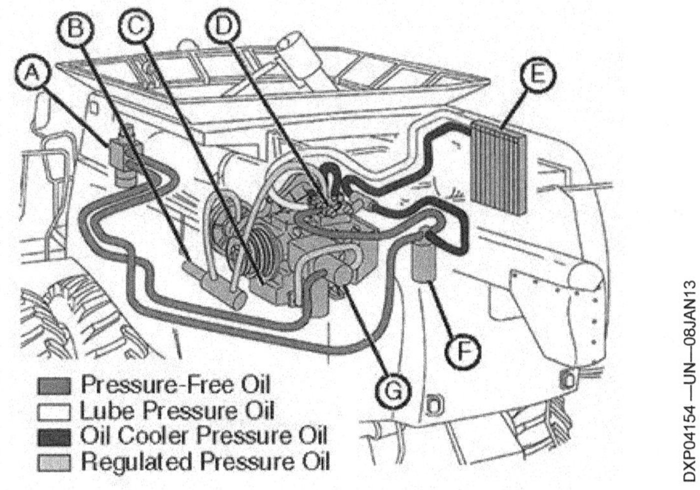


*Fig. 9 — Engine Gear Case Operation*

A—Steering Control Valve
B—Unloading Auger Engage Cylinder
C—Gear Case Sump
D—Gear Case Control Valve
E—Oil Cooler
F—Gear Case Filter
G—Gear Case/Steering Pump

KN52281,100454D -19-08JAN13-1/1

## TRANSMITTING ENGINE POWER

Most combines have a main or primary countershaft (driven by the engine) from which the various units of the combine are driven. This shaft is the prime shaft for operating the components of the combine at the proper speed. The engine must operate at the speed specified by the manufacturer to make this shaft run at the proper speed. Before making any adjustments, the countershaft speed must be checked. If the primary countershaft is not run at the designated speed, the components of the combine will not function efficiently. This may cause poor cutting, feeding, threshing, separating, or cleaning. All of these will affect the capacity of the combine.

Most combines have three basic drive systems:

- Propulsion drives
- Header drives
- Separator drives

Each system may have one or more drives to provide power for its components.

### PROPULSION DRIVES

The propulsion drive transmits power from the engine to the power train to propel the combine. Nearly all modern combines are equipped with a hydrostatic drive system that consists of a drive pump driven from the engine and a hydrostatic motor mounted on the transmission. The operator controls movement of the combine by selecting the proper gear and moving the control lever in the desired direction.

A track-type drive may be used instead of wheels for very soft or muddy fields, particularly when harvesting rice. On some combines, track and wheel drives are interchangeable, but other models must be ordered from the factory with one or the other type of drive. Power flow to either tracks or wheels is essentially the same. Later in this chapter, power trains will be discussed in more detail.

### HEADER DRIVES

The header in this example is driven by the primary drive shaft (Fig. 11). The primary drive shaft is connected at one end to the engine gear case. The opposite end of the drive shaft connects to the primary countershaft gear case. An electromagnetic clutch is mounted to one of the output shafts of the primary countershaft gear case. Power is transmitted by a belt from the electromagnetic clutch to the front of the feeder house to drive the header. The other output of the gear case is the primary countershaft. The countershaft extends to the right-hand side of the machine to power many of the separator functions.

FEEDER CONVEYOR AND HEADER DRIVE

The overall header drive system is shown in Fig. 11, while a typical feeder conveyor drive is shown in Fig. 10. The feeder conveyor in this example is driven from the header

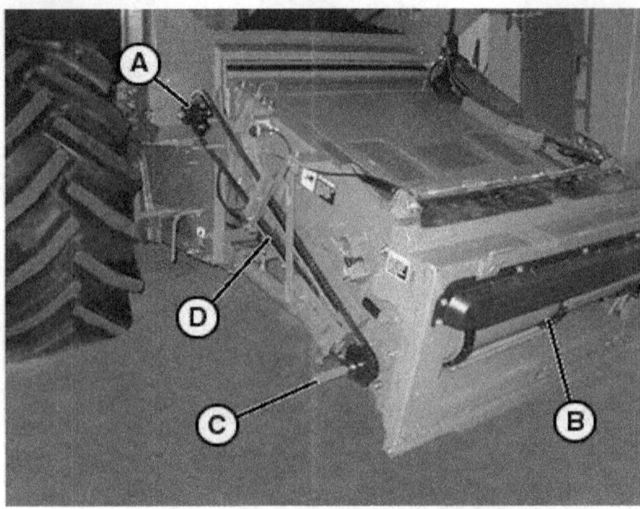

Fig. 10 — Feeder Conveyor Drive

A— Conveyor Top Shaft and Slip Clutch
B— Conveyor
C— Header Drive Shaft
D— Drive Chain

drive shaft on the bottom of the feeder house. When the header drive is engaged, power for the conveyor comes to the right-hand side of the machine from the header drive shaft. A sprocket on the drive shaft transmits power to the top shaft and slip clutch which drive the conveyor.

Continued on next page

KN52281,100454E -19-31JUL13-1/12



PN=78

In some conditions, the speed of the header drive must be changed to gather the crop at different rates, depending on crop condition and forward speed of the combine. Some manufacturers provide different sizes of sheaves to change speeds; others make variable-diameter sheaves available.

On machines where the platform or corn head is detachable from the feeder conveyor unit (Fig. 10), the header drive shaft is coupled to the drive shaft on the platform or corn head. On machines where the platform or corn head is permanently attached to the feeder conveyor unit, the shaft is usually one piece.

CUTTING PLATFORM DRIVE

**A— Feeder Conveyor**
**B— Drive Tightener**
**C— Drive from Main**

**D— Header Drive Sheave**
**E— Header Drive Shaft**

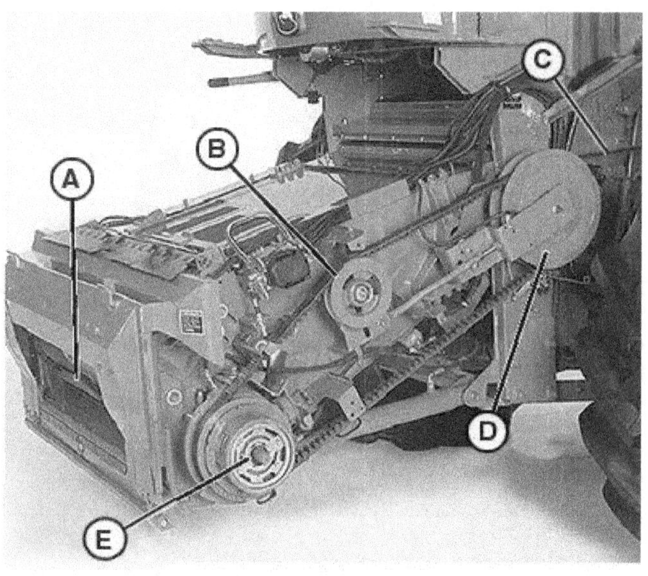

Fig. 11 — Header Drive

KN52281,100454E -19-31JUL13-2/12

As mentioned earlier, the platform is driven by the header drive shaft (Fig. 11). The cutterbar drive sheave and auger drive sprocket are mounted at the end of the header drive shaft (Fig. 12). A belt connects the cutterbar drive sheave to the cutterbar drive gearbox. The output of the cutterbar drive gearbox is connected to the platform cutterbar. The auger drive sprocket is connected to the auger-driven sprocket by a chain. The auger drive is protected by a slip clutch that is attached to the auger driven sprocket.

If the load on the auger exceeds a set limit, the slip clutch will slip and prevent damage to the auger. Different sizes of sprockets are usually available to vary the speed of the auger.

**A— Cutterbar Drive Belt**
**B— Slip Clutch**

**C— Header Drive Shaft**

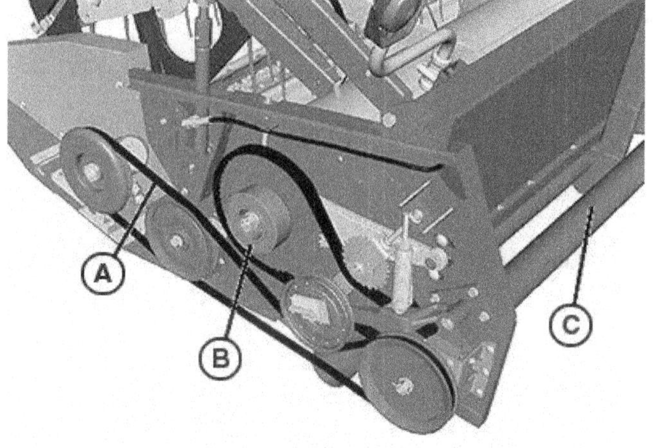

Fig. 12 — Cutterbar Drive

KN52281,100454E -19-31JUL13-3/12

The reel drive (Fig. 13) on many platforms consists of a hydraulic motor that is mounted on the reel lift arm at the end of the reel. The output shaft of the reel drive motor is connected to the reel itself by a set of spur gears. The reel speed is adjustable from inside the operator's station.

Fig. 13 — Hydraulic Reel Drive

**Continued on next page**

KN52281,100454E -19-31JUL13-4/12

The draper belt drive on rice platforms (Fig. 14) is usually driven by a hydraulic motor, with a dedicated hydraulic motor for each belt on the platform. Draper belt speed can be controlled by a switch on the armrest in the operator's station.

The cutterbar drive on the draper platform is similar to that of the cutting platform drive discussed earlier. A drive sheave is mounted on the end of a drive shaft and is connected to the cutterbar drive gearbox by a belt. The feed drum, located in the center of the draper platform, is also driven by a drive shaft. The main header drive shaft connects to a gearbox that transmits the power to two output shafts, one to drive the cutterbar and the other to power the feed drum.

The reel drive can consist of a hydraulic motor and chain drive. Another motor drives the reel from the other side of the platform. The motors are connected in series so that each is turning the reel at the same speed as the other.

## PICKUP PLATFORM DRIVE

The belts for the pickup platform are usually driven by hydraulic motors (Fig. 15). Some pickup platforms are equipped with one belt, while others may have two. Belt speed can be controlled by a switch on the armrest in the operator's station. The auger drive of a pickup platform is similar to that of the cutting platform. The auger is driven by a chain that is connected to a sprocket mounted at the end of the header drive shaft. A slip clutch is used to protect the auger from damage.

**A— Belt Pickup Drive Motor**     **B— Auger Drive**

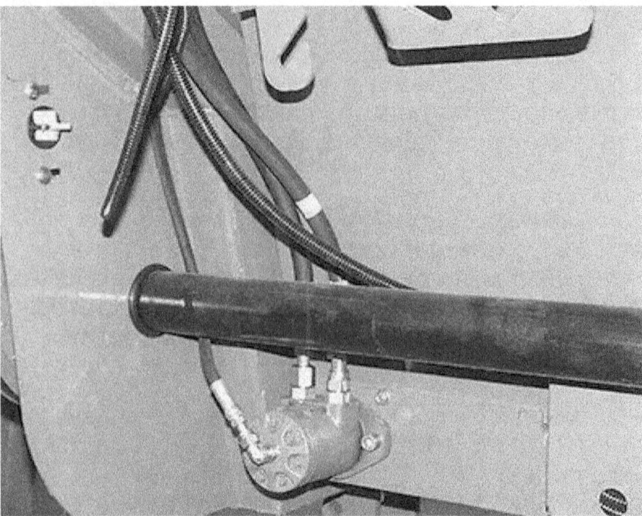

Fig. 14 — Draper Drive and Pickup Reel Drive

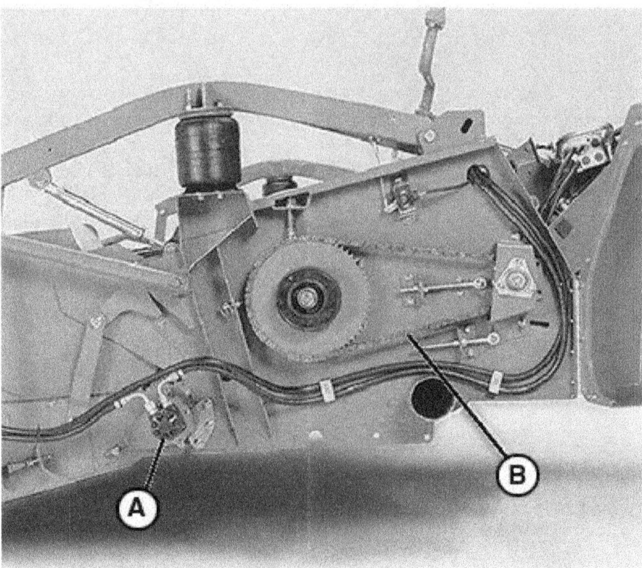

Fig. 15 — Pickup Drive System

**Continued on next page**     KN52281,100454E -19-31JUL13-5/12



050217
PN=80

## CORN HEAD DRIVE

The corn head is driven by the header drive shaft (see Figs. 10 and 16). At the outer end of the header drive shaft is a chain drive, which delivers power to the corn head main drive shaft. This main drive shaft drives the gear units for each row of the corn head. Both the gathering chains and stalk rolls are driven in turn by the gear units.

## *SEPARATOR DRIVES*

All harvesting components of the combine are driven from the primary drive shaft, primary countershaft, secondary countershaft, engine gear case, or a combination of these. The rotary combine shown in Figs. 19 and 20 is equipped with two jackshaft assemblies (front and rear) on the right-hand side of the machine. Power comes from the left-hand side of the machine to the right-hand side via the primary countershaft. The primary countershaft in turn drives each of the jackshaft assemblies, which transmit power to many of the separator functions.

On self-propelled combines, all separator drives (except the unloading auger drive) are controlled by one engage switch in the operator's station. When the switch is activated, the separator drive clutch in the engine gear case is engaged and power is distributed to the various separator functions.

Here are the harvesting drives listed by function:

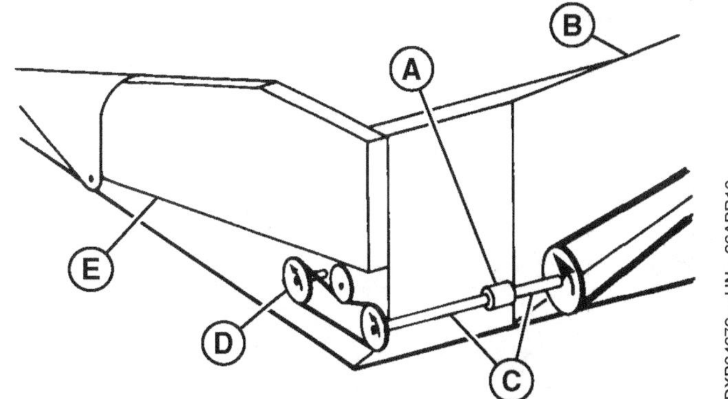

Fig. 16 — Corn Head Drive

A— **Drive Shaft Coupler**  
B— **Feeder Conveyor**  
C— **Header Drive Shaft**  
D— **Corn Head Main Drive Shaft**  
E— **Corn Head**

| Drive Type | Function |
|---|---|
| Threshing Drive | • Threshing Cylinder or Rotor |
| Separating Drives | • Beater |
| | • Straw Walkers |
| Cleaning Drives | • Shoe Supply Conveyor |
| | • Shoe |
| | • Fan |
| Grain Handling Drives | • Clean Grain Elevator and Augers |
| | • Tailings Elevator and Augers |
| | • Unloading Auger |
| Straw Disposal Drives | • Straw Chopper |
| | • Straw Spreader |

The drives shown in Figs. 17 through 20 are simplified for clarity.

Study Figs. 17 through 20 to learn the operation of these drives.

Continued on next page

KN52281,100454E -19-31JUL13-6/12

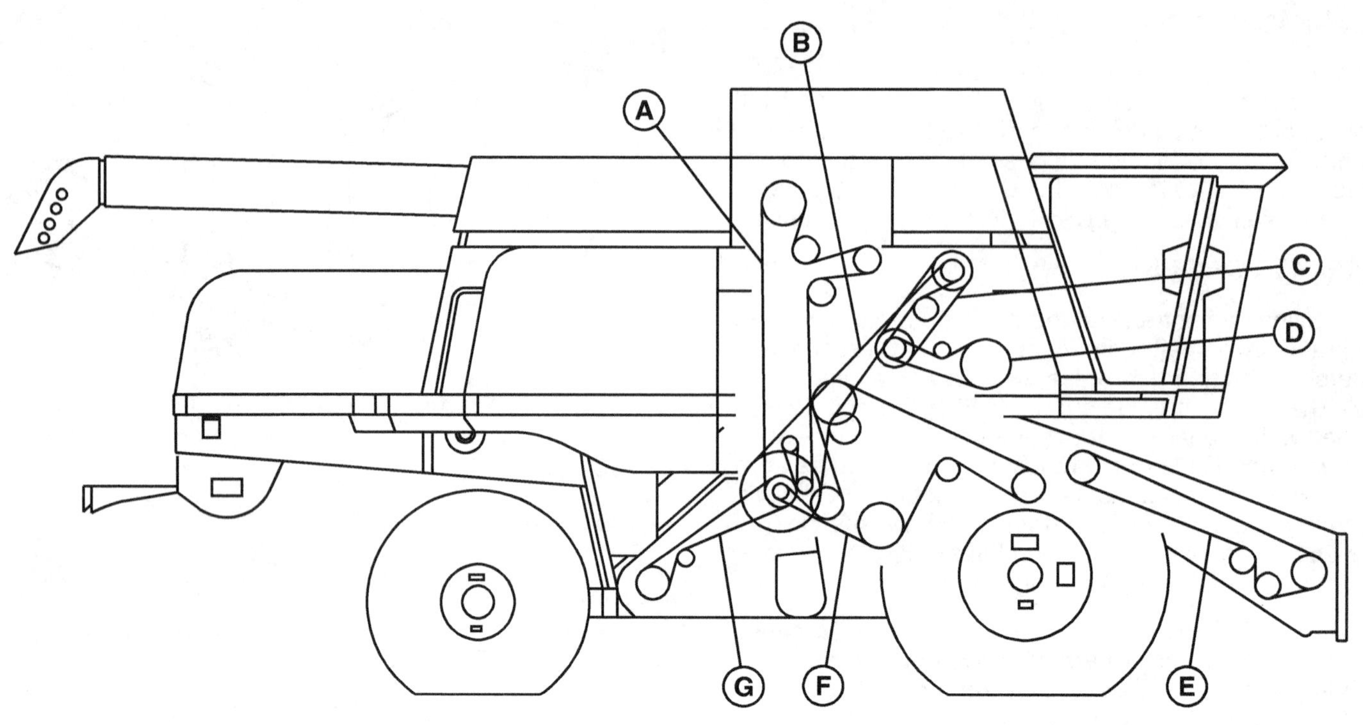

Fig. 17 — Drives on Right-Hand Side of Cylinder Walker Separation (CWS) Combine

A— Clean Grain Elevator Drive Belt
B— Secondary Countershaft Drive Belt
C— Cylinder Intermediate Drive Belt
D— Cylinder Drive Belt
E— Feeder Conveyor Drive Belt
F— Straw Walker, Shoe, and Conveyor Auger Drive
G— Lower Tailings Auger and Elevator Drive Belt

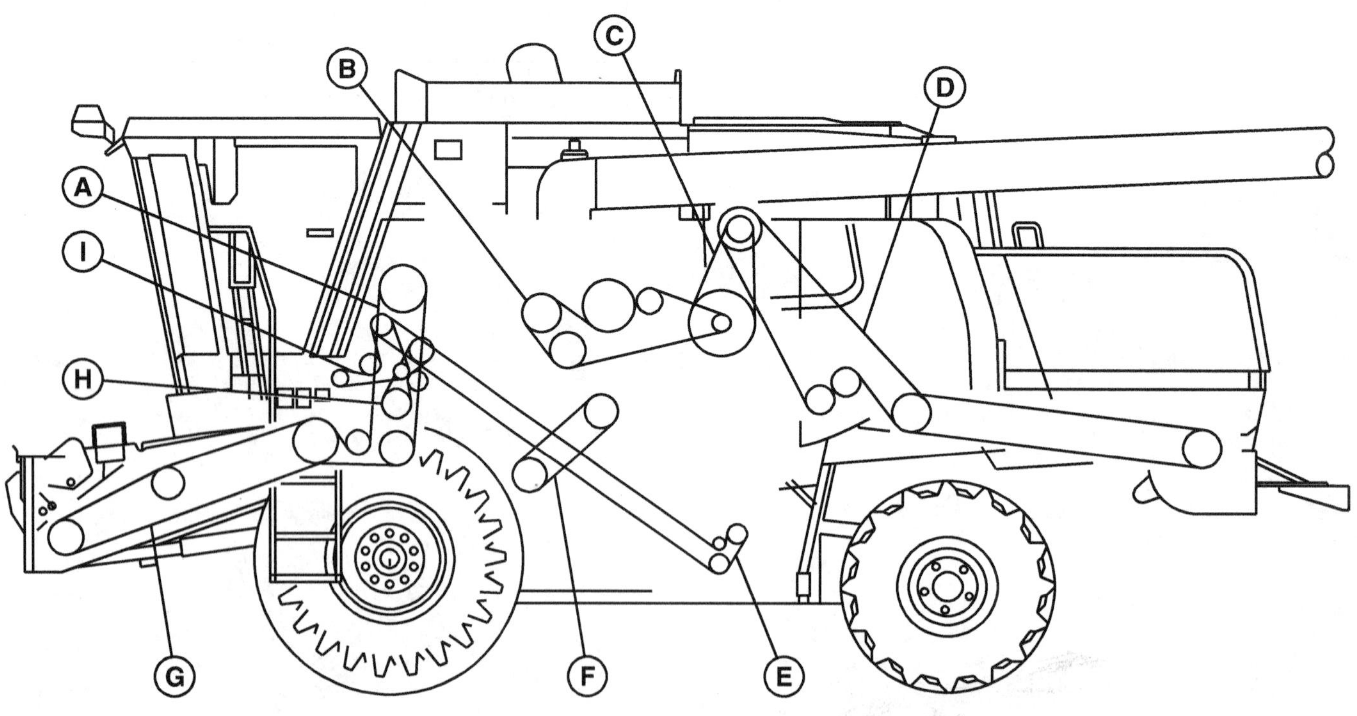

Fig. 18 — Drives on Left-Hand Side of Cylinder Walker Separation (CWS) Combine

Continued on next page

KN52281,100454E -19-31JUL13-7/12



PN=82

A— Header and Reel Pump Drive Belt
B— Unloading Auger System Drive Chain
C— Unloading Auger System Drive Belt
D— Chopper Drive Belt
E— Lower Tailings Drive Chain
F— Cleaning Fan Drive Belt
G— Header Drive Belt
H— Beater Drive Belt
I— Upper Tailings Auger Drive Chain

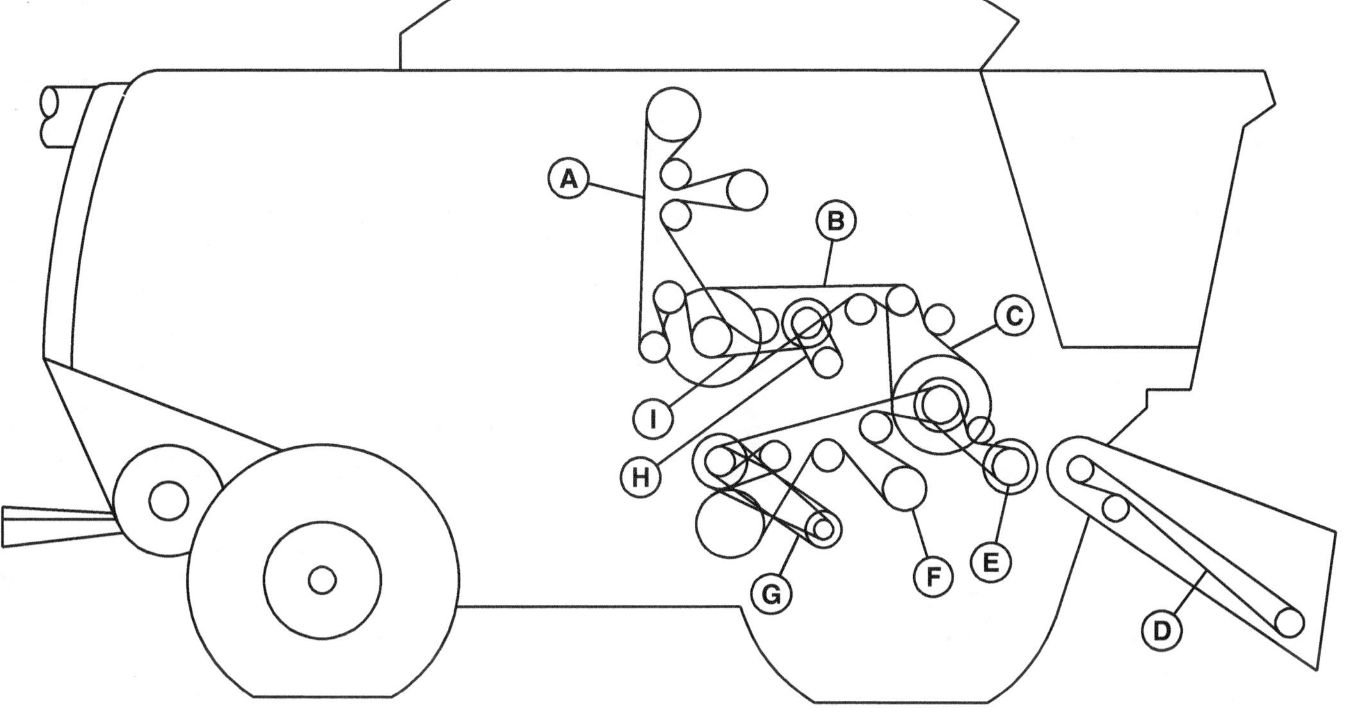

*Fig. 19 — Drives on Right-Hand Side of Typical Single Tine Separation (STS) Combine*

A— Clean Grain Elevator Belt
B— Rear Right-Hand Jackshaft Belt
C— Front Right-Hand Jackshaft Belt
D— Feeder Conveyor Chain
E— Feed Accelerator Belt
F— Shoe Fan and Conveyor Augers Belt
G— Cleaning Fan Drive Belt
H— Tailings Return Auger Chain
I— Tailings Auger and Return Belt

Continued on next page

KN52281,100454E -19-31JUL13-8/12

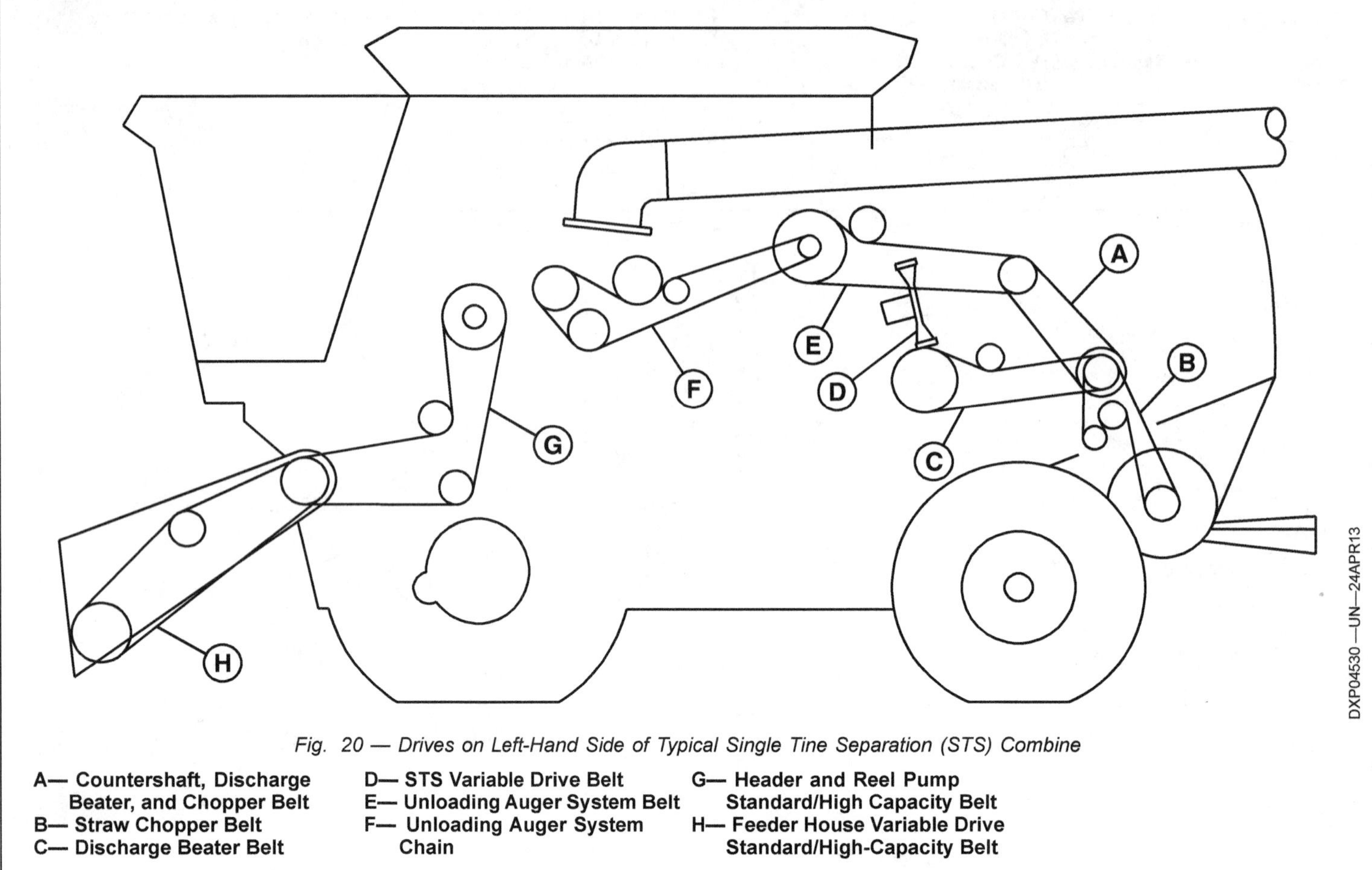

Fig. 20 — Drives on Left-Hand Side of Typical Single Tine Separation (STS) Combine

A— Countershaft, Discharge Beater, and Chopper Belt
B— Straw Chopper Belt
C— Discharge Beater Belt
D— STS Variable Drive Belt
E— Unloading Auger System Belt
F— Unloading Auger System Chain
G— Header and Reel Pump Standard/High Capacity Belt
H— Feeder House Variable Drive Standard/High-Capacity Belt

**Continued on next page**

KN52281,100454E -19-31JUL13-9/12

## THRESHING CYLINDER DRIVE

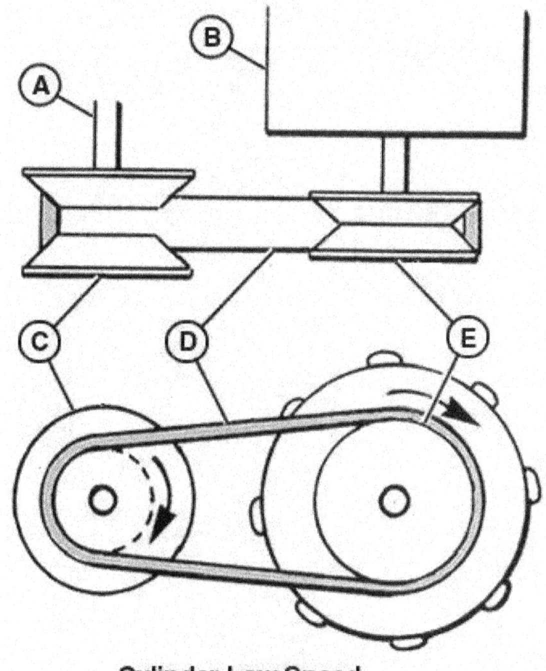


*Fig. 21 — Cylinder Drive Sheaves*

A— Primary Countershaft

B— Cylinder
C— Drive Sheaves

D— Drive Belt
E— Cylinder Sheaves

The threshing cylinder in a conventional combine (Fig. 17) is usually driven by the primary countershaft. In order to work the combine in varying crop conditions, the operator must be able to adjust the threshing cylinder speed. One method is to vary the effective diameter of the sheave that drives the cylinder. A split sheave half can be pulled toward or pushed away from a fixed half. Sheaves that are closed cause the drive belt to ride higher and create more speed and less torque on a smaller threshing cylinder sheave (Fig. 21).

A second way to control the threshing cylinder speed is to use a gear assembly on the cylinder shaft. In this example, the cylinder speed and torque depend upon the gear ratios that are chosen. Both the split sheave and gearbox methods can be used simultaneously.

**Continued on next page**

KN52281,100454E -19-31JUL13-10/12

## ROTOR DRIVE

In a rotary combine, the rotor is often driven from the engine gear case (Fig. 22). A variable drive sheave is mounted on one of the outputs of the engine gear case. The variable drive sheave is connected by a belt to the rotor driven sheave, mounted on the rotor drive gear case. The rotor gear case allows for high/low range operation. Speed is variable within each speed range of the rotor drive gear case by increasing or decreasing the hydraulic pressure to the variable drive sheave, which effectively changes the diameter of the variable drive sheave. Closing the sheave halves, bringing them closer together, effectively increases the diameter of the drive sheave and increases rotor speed. Opening the sheave halves, moving them farther apart, decreases the diameter of the drive sheave and decreases rotor speed.

## BEATER DRIVE

The beater is usually driven from the cylinder shaft so that the beater rotates either faster or slower as the speed of the cylinder is increased or decreased (Fig. 18). Thus, the speed of the beater is controlled by the speed of the cylinder.

## STRAW WALKER DRIVE

On some combines the straw walkers are driven from the secondary countershaft (Fig. 17). The speed of the walkers remains constant because the secondary countershaft is driven at a fixed speed by the primary countershaft. Straw walkers are also driven through a gearbox. This eliminates sprockets and chains.

## CLEANING SHOE SUPPLY CONVEYOR DRIVES

In the example shown of a conventional walker combine (Fig. 17), the conveyor augers are driven from the secondary countershaft. In the rotary combine example (Fig. 19), the conveyor augers are driven by the front countershaft on the right-hand side of the machine.

*Fig. 22 — Rotor Drive*

## SHOE DRIVE

On each machine (Figs. 17 and 19), the shoe arms are connected to a rotating shaft that is attached to the shoe-driven sheave. The shoe arms on each side of the machine contain an eccentric. When the shaft rotates, the eccentric in the shoe arms causes the shaking action.

**Continued on next page**

KN52281,100454E -19-31JUL13-11/12

## FAN DRIVE

Most combines are equipped with a variable-speed fan drive that provides infinite speed control to clean various crops under different conditions (Fig. 23). The variable sheaves are similar to those used on the cylinder drive.

### CLEAN GRAIN ELEVATOR AND AUGERS

In the example shown of a conventional walker combine (Fig. 17), the conveyor augers are driven from the secondary countershaft. In the rotary combine example (Fig. 19), the conveyor augers are driven by the rear countershaft on the right-hand side of the machine. Here the elevator is driven from the top, while the conveyor chain inside the elevator drives the lower auger. The upper auger is driven by a separate drive from the top elevator shaft.

## TAILINGS ELEVATOR AND AUGERS

In the example shown of a conventional walker combine (Fig. 17), the conveyor augers are driven from the secondary countershaft. In the rotary combine example (Fig. 19), the conveyor augers are driven by the rear countershaft on the right-hand side of the machine. The elevator is driven through the lower tailings auger, while the upper tailings auger is driven by the elevator conveyor chain.

## UNLOADING AUGER DRIVE

The unloading auger drive is driven from a sheave on the engine gear case. When the unloading switch in the cab

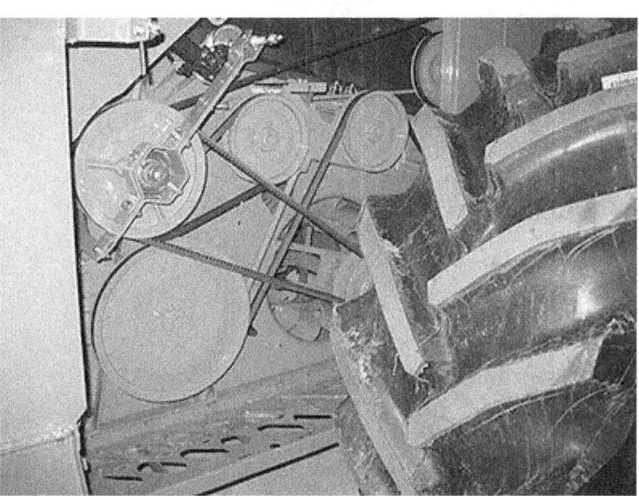

Fig. 23 — Typical Variable-Speed Fan Drive

is pressed, hydraulic oil from a control valve on the engine gear case extends a hydraulic cylinder that tensions the drive belt, which engages the unloading auger system. This permits unloading the tank without operating the entire machine.

## STRAW CHOPPER DRIVE

The straw chopper is driven from a sheave on the engine gear case.

KN52281,100454E -19-31JUL13-12/12



050217<br>PN=87

# POWER TRAINS

On self-propelled combines, engine power is transmitted to the drive wheels or tracks by the hydrostatic system and other power train components such as the transmission and final drive.

Together, the power train provides the following basic functions:

- Connects and disconnects power
- Selects speed ratios
- Provides a means of reversing
- Equalizes power to drive wheels for turning

To perform these functions, five basic parts are needed:

- Transmission — to select speed and direction
- Differential — to equalize power for turning
- Final drives — to reduce speed and increase torque to axles
- Drive wheels or tracks — to propel the machine
- Hydrostatic drive pump and motor — to transmit engine power to the transmission using hydraulic power

We will now discuss these parts in more detail.

### TRANSMISSION

The transmission has a train of gears that transfers and adapts the engine power to the drive wheels of the combine.

The transmission has two functions:

- Selects speed ratios for various travel speeds
- Reverses travel of the combine

A typical combine transmission allows ground speeds from less than 1 mph (0.6 km/h) to nearly 20 mph (32 km/h). Refer to the following chart for speed ranges in each gear for a typical combine.

| Gear | Ground Speed | |
|---|---|---|
| | **MPH** | **km/h** |
| 1st | 0.8 to 2.0 | 1.3 to 3.2 |
| 2nd | 1.7 to 4.5 | 2.7 to 7.2 |
| 3rd | 3.0 to 9.0 | 4.8 to 14.5 |
| 4th | 7.5 to 19.0 | 12.1 to 30.6 |
| Reverse | 1.5 to 3.5 | 2.4 to 5.6 |

Normally, the transmission is mounted on the rear of the drive axle (Fig. 24). The differential is housed in the same housing as the transmission gears, and the clutch housing is attached. The transmission may have three or four forward gear speeds.

Combines are usually equipped with either of these types of transmissions:

- Sliding gear
- Collar shift

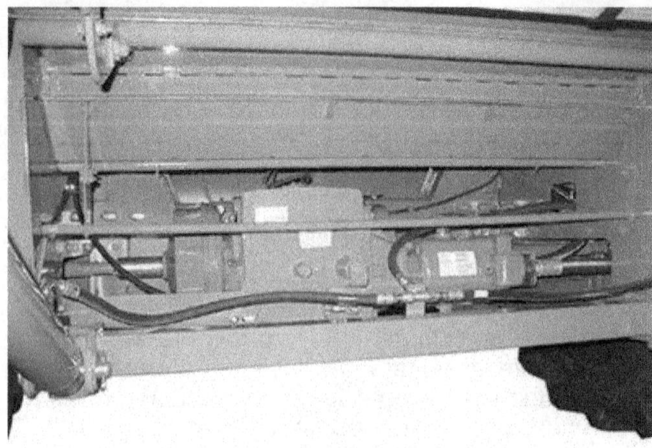

*Fig. 24 — Transmission on Typical Combine*

The sliding gear transmission has two or more shafts mounted in parallel or in line, with sliding spur gears arranged to mesh with each other and provide a change in speed or direction.

Continued on next page

KN52281,100454F -19-24JUL13-1/16



050217
PN=88

The collar shift transmission (Fig. 25) has parallel shafts with gears in constant mesh. Shifting is done by using sliding collars to lock free-running gears to their shafts. Gears with helical gear teeth are used, which gives quieter operation.

## DIFFERENTIAL

The differential has two functions:

- Transmits power "around the corner" to the drive axles.
- Allows each drive wheel to rotate at a different speed and still propel its own load.

The ring gear and bevel gears direct the power to the axles, while the bevel pinions give the differential action (Fig. 26).

A— Differential Ring Gear    D— Shifter Mechanism
B— Countershaft    E— Pinion Shaft
C— Input Shaft    F— Shifter Collars

Fig. 25 — Collar Shift Transmission in Combine

Continued on next page

KN52281,100454F -19-24JUL13-2/16

When the combine is moving straight ahead, both wheels are free to rotate as shown in Fig. 26.

Engine power comes in on the pinion gear and rotates the ring gear. The four bevel pinions and the two bevel gears are carried around by the ring gear, and all gears rotate as one unit. Each axle receives the same rotation and so each wheel turns at the same speed.

When the machine turns a sharp corner, only one wheel is free to rotate as shown in Fig. 26.

Again, engine power comes in on the pinion gear and rotates the ring gear, carrying the bevel pinions around with it. However, the right-hand axle is held stationary and so the bevel pinions are forced to rotate on their own axis and "walk around" the right-hand bevel gear.

A— Bevel Pinions  
B— Bevel Gears and Axle  
C— Ring Gear

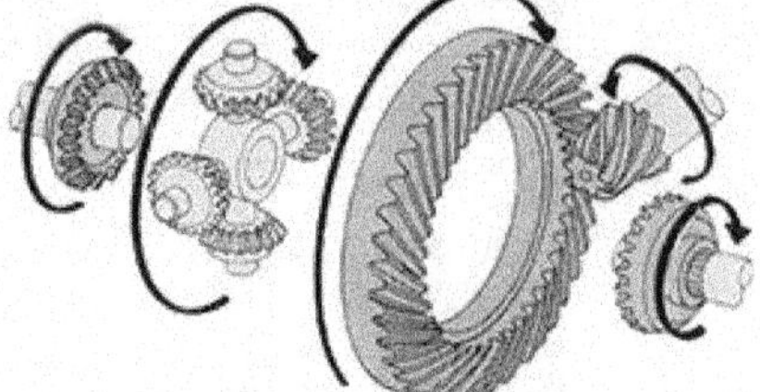

**Both Wheels Free to Rotate**

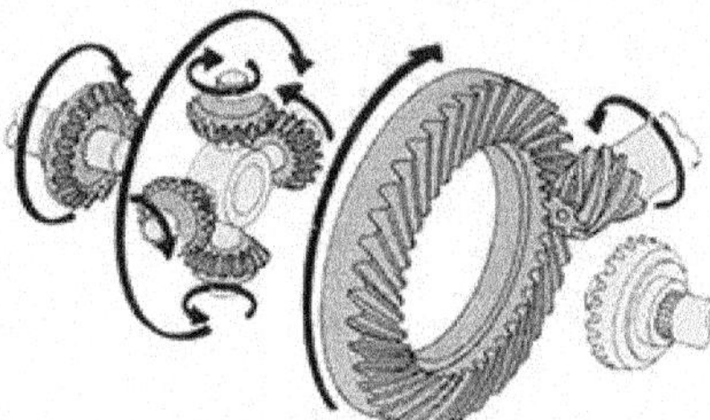

**One Wheel Free to Rotate**

*Fig. 26 — Differential in Operation*

**Continued on next page**

DXP04533 —UN—24APR13

KN52281,100454F -19-24JUL13-3/16



050217  
PN=90

## *FINAL DRIVES*

Final drives, as the last components of the power train, give the final reduction in speed and increase in torque to the drive wheels.

On most combines, the final drives are mounted near the drive wheels (Fig. 27) to avoid the stress of long axle shafts.

By reducing speeds, the final drives lower the stress and simplify the transmission, since extra gears and shafts can be eliminated.

Most final drives must support the weight of the machine as well as withstanding torque and shock loads.

There are two types of final drive axles:

- Rigid axle shaft
- Flexible axle shaft

**A— Final Drive**

*Fig. 27 — Final Drive*

KN52281,100454F -19-24JUL13-4/16

The rigid axle shaft (Fig. 28) is connected to the differential output by a splined coupling. Level-land combines use this type.

**A— Final Drive**      **C— Splined Couplings**
**B— Transmission and**
**   Differential**

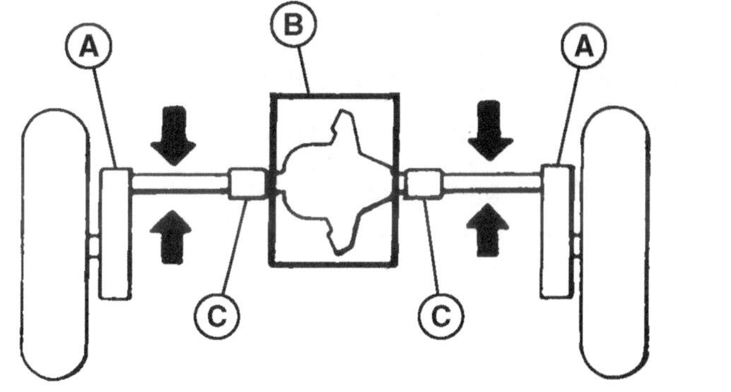


*Fig. 28 — Rigid Axle Shaft*

**Continued on next page**                    KN52281,100454F -19-24JUL13-5/16

The flexible axle shaft (Fig. 29) is used when the drive wheels are independently suspended. The axles are connected to the differential by universal joints. This type of axle is sometimes referred to as a swing axle. The drive wheel is free to move vertically without affecting the position of the differential and transmission. Flexible axles are commonly found on hillside combines.

## HYDROSTATIC DRIVES

Combines may be equipped with hydraulic or hydrostatic drives, rather than conventional transmissions, to propel the machine. The hydrostatic drive is an automatic fluid drive that uses fluid under pressure to transmit engine power to the drive wheels of the combine.

Hydrostatic drive combines are used in crop or field conditions that require considerable stop-and-go driving or downshifting, such as when combining rice. There is no clutch, and one lever provides infinite speeds from zero to top speed in each gear, whether in forward or reverse. Also, the speed remains constant as the combine travels either uphill or downhill.

To operate a hydrostatic drive combine, the operator places the speed range control lever in neutral and shifts the transmission into the desired gear. Then he or she only has to move the speed control to move the combine. As the operator moves the lever forward, the combine travels forward — the farther the lever moves, the faster the combine travels. In reverse, the opposite is true, except the lever cannot be moved as far in reverse, which prevents excessively high reverse speeds.

The mechanical power from the engine is converted to hydraulic power by a pump-motor team. This hydraulic power is then converted back to mechanical power at the transmission, which is similar to a regular transmission except it doesn't have a reverse gear because the hydrostatic drive is reversible.

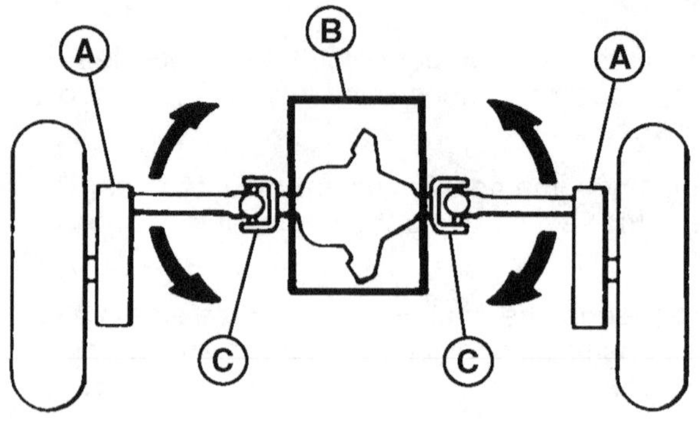

Fig. 29 — Flexible Axle Shaft

A— Final Drive
B— Transmission and Differential

C— Universal Joints

The hydrostatic drive functions as both a clutch and a transmission. The final gear train is usually simplified because the hydrostatic unit supplies infinite speed and torque ranges as well as reverse speeds.

The pump of the unit may be mounted near the engine so that it is driven directly from the engine crankshaft or powershaft. In this design, hoses connect the pump with the motor, and the motor is mounted on the transmission housing.

On other combines, the pump and motor are mounted on the transmission housing as a unit. The flow of hydraulic oil is direct between the pump and motor — no hoses are required. In this case the pump may be driven by a belt drive from the engine.

**Continued on next page**

KN52281,100454F -19-24JUL13-6/16



PN=92

The hydraulic system for the drive unit is separate from the regular combine hydraulic system. The schematic in Fig. 30 shows the basic hydrostatic system.

A— Filter
B— Reservoir
C— Cooler
D— Oil Lines
E— Motor
F— Pump

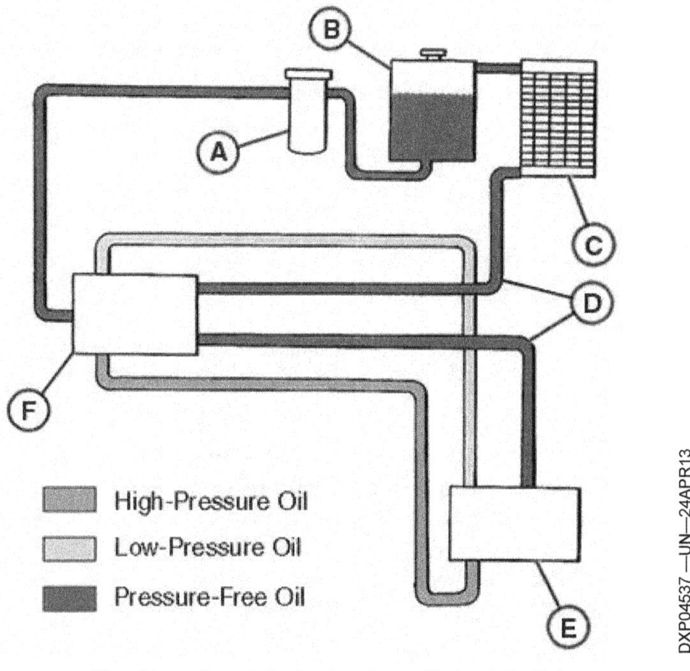


Fig. 30 — Complete System for a Hydrostatic Drive

KN52281,100454F -19-24JUL13-7/16

## OPERATION OF HYDROSTATIC DRIVE

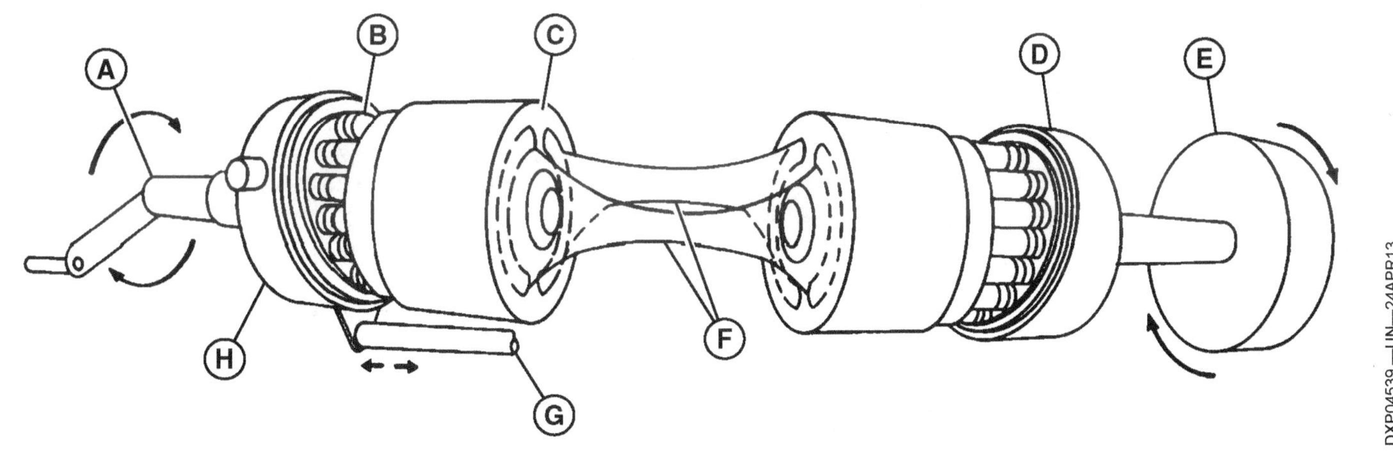

Fig. 31 — Typical Hydrostatic Pump and Motor Schematic

A— Input Shaft Drive
B— Pistons
C— Cylinder Block
D— Fixed Swash Plate
E— Output Wheel
F— Oil Circuit
G— Swash Plate Angle Control
H— Variable Swash Plate

Here is a simplified description of how a hydrostatic drive operates.

In a hydrostatic drive, several pistons are used to transmit power — one group in the pump sending power to another group in the motor (Fig. 31).

The pistons are in a cylinder block and revolve around a shaft. The pistons also move in and out of the block, parallel to the shaft, as we will see later.

Let's look at one of these cylinder arrangements.

**Continued on next page**

KN52281,100454F -19-24JUL13-8/16



PN=93

Two cylinders, each containing a piston, are connected by a line (Fig. 32). The cylinders and the line are filled with oil.

When a force is applied to the left piston as shown, that piston moves against the oil. The oil will not compress, so it acts as a solid and forces out the right piston.

A— Piston
B— Cylinder
C— Equal Force Out Here
D— Connecting Line
E— Force In Here

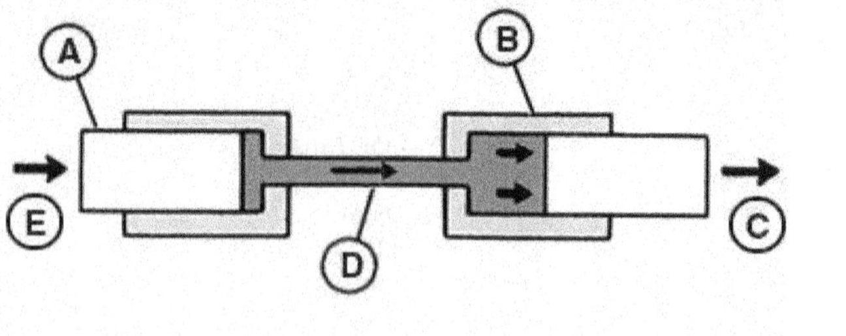


*Fig. 32 — Two Cylinders Connected*

KN52281,100454F -19-24JUL13-9/16

One piston for the pump and one for the motor are shown in Fig. 33. To provide a pumping action for the pistons, a plate called a swash plate is located in both the pump and motor. The mechanism is designed so the pistons ride against the swash plates.

The angle of the swash plate (Fig. 33) can be varied so that the volume and pressure of oil pumped by the pistons can be changed or the direction of oil flow reversed.

A pump or motor with a movable swash plate is called a variable-displacement unit. A pump or motor with a fixed swash plate is called a fixed-displacement unit. A variable-displacement pump driving a fixed-displacement motor is shown in Fig. 31.

As the pump pistons rotate, they move across the sloping face of the swash plate, sliding in and out of their cylinder bores to pump oil in and out. The more the pump swash plate is tilted, the more oil it pumps with each piston stroke and the faster it drives the motor.

The motor swash plate is at a fixed angle so that the stroke of its pistons are always the same. Thus, its speed of rotation cannot be changed except as it is driven faster or slower by the pump oil.

The point to remember now is that a given volume of oil forced out of the pump at a given pressure will cause the motor to turn at a given speed. More oil will increase the flow and speed up the motor; less oil will reduce the flow and slow it down.

The pump is driven by the machine's engine and so is linked to the speed set by the operator. It pumps a variable stream of high-pressure oil to the motor, depending on engine speed and tilt of the pump swash plate.

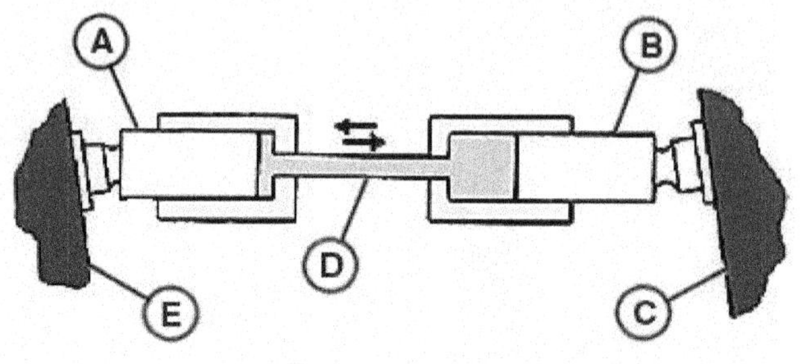


*Fig. 33 — Two Connected Cylinders with Swash Plates*

A— Pump Piston
B— Motor Piston
C— Motor Swash Plate
D— Connecting Line
E— Pump Swash Plate

Since the motor is linked to the drive wheels of the machine, it gives the machine its travel speed.

Only three factors control the operation of a hydrostatic drive:

- Rate of oil flow — gives the speed
- Direction of oil flow — gives the direction
- Pressure of oil — gives the power

Control of these three factors is infinite, giving endless selections of speed and torque in a hydrostatic drive.

The pump-motor team is the heart of the hydrostatic drive, although Fig. 30 is the complete hydrostatic system that also includes a reservoir to supply the oil, a filter to remove dirt, and a cooler to remove excess heat from the oil.

**Continued on next page**

KN52281,100454F -19-24JUL13-10/16



PN=94

Basically, however, the pump and motor are joined in a closed hydraulic loop (Fig. 34). The return line from the motor is joined directly to the intake of the pump, rather than to the reservoir. The charge pump simply supplies the oil, drawing it from the reservoir.

A— Pump
B— Motor
C— Reservoir
D— Charge Pump

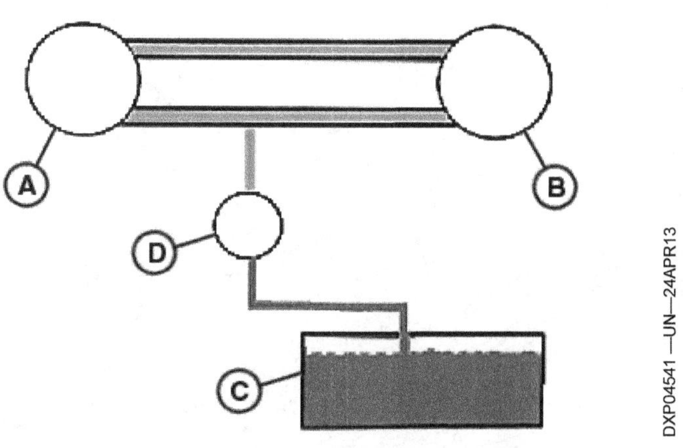


*Fig. 34 — Pump and Motor Form a Closed Hydraulic Loop*

KN52281,100454F -19-24JUL13-11/16

The direction of output shaft rotation can be reversed by shifting the pump swash plate over center (Fig. 35).

In neutral, the swash plate is vertical and no oil is pumped.

In forward, the swash plate is tilted and oil is pumped as shown at the top.

In reverse, the swash plate is tilted the opposite way and the unit pumps oil in the opposite direction.

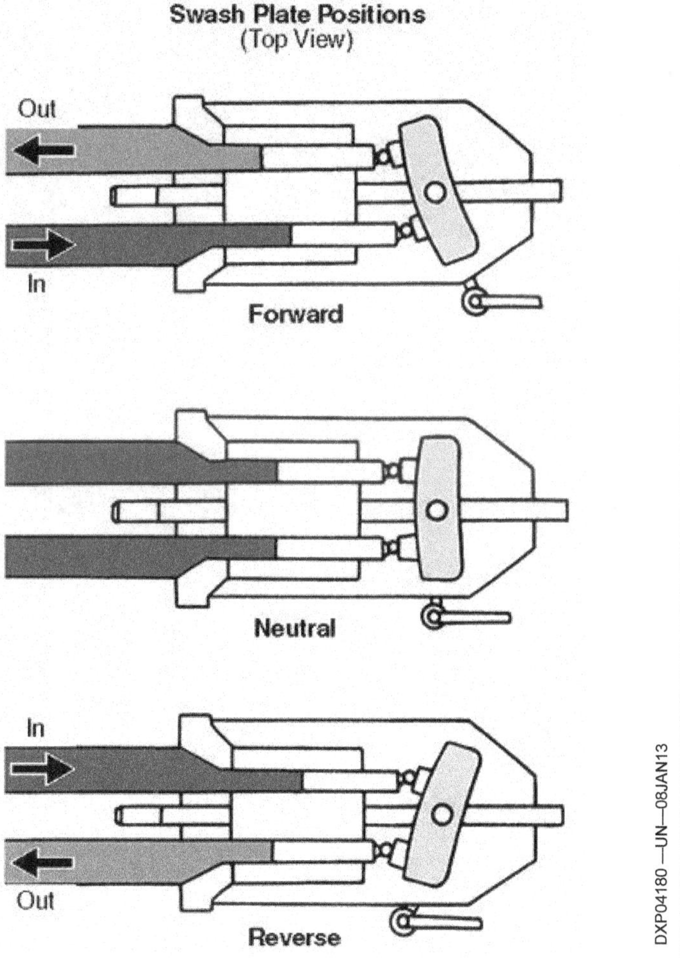


*Fig. 35 — How Reversing Is Done*

Continued on next page

KN52281,100454F -19-24JUL13-12/16

## FOUR-WHEEL DRIVE SYSTEM

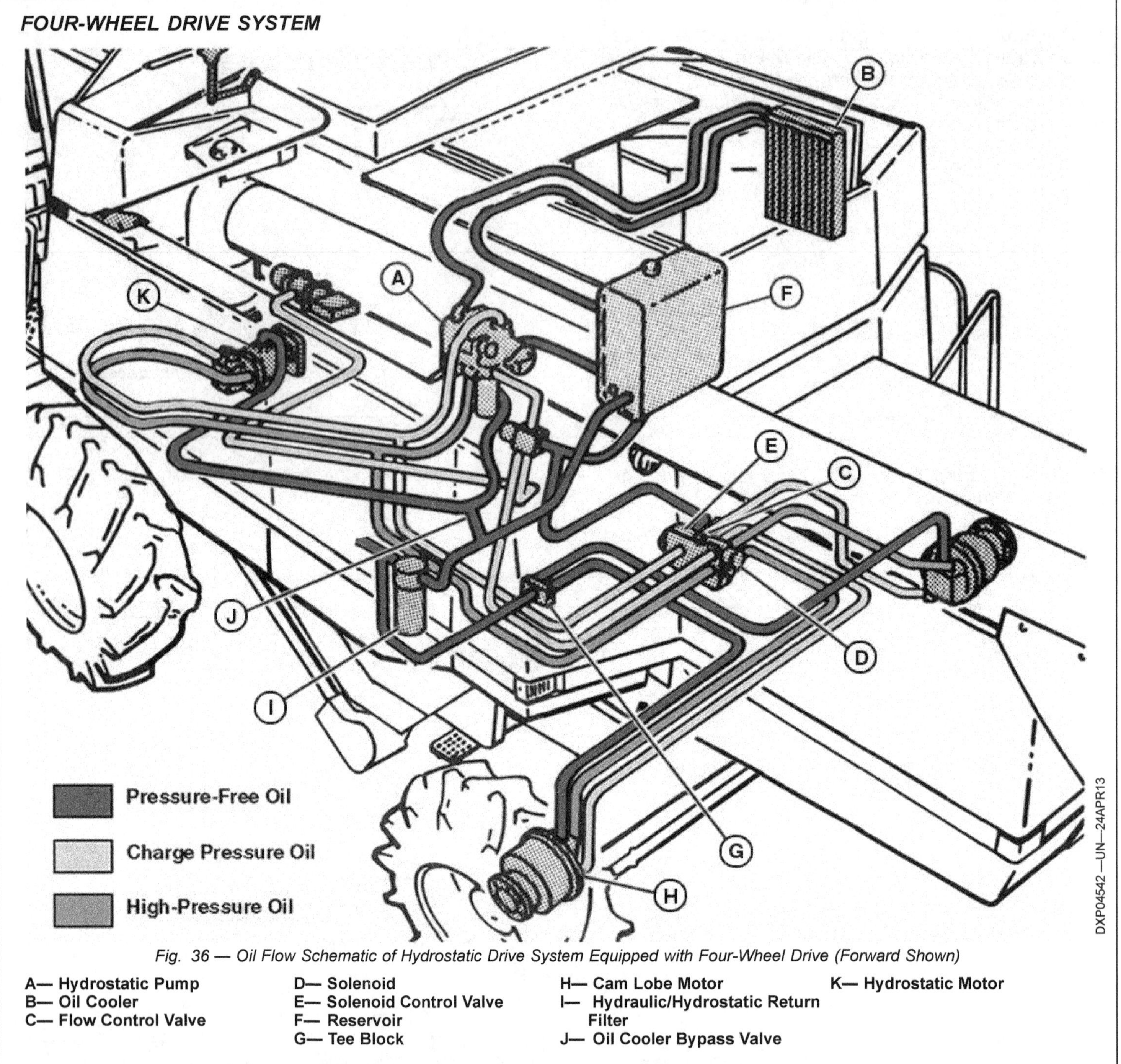


*Fig. 36 — Oil Flow Schematic of Hydrostatic Drive System Equipped with Four-Wheel Drive (Forward Shown)*

A— Hydrostatic Pump
B— Oil Cooler
C— Flow Control Valve
D— Solenoid
E— Solenoid Control Valve
F— Reservoir
G— Tee Block
H— Cam Lobe Motor
I— Hydraulic/Hydrostatic Return Filter
J— Oil Cooler Bypass Valve
K— Hydrostatic Motor

Four-wheel drive offers extra traction in muddy fields as shown in Fig. 37. When the combine needs additional traction, the operator turns on a switch that actuates this drive.

Continued on next page

KN52281,100454F -19-24JUL13-13/16

The four-wheel drive system (Fig. 36) consists of an electrohydraulic control valve, two fixed-displacement rear wheel motors of a planetary gearing or cam lobe design, an oil filter, and connecting hydraulic lines. Hydraulic power for this system is taken from the hydrostatic drive to combine front wheels.

DXP04182 —UN—08JAN13

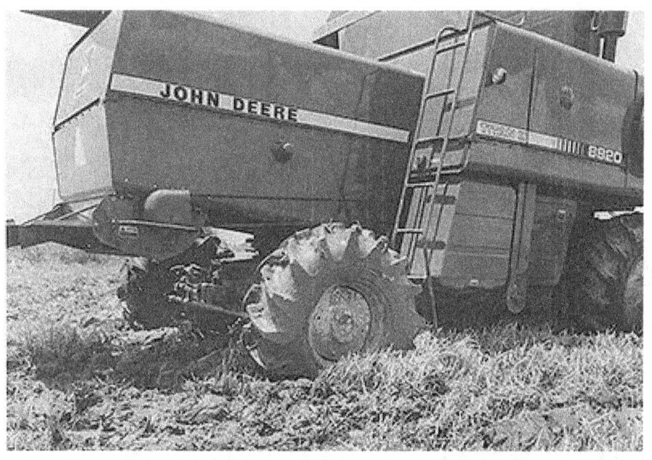

*Fig. 37 — Four-Wheel Drive Provides Extra Traction in Muddy Fields*

KN52281,100454F -19-24JUL13-14/16

To engage the four-wheel drive, the operator sends an electrical signal from the console control switch to the four-wheel drive control valve. This valve responds by opening to allow flow of hydrostatic fluid to the rear wheel motors (Fig. 36). The wheel motors convert this hydraulic energy into mechanical energy through a gear reduction to the wheel rims, propelling the combine.

DXP04183 —UN—08JAN13

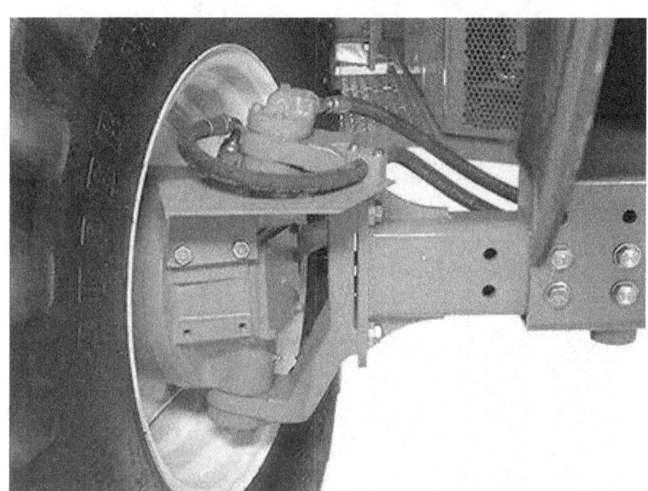

*Fig. 38 — Four-Wheel Drive on Combine*

KN52281,100454F -19-24JUL13-15/16

## HYDRAULIC SYSTEMS

On early self-propelled combines, the only hydraulic power used was for raising or lowering the header (Fig. 39). Many of today's combines use hydraulics not only to raise and lower the header, but to control the ground speed, raise and lower the reel, control the speed of the reel, drive the belt pickup, vary the speed of the feeder conveyor, control cylinder speed, swing the unloading auger in or out of position, and steady the machine (hillside combines). Also, as we discussed earlier, hydraulics may be used to propel the combine by a hydrostatic drive.

**A— Header Lift Cylinders**
**B— Control Valve**
**C— Hydraulic Pump**
**D— Hydraulic Oil Reservoir**

DXP04543 —UN—24APR13

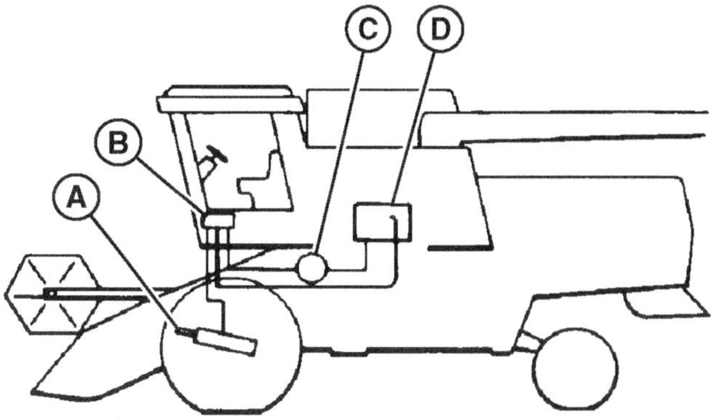

*Fig. 39 — Early Self-Propelled Combines Had a Simple Hydraulic System*

**Continued on next page**

KN52281,1004550 -19-24JUL13-1/3

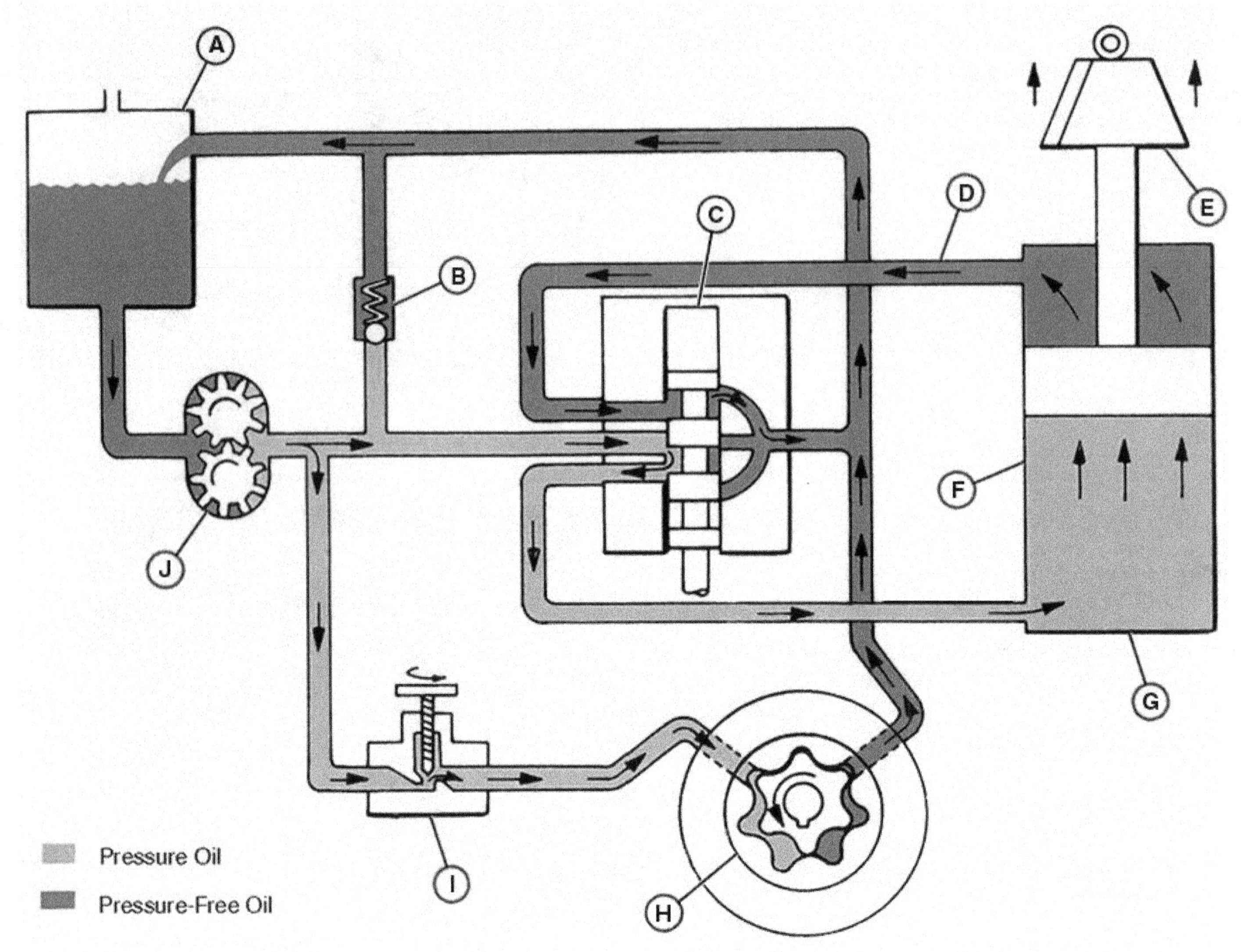



*Fig. 40 — Basic Hydraulic System*

A— Fluid Reservoir
B— Relief Valve
C— Control Valve Is Shifted, Directing Oil as Shown
D— Oil Retuning to Reservoir
E— Load
F— Pressure Oil Raises Piston and Load
G— Hydraulic Cylinder
H— Hydraulic Motor
I— Motor Control Valve
J— Hydraulic Pump

The basic hydraulic system (Fig. 40) consists of a reservoir for storing the oil, a pump for moving the oil, a control valve for directing the flow of oil, and cylinders or motors for doing the work. In this simplified system, the pump is pumping oil to the cylinder control valve and the motor control valve.

The cylinder control valve in Fig. 40 has been shifted to direct the pressure (red) oil to the cylinder to push the load up. The blue oil in the top of the cylinder is allowed to return to the reservoir.

When the cylinder control valve is shifted in the opposite direction, the oil is directed to the top of the cylinder and the load is lowered.

When the control valve is returned to the center position, the oil is trapped in the cylinder at both the top and bottom; thus, the load is held at the position selected.

The motor control valve controls the volume of oil going to the hydraulic motor. As the valve is opened farther, the speed of the motor increases. The oil passing through the motor is also returned to the reservoir. To stop the motor, the operator shuts off the control valve.

When both the control valves have shut off the flow of oil, the oil returns to the reservoir through the center passage of the valve. The relief valve opens when the pressure exceeds safe limits, such as at the end of the hydraulic cylinder stroke.

**Continued on next page**
KN52281,1004550 -19-24JUL13-2/3

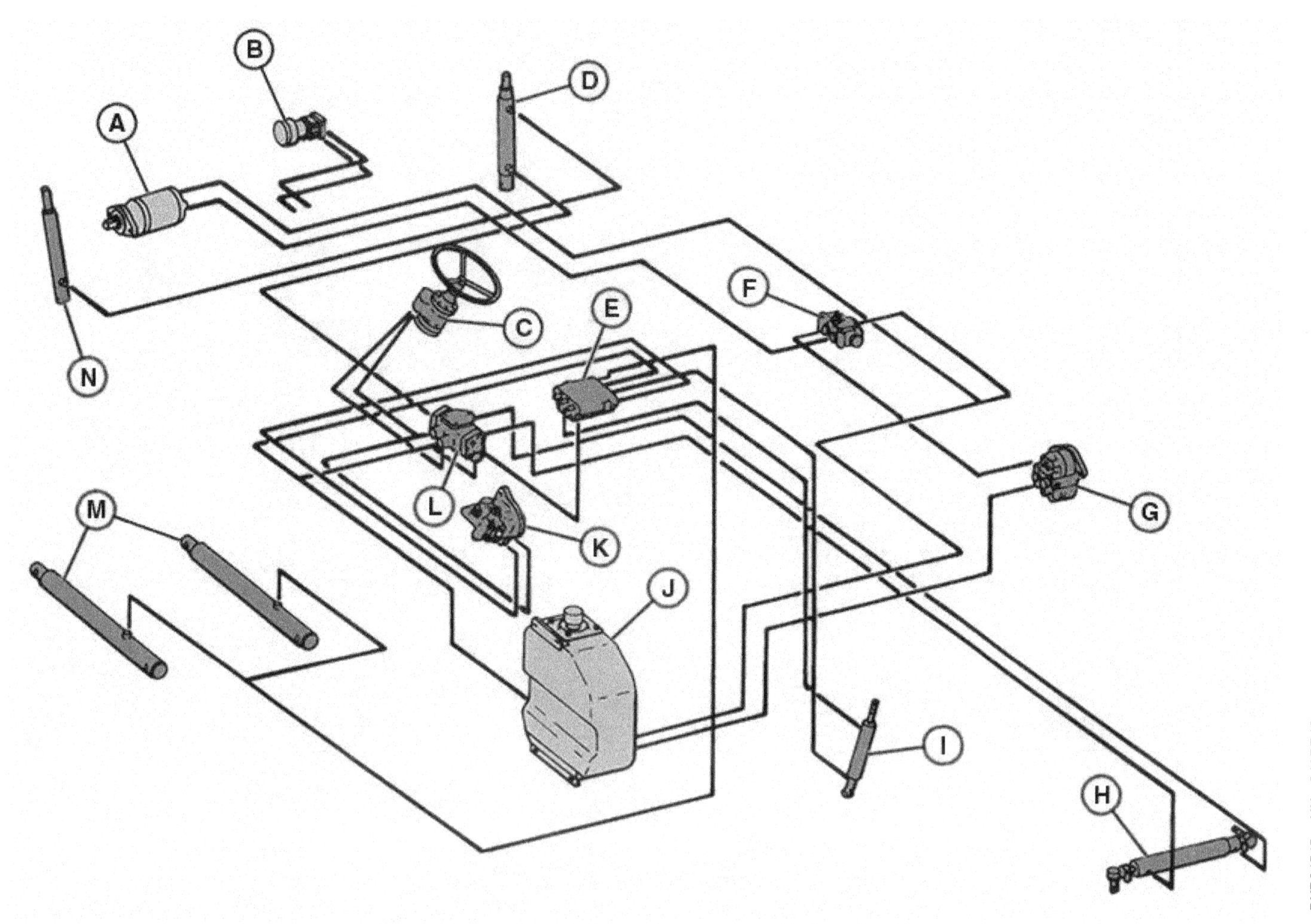

*Fig. 41 — Typical Hydraulic System on a Modern Combine*

A— Reel Drive Motor
B— Belt Pickup Drive Motor
C— Steering Pump
D— Reel Lift Master Cylinder
E— Control Valve
F— Reel Drive or Belt Pickup Drive Valve
G— Reel Drive or Belt Pickup Drive Pump
H— Steering Cylinder
I— Unloading Auger Cylinder
J— Reservoir
K— Hydraulic Pump
L— Steering Control Valve
M— Platform Lift Cylinders
N— Reel Lift Slave Cylinder

Several more hydraulic components may be added to this system, and sometimes more than one system is used on a combine. A typical complex hydraulic system is shown in Fig. 41 and discussed further.

### REEL LIFT CYLINDERS

The operator actuates the control valve to either raise or lower the reel to the desired height. In this design, one cylinder is the master cylinder, which controls the slave cylinder so that both move the same distance to keep the reel level.

When the proper height is reached, the control valve is allowed to return to center and the cylinders are held in position by trapped oil.

### DRIVE MOTORS

The operator controls the speed of the reel drive or belt pickup drive by opening or closing the motor drive valve. The operator can adjust the reel or belt pickup speed to the ground speed to keep the proper speed ratio. Attachments are available for some combines that automatically adjust reel or pickup speed in proportion to ground speed.

### HEADER LIFT CYLINDERS

Again, by operating the control valve, the operator can raise or lower the header to the desired operating height. When the control valve is returned to center, the header is held in position by trapped oil.

### STEERING PUMP

The steering system in Fig. 41 uses a steering pump that is attached to the steering wheel. When the operator turns the steering wheel to the right or left, the steering pump delivers oil to the steering control valve that directs the pressurized oil from the main hydraulic pump to the proper end of the steering cylinder. This causes the rear wheels to turn in the desired direction.

KN52281,1004550 -19-24JUL13-3/3

# LEVELING SYSTEM

Leveling systems are used on some combines to help keep the separator components on a level plane. If the separator is allowed to tilt too far to one side, as when operating on the side of a hill, chaff and grain will build up on the downhill side of the separator components, resulting in poor separating performance. In the sample leveling system shown (Fig. 42), the combine final drives are mounted on large, pivoting castings on each side of the machine.

The pivot castings are connected to the combine main frame by two large hydraulic cylinders. An electronic control unit is mounted to the machine to constantly monitor the angle of the separator as the machine moves through the field. To make the necessary changes to keep the machine level, the control unit energizes hydraulic solenoid valves to direct hydraulic oil to the hydraulic cylinders mounted to the pivot castings. Extending or retracting the cylinders allows these castings to pivot up or down to tilt the machine as needed to remain level on slopes up to 15 percent.

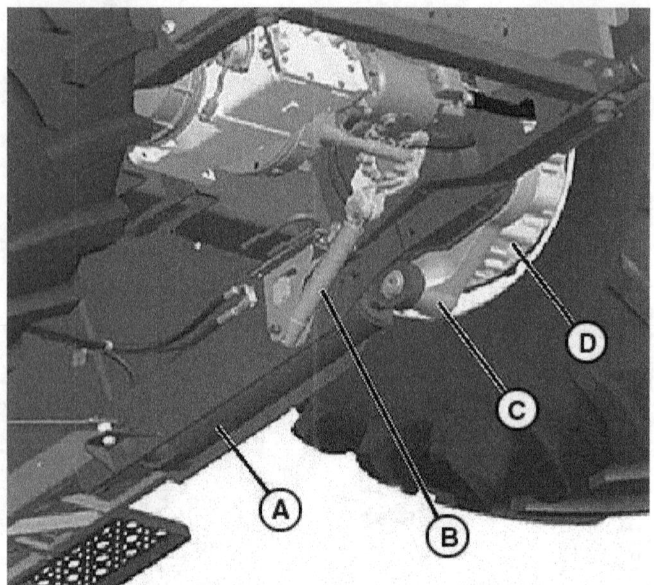


*Fig. 42 — Leveling System*

A— Leveling Cylinder  
B— Header Tilt Master Cylinder  
C— Pivot Casting  
D— Final Drive

KN52281,1004551 -19-31JUL13-1/2

To ensure that the header stays parallel to the ground a master-slave cylinder arrangement is used to tilt the header in respect to the chassis (Fig. 43). The header master cylinder is mounted on one of the pivot castings. As the casting is pivoted up or down to keep the machine level, the master cylinder directs oil to the slave cylinder that is mounted to a tilting frame on the front of the feeder conveyor to tilt the header as required to move parallel with the ground. Height sensors at the left- and right-hand ends of the platform are also used to monitor header position.

The system can be used in automatic mode or may be manually operated using the controls on the combine armrest.

A— Header Tilt Slave Cylinder

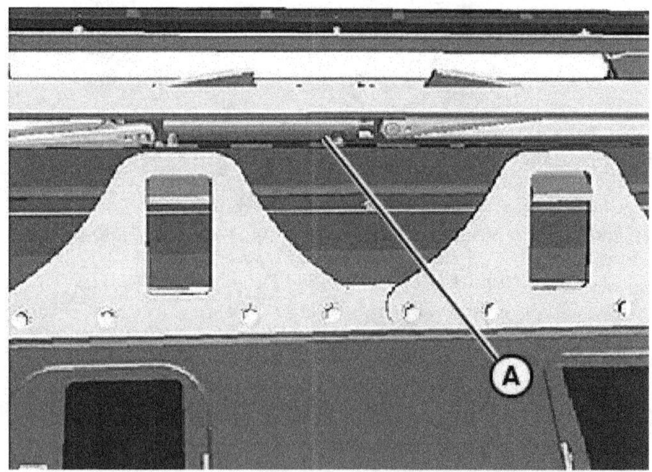


*Fig. 43 — Header Tilt Slave Cylinder*

KN52281,1004551 -19-31JUL13-2/2

## TEST YOURSELF

### Questions

1. What are the three power systems required to operate a modern combine?

2. List the following engine strokes in proper sequence:

   - Compression stroke
   - Exhaust stroke
   - Intake stroke
   - Power stroke

3. Name five of the six basic engine systems.

4. (True or False) The engine lubricating system has four major parts.

5. (Fill in the blanks.) The engine's _______   _______ system absorbs heat created by combustion and friction.

6. What are the three basic drive systems found on most combines?

7. What are four components that can be controlled by the hydraulic system on modern combines?

KN52281,1004552 -19-06SEP13-1/1

## INTRODUCTION

Almost all the operating controls and instruments are located in the operator station within easy reach of the operator seat. The operator station is located high on the front of the combine to give the operator a good view of the header. Here the operator can control all of the combine functions.

OUO1082,000624C -19-28AUG13-1/2

Most combines have essentially the same controls and instruments (Fig. 1). Their names and locations may be different, but they operate the same functions. For example, the control to raise or lower the header may be called either a platform lift control, a table lift control, or a header lift control.

Levers, knobs, or switches may be used to control the various components. Some manufacturers have controls of different colors and shapes to help the operator quickly identify the controls while operating the combine.

Here are some of the color codes one manufacturer uses for the controls:

- BLACK — Operating adjustments and controls
- ORANGE — Ground drive and engine speed
- YELLOW — Drive engagement

Fig. 1 — Combine Operating Controls

OUO1082,000624C -19-28AUG13-2/2



PN=102

# IDENTIFYING CONTROLS

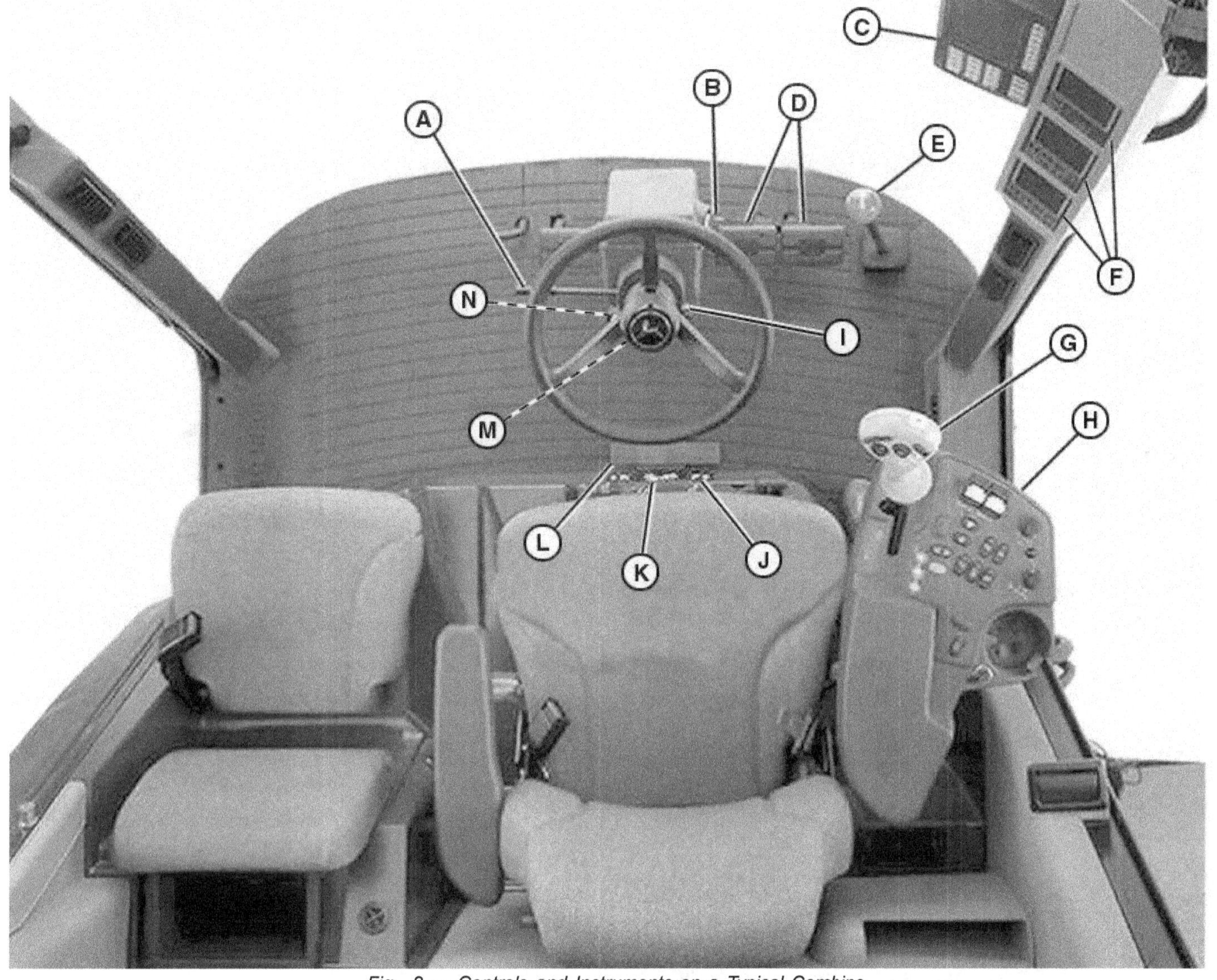

*Fig. 2 — Controls and Instruments on a Typical Combine*

A— Turn Signal Lever
B— Start Switch
C— GPS Unit
D— Brake Pedals
E— Gearshift Lever
F— Cornerpost Multi-Function Displays
G— Forward-Reverse Control Handle
H— Armrest Console Controls
I— Horn Button
J— Seat Height Lever
K— Seat Position Lever
L— Reverser Control Pedal
M— Steering Column Tilt Pedal
N— Starting Aid Button

Each combine operator's manual has a section on controls and instruments to help the operator identify and operate the controls.

Because each manufacturer locates these controls differently, the operator must become familiar with the controls before attempting to operate the combine. Typical controls and instruments on a modern combine are shown in Figs. 2 through 5. Some combines have only a few of these, and others may have different ones. Study these illustrations to become familiar with how many controls there are and what their functions are.

Continued on next page

KN52281,1004553 -19-28AUG13-1/3

Fig. 3 — Armrest Console Controls

A—Header Engage and Feeder House Reverser Switch
B—Separator Engage Switch
C—Road Transport Disconnect Switch
D—Draper Belt Speed Switch
E—Header Control Dial
F—Reel Speed Switch (Manual)
G—Concave Clearance Switch
H—Threshing Speed Switch
I— Cleaning Fan Speed Switch
J— Reel Speed Dial (Preset)
K—Cup Holder
L—12V DC Power Outlet
M—Spreader Speed Switch
N—Sieve Adjust Switch
O—Chaffer Adjust Switch
P—Four-Wheel Drive Switches
Q—Engine Speed Switches

Before discussing controlling combine movement and operation, combine and engine break-in should be mentioned. Most manufacturers recommend that certain procedures be followed to ensure proper break-in of a new combine. These procedures may include operating the combine at reduced loads for a time to seat the piston rings and give the other parts a chance to adjust to loads. Other things, such as checking torque on important bolts, checking stretch of drive belts, and changing oil in the hydraulic system and engine, may be recommended to be made after the combine has operated for several hours.

A—Forward/Reverse Control Handle
B—Quick Stop Switch
C—Unloading Auger Swing In/Out Switch
D—Unloading Auger Drive Engage/Disengage Switch and Indicator Light
E—Header Raise/Lower Switch
F—Reel Raise/Lower Switch and Reel Fore/Aft Switch

F—Front End Speed Control Switch (if Equipped with Corn Head or Row-Crop Head)
F—Corn Head Deck Plate Adjustment Switch
G—Activation Button 3
H—Activation Button 2
I— Activation Button 1

A— Light Switches
B— Climate Control Switches

C— Windshield Wiper and Washer Switch
D— Overhead Warning Display Panel

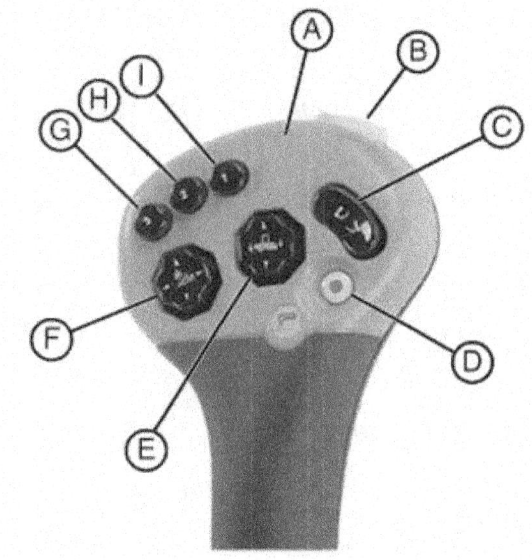

*Fig. 4 — Multi-Function Control Handle (Orange)*

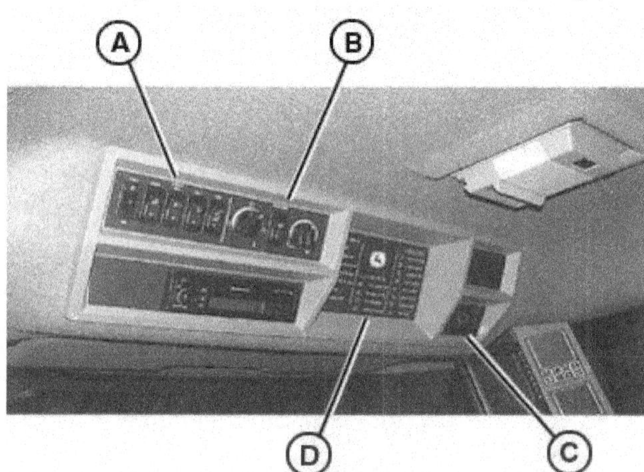

*Fig. 5 — Overhead Monitor and Control Panel*

KN52281,1004553 -19-28AUG13-3/3

## CONTROLLING COMBINE MOVEMENT

⚠ **CAUTION: For your own safety, always keep in mind — a combine is a big, heavy machine that requires careful handling at all times. To control combine movement, the operator must know how to:**

• Operate the engine

• Drive the combine

The operator must also know how to transport the combine, which we will cover later.

### *OPERATING THE ENGINE*

Before attempting to operate the engine, read the operator's manual to become aware of the controls and instruments related to starting and stopping the engine, and the safety instructions. Following are some common engine controls and instruments found on modern combines.

### ENGINE STARTING KEY SWITCH

The key switch on most combines is similar to those used on automobiles. Three or four positions may be available: OFF, ON, START, and ACCESSORIES (Fig. 6). Most manufacturers recommend that you turn the key to the ON position before attempting to start the engine to check the operation of the alternator and oil pressure indicator lights. Then turn the switch to START to engage the starter to crank the engine. After the engine has started, release the key. The alternator light should go out. If it doesn't go out after 10 seconds, shut off the engine at once and determine the cause.

The operation of other warning lights may also be checked by turning the switch to the ON position. They are the parking brake indicator light, oil pressure indicator light, water temperature indicator light, etc. Refer to your combine operator's manual.

### ENGINE SPEED CONTROL LEVER OR SWITCHES

Engine speed control lever or switches should be set at slow idle when starting engine. Place the speed control in the full-open position when operating the combine in the field so that the units of the combine are operating at the proper speed, as regulated by the governor.

### FUEL SHUTOFF (DIESEL ENGINES ONLY)

Some diesel engines are equipped with a fuel shutoff control that shuts off the fuel supply to the fuel injection pump. Usually, the procedure used to stop a diesel engine equipped with a fuel shutoff control is to:

1. Run the engine at slow speed until the engine cools.

2. Close the speed control.

3. Shut off the fuel supply.

4. Turn off the key switch.

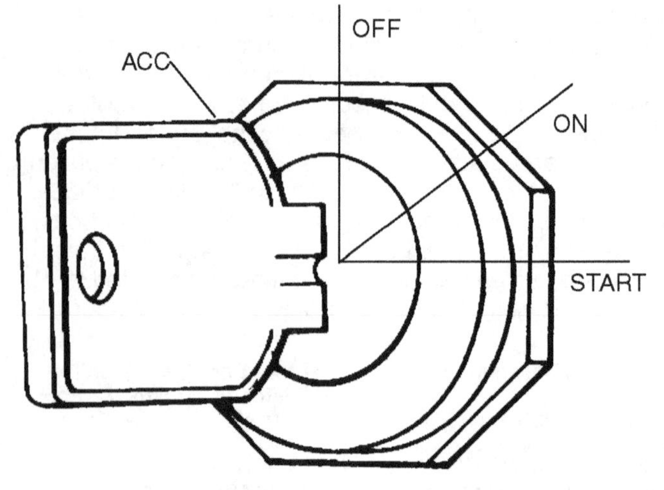


*Fig. 6 — Typical Key Switch Positions*

### ALTERNATOR INDICATOR LIGHT

This light glows when the alternator is not charging the battery. If the light goes on while the engine is running, stop the engine and determine the cause without delay. Some combines have an ammeter gauge rather than a warning light. This gauge indicates the amount of charge or discharge. If, while the engine is running, the gauge indicates that the system is discharging, stop the engine immediately and determine the cause.

### ENGINE COOLANT TEMPERATURE GAUGE

This gauge indicates the coolant temperature in the engine cooling system, not the quantity. The gauge usually has a warning zone to indicate when the coolant is above normal operating temperature. If the gauge indicates that the coolant is in the danger zone, stop the engine and determine the cause.

### AIR RESTRICTION INDICATOR

The air restriction indicator is usually located on the air intake tube to the engine. The red signal in the indicator appears whenever the air cleaner element is dirty and needs servicing. Most manufacturers recommend that this indicator be checked daily.

### ENGINE OIL PRESSURE GAUGE

This gauge (indicator light on some combines) indicates the pressure of the engine lubricating oil — not the amount of oil in the crankcase. When the engine is running, the gauge should show oil pressure in the normal range (the indicator light should not be on). If the oil pressure drops below the normal range and into the danger zone (indicator light on), stop the engine immediately and determine the cause.

### FUEL GAUGE

The fuel gauge simply tells approximately how much engine fuel is in the fuel tank.

Continued on next page

KN52281,1004554 -19-31JUL13-1/2

## ENGINE TACHOMETER

The tachometer usually shows the engine speed in hundreds of rpm. Many operators use the tachometer to tell if the engine is running at the proper working speed, especially when operating in the field. If the engine speed drops appreciably, you must reduce the load by reducing ground speed. This gauge may also contain the hour meter.

## HOUR METER

The hour meter indicates the accumulated engine service hours. This is usually based on the normal operating speed of the engine (2200 to 2900 rpm). The hour meter provides a means of determining when lubrication and periodic services are needed.

KN52281,1004554 -19-31JUL13-2/2

## STARTING AND STOPPING THE ENGINE

Here we will cover typical starting and stopping procedures for diesel engines and turbocharged diesel engines. Read the starting tips.

 **CAUTION: Follow these procedures to avoid safety hazards and possible damage to the combine.**

These instructions are of a general nature. Always follow specific instructions in the operator's manual for the combine being operated.

*NOTE: See instructions describing general operation of controls and instruments later in this chapter.*

### BEFORE STARTING THE ENGINE

- Check crankcase oil level.
- Check cooling system fluid level.
- Check hydraulic system oil level.
- Be sure fuel tank is full.

 **CAUTION: Before attempting to start the engine, be sure no one is near the combine because a bystander might be injured by moving parts. Be especially concerned for the safety of children.**

**Never start or operate the combine in a closed building. Always be sure there is plenty of ventilation when running a combine because the exhaust fumes are poisonous.**

When starting the engine, never hold the key in start position for more than 30 seconds at a time, or you may damage the starter by overheating internal electrical parts. If the engine does not start within 30 seconds, allow at least 2 minutes for proper cooling of the starter before attempting to start the engine again.

Be sure to pause long enough after a false start to make certain the starter has stopped completely before attempting to start the engine again. Otherwise, you may damage the gears on the starter or flywheel.

On diesel engines, do not let the fuel tank run dry. If it does, you must bleed the entire fuel system to remove air bubbles. Follow instructions in your operator's manual for this procedure.

 **CAUTION: Do not attempt to start a combine by towing it. It is an unsafe practice which could result in injury to you or others. Use booster batteries if the engine is hard to start, and follow the directions in the operator's manual for the proper procedure.**

On hydrostatic drive combines, the pump and motor may be damaged if towing is used in an attempt to start the engine.

### STARTING THE DIESEL ENGINE

**CAUTION: Before starting the engine, make sure no one is standing near the combine.**

1. Disengage header drive, separator drive, and grain tank unloading auger drive.
2. Place the gearshift lever in neutral.
3. Place the hydrostatic control lever in neutral.
4. Move the speed control lever to slow idle position and push in the fuel shutoff.
5. Turn the key switch to ON. Check operation of the warning lights according to the operator's manual.
6. Turn the key switch to START. After the engine starts, release the key.
7. Be sure the oil pressure gauge registers normal pressure and the warning lights are not on. **If either of these conditions is not correct, stop the engine and determine the cause.**
8. Allow the engine to run for several minutes at slow idle without load until the engine and transmission have reached a safe operating temperature.

### STOPPING THE DIESEL ENGINE

1. Place the hydrostatic control lever in neutral and move the gearshift lever into neutral.
2. Set the speed control at half speed and allow the engine to run at this speed for several minutes before stopping. This cools the engine and prevents possible damage to parts because of overheating.
3. Move the speed control to slow idle.
4. Pull out the fuel shutoff, if equipped, and turn the key to OFF.
5. Remove the key from the switch to prevent tampering and unauthorized operation.
6. Engage the parking brake and follow any other parking procedures that may be recommended in the operator's manual.

### STARTING THE TURBOCHARGED DIESEL ENGINE

**IMPORTANT: When starting the engine after the combine has not been used for an extended period of time or after the oil filter has been changed, pull the fuel shutoff all the way out and crank the engine until the engine oil pressure light goes out. This ensures that the turbocharger has a supply of lubricating and cooling oil. Do not operate the starter for more than 30 seconds at a time. After the indicator light goes out, move the throttle to the slow idle position. Make sure the fuel shutoff is all the way in and then start the engine.**

1. Repeat steps 1 through 7 as described in Starting the Diesel Engine.
2. Idle the engine for several minutes at slow speed (see operator's manual) to ensure turbocharger lubrication before accelerating or applying load.

Continued on next page

KN52281,1004555 -19-25JUL13-1/2

**IMPORTANT: If the engine quits running when operating under load, immediately restart the engine to prevent overheating of turbocharger parts, which is caused when the flow of oil for cooling and lubrication is stopped.**

## *STOPPING THE TURBOCHARGED DIESEL ENGINE*

1. Place the hydrostatic control lever in neutral and move the gearshift lever into neutral.
2. Allow the engine to idle for a few minutes to cool the engine and turbocharger. (Lubrication and cooling of the turbocharger and some engine parts is provided by the engine lubrication oil. Therefore, sudden stopping of a hot engine may allow some parts to overheat and cause possible damage.) Allow the temperature gauge to drop well into the normal range on the dial.
3. Pull out the fuel shutoff, if equipped, and turn the key to OFF.
4. Remove the key from the switch to prevent tampering and unauthorized operation.
5. Engage the parking brake and follow any other parking procedures that may be recommended in the operator's manual.

KN52281,1004555 -19-25JUL13-2/2

# DRIVING THE COMBINE

To drive the combine safely, the operator must be aware of all the functions of the combine and how they operate. The operator must also know how to operate and adjust the following:

- Operator seat
- Steering
- Brakes
- Propulsion units
- Ladder

In addition to these, the operator must know how to transport the combine. This will be covered later.

## *OPERATOR SEAT*

The operator must be comfortable and within easy reach of the controls of the combine to drive it properly. Before attempting to operate the combine, the seat should be adjusted to the operator's height and reach.

All combines have adjustable seats that can be raised and lowered or moved forward and rearward. Some may be folded up or moved back out of the way so that the operator can stand while driving. A typical seat mounting is shown in Fig. 7.

**To adjust the seat forward or rearward:**

1. Sit down and then disengage the seat lock.

Fig. 7 — Operator and Passenger Seats

2. Put the seat forward or rearward.
3. When the desired position is reached, release the lock and the seat is held in position.

**To raise or lower the seat:**

1. Remove the spring locking pins.
2. Raise or lower the seat to the desired position.
3. Reinsert the pins to hold the seat in place.

**Continued on next page**

KN52281,1004556 -19-31JUL13-1/7



050217<br>PN=109

## STEERING

Most combines also have an adjustable steering column for individual arm lengths or for standing position. Adjust the steering column by depressing the lock pedal and moving the column to the desired setting (Fig. 8).

The wheels that steer a combine are mounted on the rear of the machine. Be careful when making a turn near obstacles, because the rear of the combine may swing around and strike something.

Most steering on modern combines is power steering, which makes driving of the combine down rows or fields an easy task.

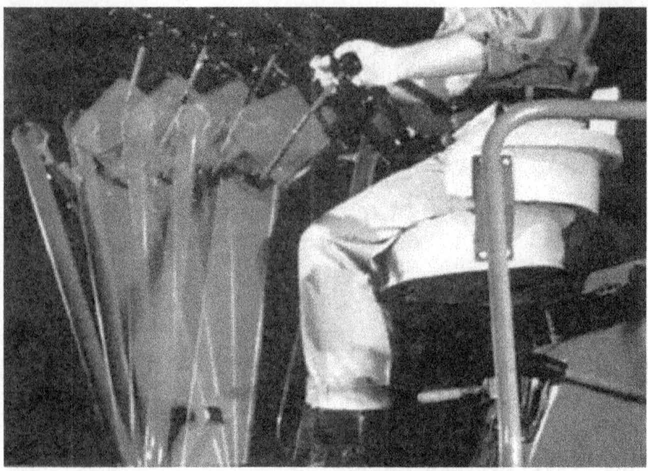

Fig. 8 — Adjusting Steering Column

KN52281,1004556 -19-31JUL13-2/7

## BRAKES

**CAUTION: Always reduce travel speed before applying brakes. Quick stops can result in the combine nosing forward.**

The brakes on most combines are the hydraulic type; however, some combines still have mechanical brakes. Usually, two brake pedals are provided that control individual drive wheels (Fig. 9). When used separately, these pedals can be used to assist in turning. When used together, a quick stop is ensured. When stopping the combine, press on both brake pedals. Uneven application of brakes will cause the combine to swerve to one side at high speeds. This could result in the combine turning over.

When using the brakes to make sharp turns, slow down to a safe speed. Begin turning the steering wheel before applying the brake to assist turning. Otherwise the rear wheels will skid sideways and the turn will be more difficult. Be careful when making a turn near obstacles, because the rear of the combine may swing around and strike something.

Some combines have a brake pedal lock, which couples the pedals together to aid in even braking when transporting the machine.

Fig. 9 — Brake Pedals

A— Brake Pedals

Continued on next page

KN52281,1004556 -19-31JUL13-3/7

## PARKING BRAKE

The parking brake locks the wheel brakes so the combine cannot move when left unattended. Never attempt to move the combine with the parking brake engaged. Some combines have an indicator light to notify the operator when the parking brake is engaged.

To engage the parking brake shown in Fig. 10, pull the lever upward; to disengage, push the lever downward.

A— Parking Brake Lever

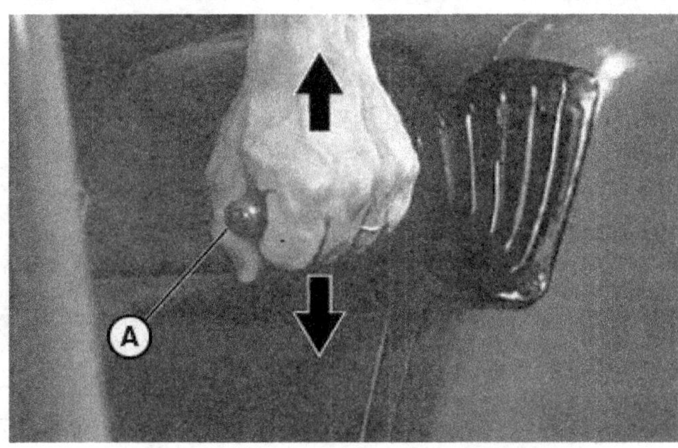

*Fig. 10 — Operating Parking Brake Lever*

KN52281,1004556 -19-31JUL13-4/7

Many machines are equipped with a parking brake pedal and a release pedal located on the floor near the steering column (Fig. 11). Pressing the pedal down engages the parking brake. Pressing the release pedal disengages the parking brake.

A— Parking Brake Release    B— Parking Brake Pedal
     Pedal

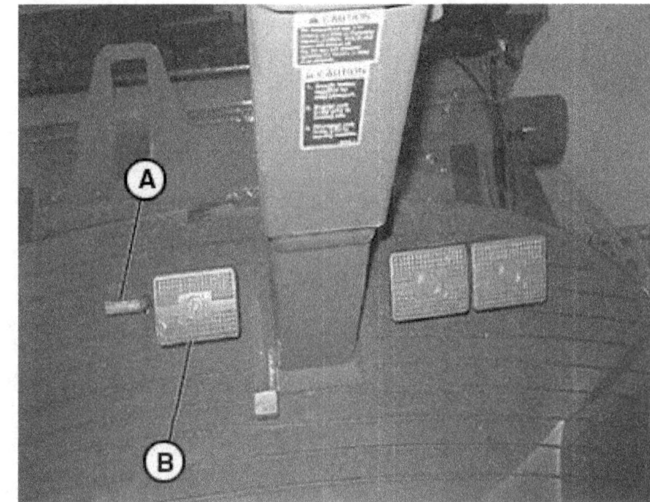

*Fig. 11 — Parking Brake Pedal System*

**Continued on next page**      KN52281,1004556 -19-31JUL13-5/7

## PROPULSION UNITS

Controlling the propulsion units on hydrostatic drive combines consists of operating the gearshift lever and the hydrostatic speed range lever (Fig. 12).

## HYDROSTATIC DRIVE COMBINES

The hydrostatic unit (pump and motor) transmits engine power to the transmission. When the speed range lever is moved, oil moves through the hydrostatic pump and motor, causing the ground travel speed to increase or decrease within a selected transmission gear.

To move the combine forward, place the speed range lever in neutral (Fig. 12). Move the gearshift lever to the desired gear. Push the speed range lever forward. To increase forward travel speed in the gear selected, move the lever farther forward.

To move the combine in reverse, place the speed range lever in neutral (Fig. 12). Move the gearshift lever to the desired gear. Pull the speed range lever rearward.

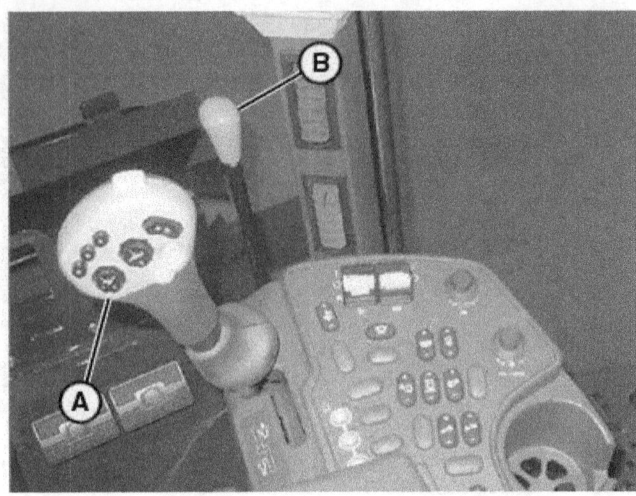

*Fig. 12 — Propulsion Unit Controls (Armrest Console Controls)*

**A— Speed Range Lever**　　　**B— Gearshift Lever**

KN52281,1004556 -19-31JUL13-6/7

## Ladder

Some combines have a movable ladder that can be moved out of the way of uncut grain to avoid grain loss by ladder impact. The ladder may be folded or pivoted out of the way. The ladder shown in Fig. 13 is a pivoting ladder.

To move the ladder, pull the lock release lever down.

⚠ **CAUTION: Keep steps and walking surfaces free of grease and dirt. Use handrails for safe mounting and dismounting. Be careful on frosty mornings. Do not store tools and equipment on operator platform.**

*Fig. 13 — Operating Ladder*

KN52281,1004556 -19-31JUL13-7/7

## TRANSPORTING THE COMBINE

A combine can be transported by driving it under its own power, carrying it on a truck, or towing it. Often the manufacturer recommends towing in an emergency situation only, and then only over a short distance at a slow speed. Refer to machine operator's manual for towing instructions.

### HOW TO TRANSPORT

*NOTE: See instructions describing operation of controls elsewhere in this chapter.*

1. When transporting the combine under its own power, couple the brake pedals together if a lock is provided.
2. Reduce the width of the combine by folding the unloading auger back along the separator.
3. Remove the header if necessary.
4. If the header and feeder house are removed, wire the hydraulic cylinders up or support them with chains.
5. If the header is left on, raise it to a position that allows good visibility for the operator, yet provides ample ground clearance.
6. To reduce the spread of noxious weed seeds, thoroughly clean the combine before leaving one field and going to the next. Sweep trash and straw from the outside of the combine. Open doors at the bottom of the elevators. Remove the grain tank drain hole cover and run the combine until all straw, trash, and grain are removed from inside. Shut off the engine and clean out the interior of the combine.
7. Use appropriate warning devices when traveling on the road or highway, such as a slow moving vehicle (SMV) emblem, reflectors, operating lights, and warning lights. Keep these items clean so that they are visible to other traffic.
8. Have a car or truck drive in front of and behind the combine to warn traffic on public roads. Stay in your lane. Pull off, stop, and let traffic pass. Do not drive on the shoulder.
9. Because wheels for steering are in the back, self-propelled machines often fishtail when turned quickly at transport speeds. Steering to the right will whip the rear to the left, and vice versa. The back of the combine will swing out into the path of oncoming traffic.
10. Slowing or braking the combine too rapidly could cause loss of some steering control (weight on rear wheels). This is most noticeable when driving with a corn head or some other heavy header, or with the header raised too high. In this case, most of the weight will be on the drive wheels. Use rear weights. Keep the header as low as possible. Use variable speed drive or engine speed control to slow the machine. Reduce speed before applying brakes, and always transport with brake pedals locked together.
11. Check your local governmental regulations for required warning devices.

KN52281,1004557 -19-25JUL13-1/1

## CONTROLLING COMBINE FIELD OPERATION

Here we will discuss the controls necessary to operate the combine in the field. Field adjustments will be covered in the next chapter.

The controls for the following units will be discussed:

- Header
- Separator
- Threshing
- Cleaning
- Grain tank unloading
- Electrical

*Fig. 14 — Controlling Field Operation*

Continued on next page　　KN52281,1004558 -19-31JUL13-1/18



050217
PN=113

## HEADER CONTROLS

The combine may be equipped with either a cutting platform, pickup platform (Fig. 15), or a corn head. Some of the same controls are used to operate these units.

These controls are necessary to allow the operator to adjust the header for crop and field conditions. It is important to control the header for proper cutting, gathering, and feeding the crop to the combine separator for efficient harvesting.

Here are the two main header controls.

Fig. 15 — Pickup Platform

KN52281,1004558 -19-31JUL13-2/18

### HEADER DRIVE CONTROL

In chapter 3 we discussed the drive for the header. The header drive may be controlled either mechanically with a lever, or electrically with a switch (Fig. 16) and electromagnetic clutch. In either case the operator can engage or disengage the drive easily from the operator seat.

To engage the header drive with the switch control, simply actuate the switch. To disengage, turn the switch off.

### HEADER HEIGHT CONTROL

The operator can control the height of the header. Depending on the combine, the hydraulic cylinders to raise or lower the header are controlled with either a lever or switch linked to a hydraulic valve.

On the machine shown in Fig. 17, the header can be raised and lowered manually by pressing the header raise/lower switch located on the control lever.

The header control dial on the armrest allows the operator to preset the position of the feeder conveyor relative to the combine or relative to the ground, and return to that preset position automatically. The system response rate can also be adjusted using additional control knobs on the armrest.

A— Header Raise/Lower Switch    B— Header Control Dial

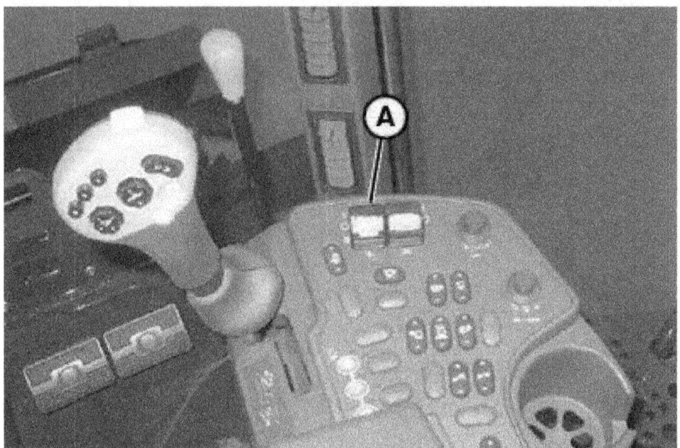

Fig. 16 — Header Engage and Feeder House Reverser Switch

A— Header Engage and Feeder
House Reverser Switch

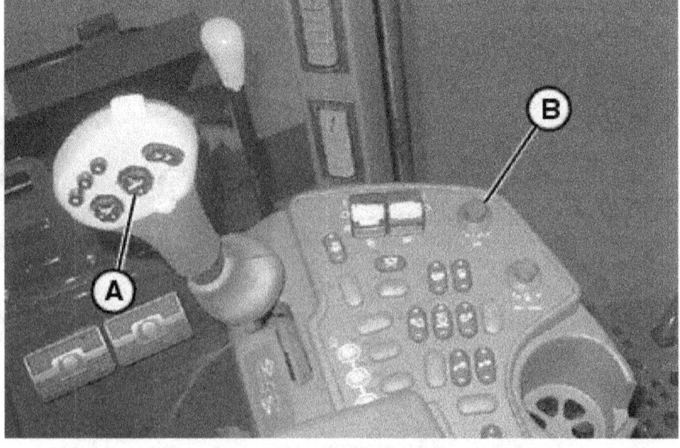

Fig. 17 — Header Raise/Lower Controls

**Continued on next page**

KN52281,1004558 -19-31JUL13-3/18



PN=114

## ACCUMULATOR

Accumulators are used in the hydraulic system of headers that are both automatic height controlled and those that ride on skid plates. Accumulators help protect the hydraulic system from internal damage by absorbing load and shock as the headers are operated.

An accumulator for a combine is usually a container that is charged with nitrogen gas on one side of a rubber bladder or piston, and hydraulic oil on the other side of the bladder or piston (Fig. 18).

The nitrogen gas compresses as pressure from the hydraulic fluid is applied to the bladder or piston. By doing this, the nitrogen absorbs the pressure that is otherwise absorbed by the other parts of the hydraulic system.

Accumulators can be adjusted in two ways. The header movement and free play can be stiffened by adding nitrogen gas to the container. (Do not overfill the container.) Another adjustment is at the valve that controls the hydraulic fluid flow into the container. A fully open valve will allow looser header movement than one that is closed.

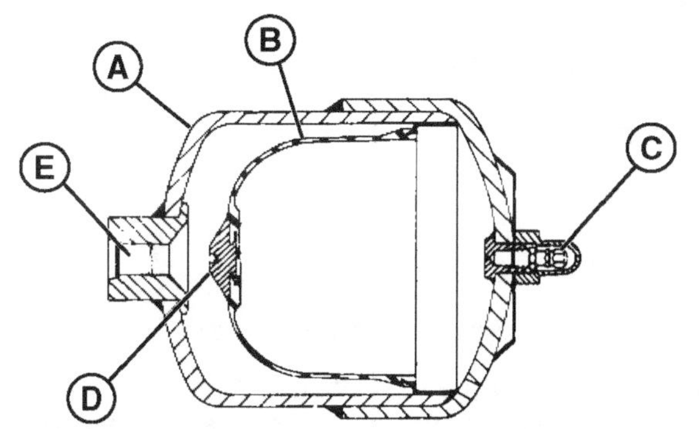

Fig. 18 — Cross Section of Typical Gas-Filled, Bladder-Type Combine Header System Accumulator

A— Housing
B— Bladder
C— Gas Valve
D— Button
E— Oil Port

KN52281,1004558 -19-31JUL13-4/18

## CUTTING PLATFORM

Controls for the cutting platform consist of the above controls in addition to hydraulic lift reel and, on some combines, variable-speed reel controls. The hydraulic lift reel switch (Fig. 19) actuates a control valve that causes the hydraulic cylinders for the reel to raise or lower the reel to the desired position. The variable-speed reel control knob allows the operator to adjust the reel speed to the ground speed. The switch also controls the reel's fore-and-aft motion.

To change reel speed, turn the knob in the proper direction to either increase or decrease speed. The reel speed can also be manually controlled using the reel speed switch on the armrest.

## PICKUP PLATFORM

Controls for the pickup platform consist of the header drive control and height control as described above. In addition to these, some combines have a hydraulic drive variable-speed belt pickup. In some cases the control that is used to control the variable-speed reel is used to operate this drive. Again, the operator can choose the exact speed at which the pickup belt revolves by turning the knob in the proper direction.

## CORN HEAD

The corn head controls also consist of a header drive control and a header height control. Many corn heads have variable-speed drives that are hydraulically controlled. On some combines, the same control lever used to raise and lower the reel is used to actuate the

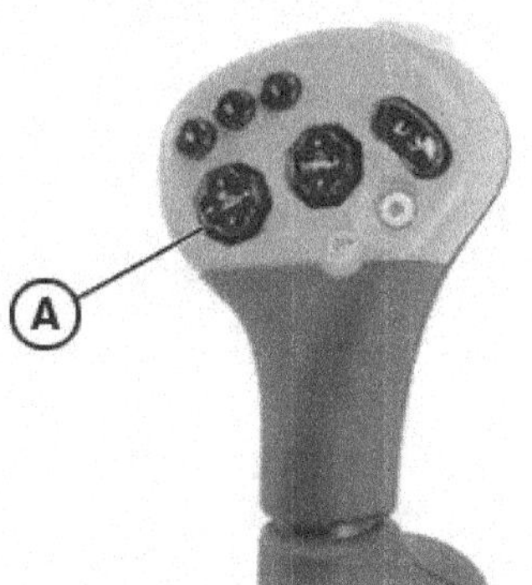

Fig. 19 — Lift Reel Control Switch

A— Lift Reel Control

hydraulic cylinder that controls the variable drive on a corn head. The operator can adjust the corn head speed to the forward travel of the combine.

Proper operation of snapping rolls, snapping bars, and gatherer chains is important for the smooth flow of materials into the combine. Improper functioning of these parts can cause uneven feeding, plugging, more downtime, and increased risk to the operator.

**CAUTION: Trying to unclog snapping rolls without shutting off the machine is a leading cause of serious injuries during corn harvesting.**

Continued on next page

KN52281,1004558 -19-31JUL13-5/18



050217
PN=115

## SEPARATOR CONTROL

The separator engage switch controls all the drives of the separator. The separator drive must be engaged to operate the following drives:

- Header drives
- Threshing drives
- Separating drives
- Cleaning drives
- Grain handling drives (except unloading auger drive)

Only a few of the above drives have individual controls. Most combines have a separator engage switch as shown in Fig. 20. To engage the separator, push the separator engage switch down and forward. To disengage, pull the switch back.

*Fig. 20 — Separator Engage Switch*

**A— Separator Engage Switch**



## THRESHING CONTROLS

These are probably the most important controls on the combine. The operator must adjust the cylinder or rotor speed and concave spacing to match crop and harvesting conditions. On most combines, these adjustments can be made by pushing a switch on the armrest console.

### CYLINDER OR ROTOR SPEED

Cylinder or rotor speed is controlled by a switch at the armrest console (Fig. 21). Cylinder speed may be adjusted from 150 rpm to 1500 rpm. The switch controls a hydraulic valve that directs hydraulic oil to the cylinder variable-drive sheave to effectively increase or decrease the diameter of the sheave by moving the sheave halves closer or farther apart. This increases or decreases the speed of the cylinder. Cylinder or rotor speed may be viewed on a multi-function display, usually located on the cornerpost of the cab (Fig. 22).

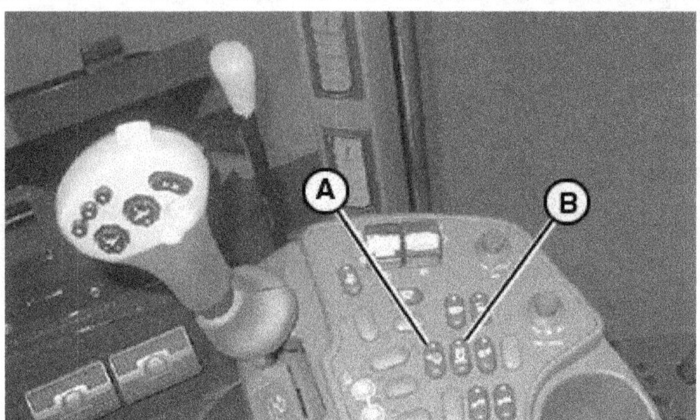

*Fig. 21 — Threshing Controls*

**A— Concave Spacing Switch**     **B— Cylinder Speed Switch**



### CONCAVE SPACING

On most combines, the concave spacing is adjustable from the operator seat. In Fig. 21, a concave clearance switch is located on the armrest console. Pressing the switch to increase or decrease concave clearance engages an electric actuator attached to the concave. A sensor mounted to a shaft or linkage is used to sense the concave position. On most machines the concave position is viewable, as a number, on a multi-function display (Fig. 22).

**A— Cylinder Tachometer**     **B— Concave Spacing**

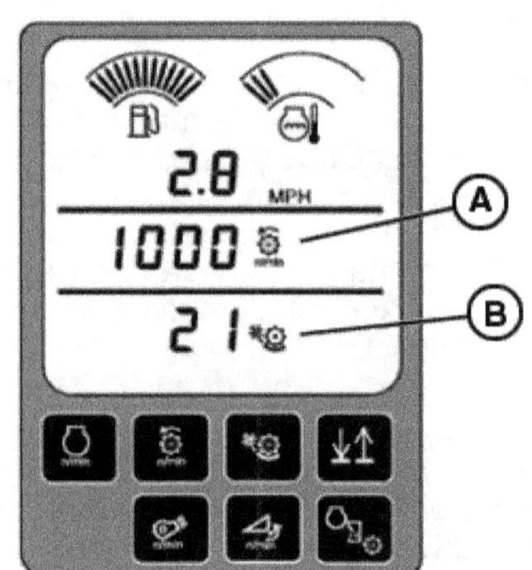

*Fig. 22 — Multi-Function Display — Cylinder Tachometer and Concave Spacing*

Continued on next page     

## CLEANING CONTROLS

The cleaning controls adjust the fan speed, chaffer, and sieve. The fan must be adjusted to the crop and feed rate. The chaffer is adjusted to allow the fan blast to separate the chaff from the grain and not allow too much coarse material through. The sieve is adjusted to allow only grain through.

Chaffer and sieve adjustments vary from machine to machine. In Fig. 23, chaffer and sieve adjustments are made by moving the adjusting lever at the rear of the machine to the left or right. A gauge underneath the adjustment lever shows the chaffer and sieve opening measurements.

*Fig. 23 — Chaffer and Sieve Controls*

A— Sieve                    B— Chaffer

KN52281,1004558 -19-31JUL13-9/18

In Fig. 24, the adjustment is made from the side of the combine using a hand crank. A gauge extending through the side of the combine shows the opening measurements.

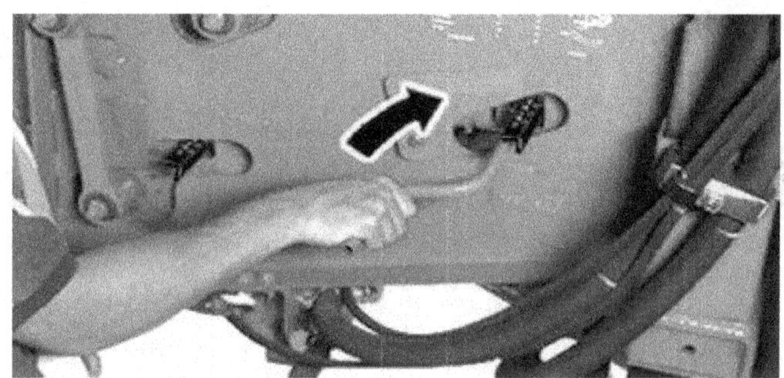

*Fig. 24 — Chaffer Adjustment*

KN52281,1004558 -19-31JUL13-10/18

On some combines, chaffer and sieve adjustments can be made from the operator seat (Fig. 25). Pressing the chaffer or sieve adjustment switch on the armrest console activates an electric motor mounted to the chaffer or sieve. Sensors mounted to the chaffer or sieve monitor the adjustments, and the values can be displayed in the cab.

Cleaning fan speed is also controlled from the operator seat. Pressing the cleaning fan speed switch activates an electric actuator. The actuator controls a variable-drive sheave. The actuator moves the sheave halves closer together or farther apart, altering the diameter of the drive sheave and changing fan speed. A speed sensor mounted on the fan shaft monitors fan speed, and fan speed can be displayed on the multi-function display.

A— Sieve Switch             C— Chaffer Switch
B— Fan Speed Switch

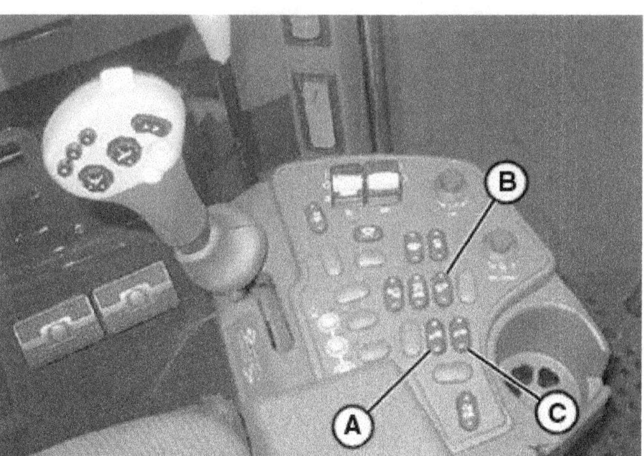

*Fig. 25 — Cleaning Controls*

Continued on next page

KN52281,1004558 -19-31JUL13-11/18

## GRAIN TANK UNLOADING AUGER CONTROLS

The grain tank unloading auger has two operations that are controlled by the operator:

- Swinging the auger into operating position or transport position
- Emptying the grain tank

The unloading auger swing is controlled by a switch in the cab, located on the control lever (Fig. 27). Pressing the extend portion of the switch energizes a hydraulic valve that directs hydraulic oil to the unloading auger swing cylinder, which will swing the auger into the unloading position. When the grain tank has emptied, pressing the retract portion of the switch will swing the auger back toward the machine. When transporting the combine, install the unloading auger locking pin to secure the auger in place.

An unloading auger extension is required on some combines when using very wide headers to permit easier unloading.

### UNLOADING GRAIN TANK

To empty the grain tank, push the unloading auger drive engage/disengage switch on the control lever (Fig. 27). When the grain tank is empty, push the button again to disengage the drive.

A— **Unloading Auger Swing In and Out**

B— **Unloading Auger Drive Engage/Disengage Switch with Indicator Light**

Fig. 26 — *Placing Auger in Unloading Position*

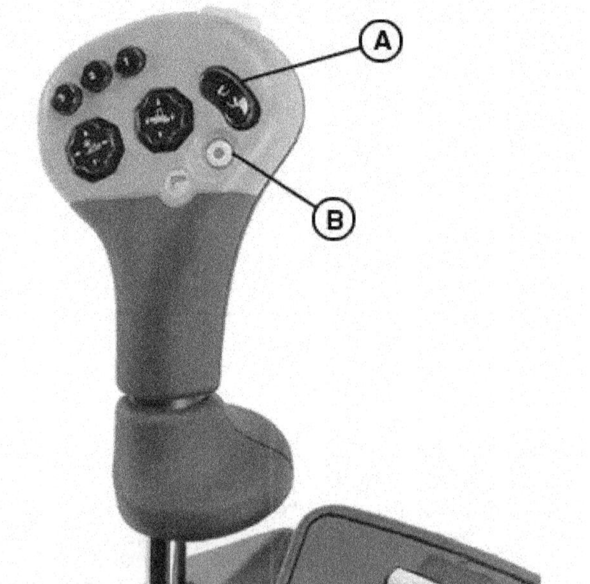


Fig. 27 — *Unloading Auger Controls (Multi-Function Lever)*

**Continued on next page**          KN52281,1004558 -19-31JUL13-12/18

## ELECTRICAL ACCESSORIES

The combine may be equipped with the following electrical devices for operating the combine:

- Field and highway safety lighting
- Separator warning device
- Slow shaft speed monitor
- Electronic grain loss monitor
- Feed rate control system
- Moisture sensors, yield monitor, and yield mapping
- GPS-assisted steering systems

We will now look at these accessories in detail.

### FIELD AND HIGHWAY SAFETY LIGHTING

A modern combine is equipped with many different lighting options. Fig. 29 shows an example of light switches located on the overhead panel of a combine. Road and hazard lights (Fig. 28) ensure proper visibility when transporting the combine. Field lights provide necessary light when harvesting at night, including lights on the front of the combine, grain tank, and at the end of the unloading auger. Work lights illuminate the front, rear, and sides of the machine, as well as the cleaning shoe area when servicing the machine. Refer to machine operator's manual for specific lighting details and operation.

Fig. 28 — Field and Highway Safety Lighting

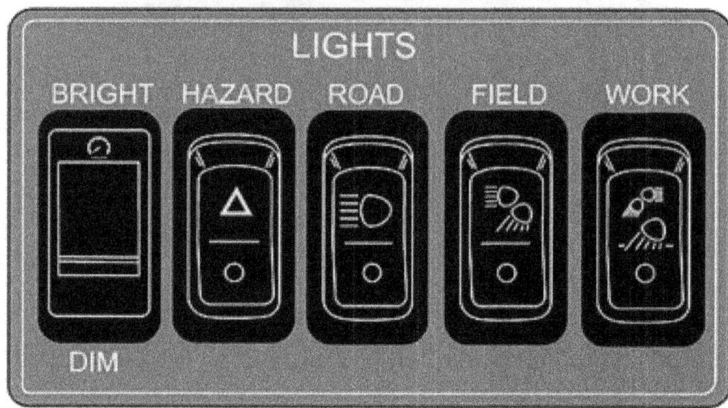


Fig. 29 — Light Switch Panel

KN52281,1004558 -19-31JUL13-13/18

### SEPARATOR WARNING DEVICE

Most combines are equipped with systems to monitor the operating conditions of the separator. Straw walker combines are usually equipped with sensors that check the walker area for plugging. If the straw walkers become plugged, a warning light on the display panel will illuminate and a warning horn will be sounded. When the horn sounds, stop the combine and turn off the engine. Inspect the straw walkers and remove accumulated straw.

Rotary machines may sense the load on the rotor and rotor drive components, and warn when operating loads exceed a set point. If the operating load on the drive components becomes excessive, a warning light on the display panel will illuminate and a warning horn will be sounded. Reduce ground speed or revise separator settings to reduce power requirements.

### SLOW SHAFT SPEED MONITOR

This device indicates (with warning lights) when certain essential shafts of the combine are running below their designed speed (Fig. 30). Units such as tailings elevator, grain conveyor, clean grain elevator, straw walkers, and straw choppers can be monitored.

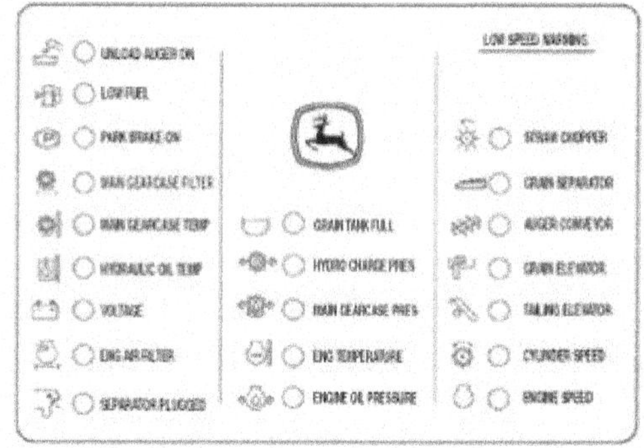

Fig. 30 — Overhead Warning Display Panel

When one or more of the lights glow, the operator must shut off the combine engine and investigate and remedy the problem before continuing to operate.

### ELECTRONIC GRAIN LOSS MONITOR

Electronic measuring and sensing devices are becoming increasingly popular with combine operators. Not only can remote sensing devices help increase yield, they can help save harvest time and make the job safer.

Continued on next page

KN52281,1004558 -19-31JUL13-14/18

The most common type of monitoring system is the grain loss monitor. It measures impacts from grain as grain comes off the cleaning shoe, separator, and tailings elevator. These grain impacts on the monitor create impulses that are transmitted to a preamplifier (Fig. 31). There the impulses are averaged into a DC current that is connected to a display in the operator cab.

A— Control Panel
B— Straw Walker Sensors
C— Cleaning Shoe Sensors
D— Preamplifier Circuit Box

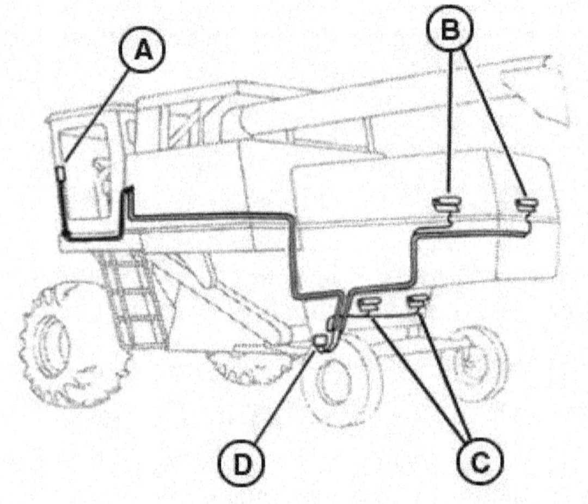

*Fig. 31 — Grain Loss Sensing System in a Typical Combine*

KN52281,1004558 -19-31JUL13-15/18

By presetting the grain size, and acceptable loss of grain at the shoe and separator, the combine operator can determine grain loss and modify the combine to attain the best operating results (Fig. 32).

FEED RATE CONTROL SYSTEMS

Some newer combines are equipped with systems that automatically vary the ground speed to maintain a constant flow of material through the machine, maximizing productivity and reducing operator stress. As the crop material gets lighter, machine ground speed will increase, and as crop material gets heavier, machine ground speed will decrease. The system will continually monitor engine speed, separator load, and grain losses, and will vary the ground speed accordingly.

Essentially, the machine is driven at the fastest possible speed to prevent the separator from plugging or the engine speed from dropping below an acceptable rpm. An electronic control unit monitors machine functions and makes adjustments to ground speed by energizing proportional control valves mounted on the hydrostatic

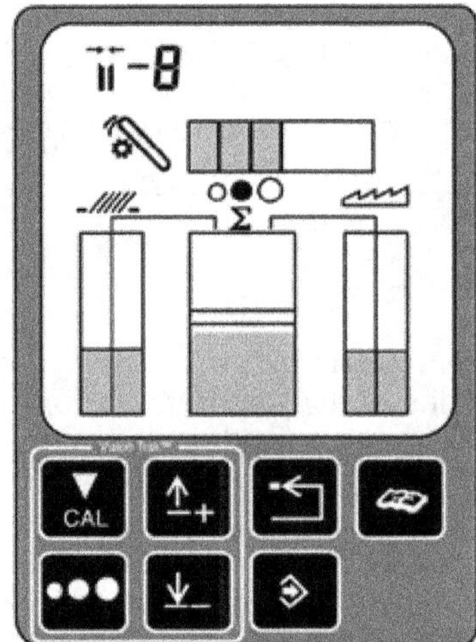

*Fig. 32 — Grain Loss Monitor*

drive pump. The proportional control valves control pump swash plate angle, which varies oil flow to the hydrostatic drive motor mounted on the transmission.

**Continued on next page**

KN52281,1004558 -19-31JUL13-16/18



PN=120

## MOISTURE SENSING, YIELD MONITORING, YIELD MAPPING

Many combines are now equipped with the latest technology, all used to increase machine efficiency, productivity, and overall profitability. Moisture sensors (Fig. 33) are used to take moisture readings of the crop while it is being harvested. A small portion of grain is taken from the clean grain elevator and passed through the moisture sensor, where moisture content is measured, and then returned to the grain tank. Information from the moisture sensor is received by an electronic control unit and then displayed on a readout in the operator station.

The mass flow sensor in Fig. 33 is used to determine crop yield as it is being harvested. The sensor is mounted to the top of the clean grain elevator. As the grain reaches the top of the elevator, it contacts the sensor. This information is sent to an electronic control unit, which displays the yield information for the operator to view.

Fig. 33 — Moisture and Mass Flow Sensors

KN52281,1004558 -19-31JUL13-17/18

Many combines are also equipped with a GPS (global positioning system) receiver (Fig. 34), which uses information from satellites to precisely calculate the machine's position in the field. With the use of GPS, the moisture and yield data, along with productivity, can be recorded during harvest and later downloaded to a personal computer.

The farmer can then create detailed maps of each field harvested, showing moisture content and yield information at each specific point in the field. This information can be used to help make better management decisions.

## GPS-ASSISTED STEERING SYSTEMS

Using the same GPS systems previously discussed, assisted steering systems are now available on some combines to permit GPS-guided steering. Using signals from the GPS satellites to determine machine location, the assisted steering system can steer the combine in straight, parallel paths through the field.

Machine efficiency is increased by providing a consistent cut using the full width of the platform at every pass, reduced grain loss, and the ability to harvest more acres per hour.

Operator fatigue is also reduced by enhancing operator comfort and providing easier operation in dusty, low-visibility conditions. An electronic control unit uses inputs from the GPS, steering wheel position, and rear wheel position to make the necessary corrections to keep the machine on a straight, parallel path. The electronic control unit controls a hydraulic steering valve that directs hydraulic oil flow to the steering cylinder at the rear of the

Fig. 34 — Global Positioning System

A— GPS

combine. When the operator positions the combine in the field, the system can be activated by pressing a button. The operator may resume manual steering control at any time while operating the machine.

With the latest advancements in technology, some companies have developed a system in which a combine operator can autonomously control an operator-less tractor and grain cart from inside the cab of the combine. The controlled tractor will follow the combine at a safe distance until the combine operator gives a command to the tractor that the clean grain tank is full, and the tractor and grain cart pull up under the unloading auger. When done unloading, the combine operator sends another command to park the tractor, which sends it to a predetermined location for manual unloading of the grain cart.

KN52281,1004558 -19-31JUL13-18/18

## SYMBOLS FOR OPERATING CONTROLS

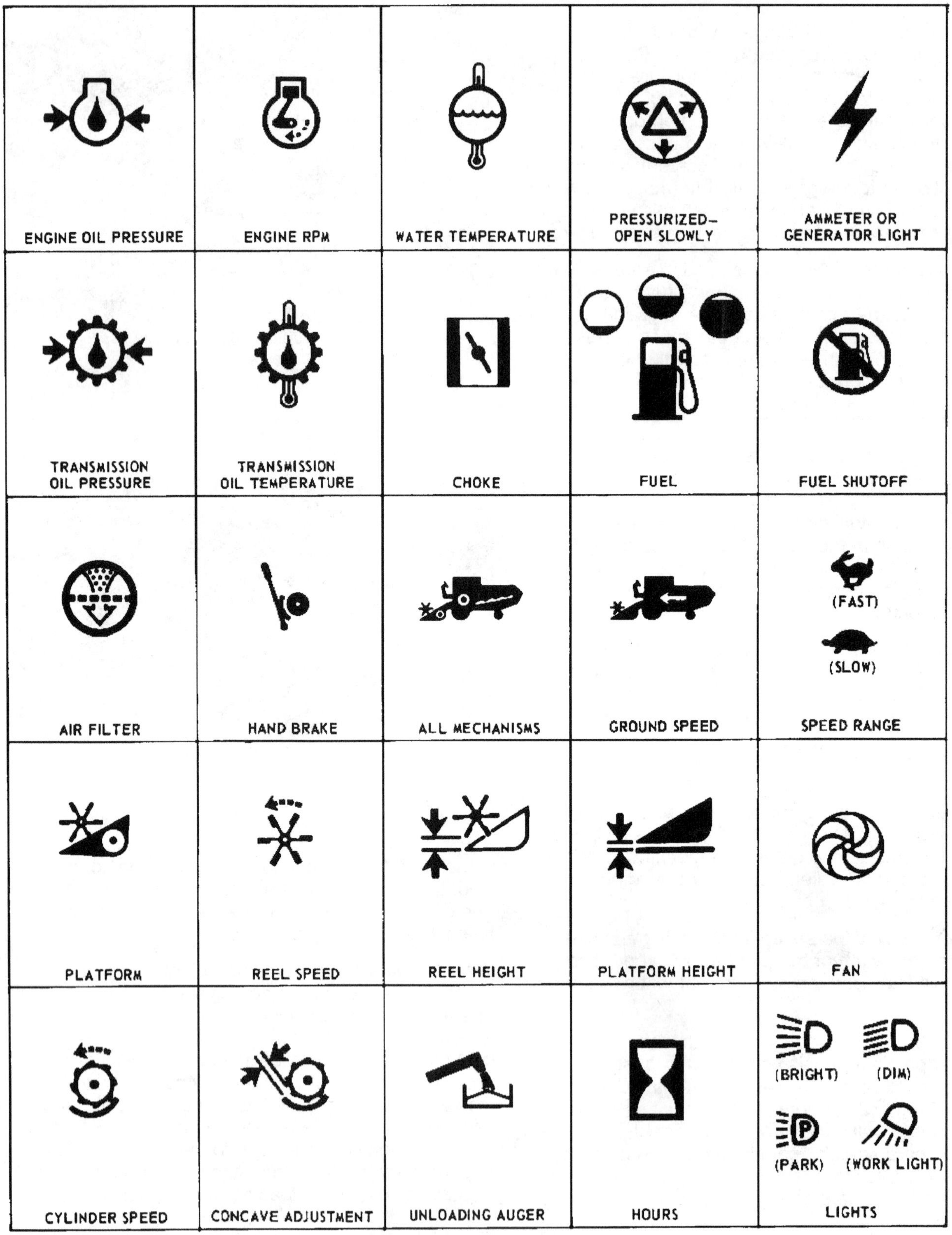

| | | | | |
|---|---|---|---|---|
| ENGINE OIL PRESSURE | ENGINE RPM | WATER TEMPERATURE | PRESSURIZED— OPEN SLOWLY | AMMETER OR GENERATOR LIGHT |
| TRANSMISSION OIL PRESSURE | TRANSMISSION OIL TEMPERATURE | CHOKE | FUEL | FUEL SHUTOFF |
| AIR FILTER | HAND BRAKE | ALL MECHANISMS | GROUND SPEED | (FAST) (SLOW) SPEED RANGE |
| PLATFORM | REEL SPEED | REEL HEIGHT | PLATFORM HEIGHT | FAN |
| CYLINDER SPEED | CONCAVE ADJUSTMENT | UNLOADING AUGER | HOURS | (BRIGHT) (DIM) (PARK) (WORK LIGHT) LIGHTS |

*Fig. 35 — Universal Symbols*

DXP04223 —UN—08JAN13

Continued on next page

KN52281,1004559 -19-28AUG13-1/2

Some manufacturers use symbols instead of words to indicate the function of each operating control. Many of these symbols are shown in Fig. 35. Study these universal symbols and learn to recognize them at a glance.

KN52281,1004559 -19-28AUG13-2/2

# TEST YOURSELF

## *QUESTIONS*

1. (Fill in the blanks) Before starting the engine, turn the key to the ON position and always check the operation of the _________________ _________________ according to the operator's manual.
2. (Fill in the blank) The _________________ control should be placed in the full open position when operating the combine in the field.
3. Describe the proper procedure for shutting down a turbocharged diesel engine.
4. Before starting a combine, what four fluid levels should be checked?
5. (True or False) The combine should always be operated at maximum ground speed.
6. What three ways are suggested for transporting a combine?
7. (True or False) The threshing cylinder or rotor speed may be changed at any time while in operation.
8. What are three of the cleaning unit controls that may need adjustment during operation?
9. (True or False) A warning device may warn the operator if the straw walkers become plugged.
10. What is the main purpose of the leveling systems used on some combines?

KN52281,100455A -19-25JUL13-1/1

## INTRODUCTION

DXP02705 —UN—23FEB11

**5**

OUO1082,000624D -19-25JUL13-1/4

No one can qualify as a good combine operator by simply reading the operator's manual; a competent operator must have experience. Even those who have had experience can always learn new "tricks." Here we will discuss the basic operation and adjustments that can help the operator do a good job of combine harvesting.

To be a good combine operator, you must know:

- Functional design of the combine
- Basic principles of operation
- How to make proper adjustments
- How to identify harvesting losses
- How to maintain efficient operation

In previous chapters we have covered the first two items; in this chapter we will discuss the others.

*Fig. 1 — Even an Experienced Operator Can Learn New Tricks*

**Continued on next page**

OUO1082,000624D -19-25JUL13-2/4

## *PROPER OPERATION AND ADJUSTMENT ARE IMPORTANT*

Combine harvesting can be profitable only if the operator knows how to adjust the combine properly and operate it efficiently with a minimum of losses.

In corn, for example, recent studies have shown that profit losses of 9 percent or more can result if harvesting efficiency is not ideal (Fig. 2). In soybeans, the profit loss can be 13 percent or more.

What is ideal efficiency? In some studies the average harvesting efficiency for 100-bushel-per-acre (5 t/ha) corn was 92 percent, while the ideal was 97 percent. In 30-bushel-per-acre (1.6 t/ha) soybeans, 87 percent efficiency was average, while ideal was 97 percent.

How do you get ideal efficiency? This chapter helps solve that complex problem, which is basically a question of proper operation and adjustment of the combine.

Harvest losses directly result in decreased profits. These losses are caused by poor adjustment and operation

Fig. 2 — *Poor Harvesting Efficiency in Corn Can Result in a Profit Loss of 9 Percent or More*

of the combine. In corn yielding 100 bushels per acre (5 tonnes/ha), for example, if losses were 4 bushels per acre (200 kg/ha), value lost per acre would be 10 dollars at a price of $2.50 per bushel (10 cents per kg). In addition, marketing losses include penalties for low test weight, moisture, damages, and excessive trash or foreign material, all of which can add significantly to the profit losses.

**Continued on next page**

OUO1082,000624D -19-25JUL13-3/4

## WHAT ARE PROPER OPERATION AND ADJUSTMENT

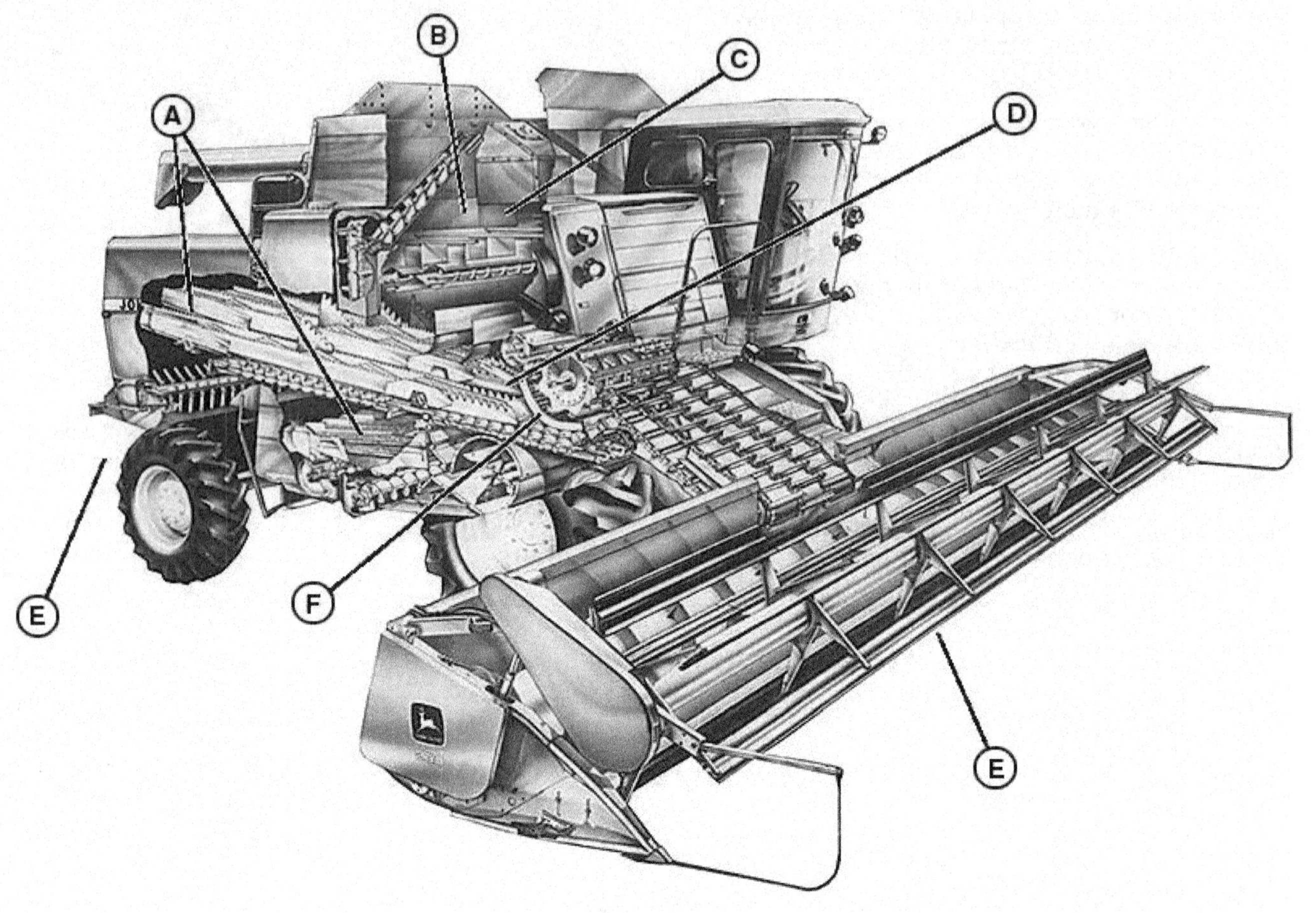

*Fig. 3 — Signs of Poor Combine Harvesting*

A— Grain Lost from Shoe
B— Chaff and Trash in Grain Tank
C— Cracked Grain in Grain Tank
D— Straw Chewed Up Excessively
E— Grain Losses on the Ground
F— Unthreshed Kernels

The operator must recognize the effects of both good and poor operation. Illustrated in Fig. 3 are some of the signs of poor combine harvesting:

- Grain losses on the ground
- Unthreshed kernels on the straw, cob, or in the pod
- Straw chewed up excessively
- Grain lost from the straw walkers or shoe
- Excessive tailings in the tailings elevator
- Cracked grain in the grain tank
- Chaff or trash in the grain tank
- Marketing penalties for low-quality grain due to harvesting damage or crop condition

### FACTORS THAT AFFECT COMBINE HARVESTING

Several important factors affect the harvesting of crops with a combine. The following factors must be considered in every instance of combining:

- Skill of the operator
- Condition of the crop and field
- Adjustment of the combine
- Proper operating speed of harvesting components
- Ground speed of the combine
- Width of the header

### COVERAGE IN THIS CHAPTER

This chapter is not meant as a guide to combine harvesting, but it is meant to explain the principles of combine harvesting, which are invaluable to any combine operator.

Here are the subjects that will be covered in this chapter:

- Planning and preparation
- Preliminary settings
- Field operation and adjustments
- Field problems

OUO1082,000624D -19-25JUL13-4/4

## PLANNING AND PREPARATION

Prior to combining, the operator must make plans and prepare for the harvesting operation. It is not enough to know how to operate a combine; the operator must also consider the following factors:

- Harvesting methods
- When to combine
- Consequences of early and late harvesting
- Production capacity of the combine
- Mechanical condition of the combine

Let's look at these subjects to learn their importance.

### HARVESTING METHODS

Usually the crop determines the harvesting method (Fig. 4). For example, when harvesting corn with a combine, a corn head must be used; when harvesting other crops, a cutting platform is used. A row-crop head may be used for soybeans, sunflowers, or sorghum. In addition, some crops are harvested with a windrow-pickup platform. Many combine owners raise one or two different crops that may require different headers (discussed in chapter 2) and often different combines, such as a corn-grain-soybean combine, an edible-bean combine, or a rice combine. These combines may be converted to other crops by changing the header or threshing cylinder or both. Of course, other minor changes may be made.

One harvesting method that needs further explanation is the windrow method. We will discuss the combining of the windrow later in this chapter. However, before a combine can be used to harvest the crop, the crop must be cut and windrowed by a grain windrower.

### GRAIN WINDROWERS

Grain windrowers (sometimes called swathers) are often used when there are many weeds in the grain, when there is considerable moisture at harvest time, when the crop ripens unevenly, or when it does not ripen in time for harvest before bad weather begins (Fig. 5). In these cases it is usually desirable to cut the grain with a windrower and thresh it later with a combine equipped with a windrow-pickup attachment (pickup platform or belt pickup).

The grain windrowers place the grain in a windrow. Advantages provided by a windrow are:

- Crop is less susceptible to bad weather
- Allows even drying of grain
- Allows full use of the combine by placing up to 42 feet (12.8 m) of grain in narrow swath

*Fig. 4 — Four Harvesting Methods*

**Continued on next page**

DK75838,0000206 -19-26AUG13-1/17

## TYPES AND SIZES

The grain windrower may be either a self-propelled (Fig. 5) or a pull-type (Fig. 6). The self-propelled machines are gradually replacing the pull-types because they are more maneuverable and they free the tractor to perform other functions.

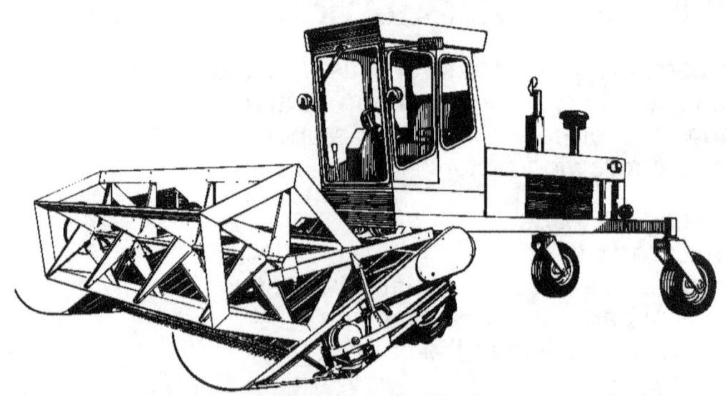

Fig. 5 — Self-Propelled Grain Windrower

Fig. 6 — Typical Pull-Type Windrower Planning and Preparation

**Continued on next page**

DK75838,0000206 -19-26AUG13-2/17

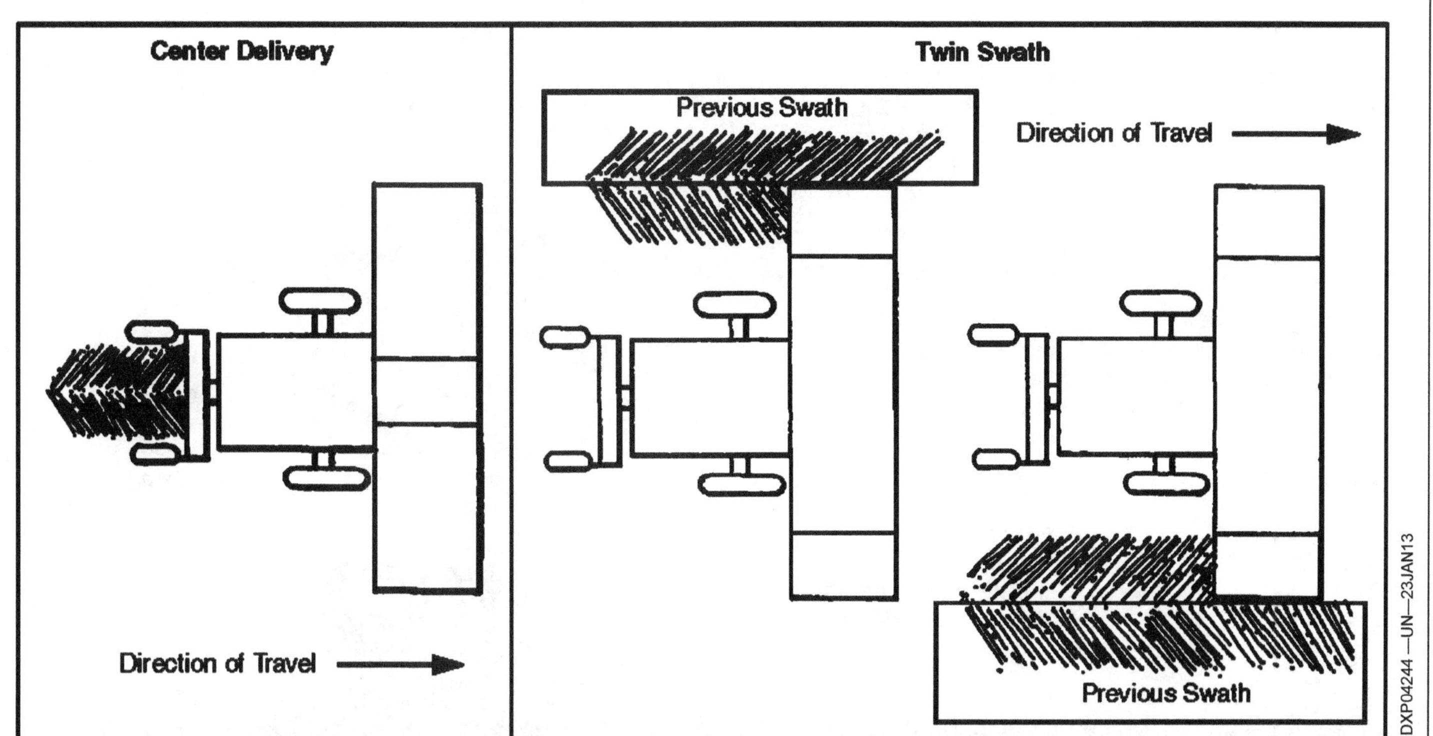


*Fig. 7 — Two Types of Deliveries Are Possible*

With both the self-propelled and pull-types, two kinds of deliveries are possible. The most common is the center or single delivery. The other type is the twin or double swath (Fig. 7). When the twin swath is used, windrows are placed side by side. The first swath is shown in the box. With the double swath, one windrow is placed on top of the other.

**Continued on next page**

DK75838,0000206 -19-26AUG13-3/17

## CONSTRUCTION

The grain windrower is constructed with a bat-type reel at the front (Fig. 8). The reel makes the first contact with the grain. A cutterbar operates beneath the reel to cut the crop. Behind the cutterbar is a pair of canvases. They travel from the outside edge of the platform to an opening in the center of the single swath platform.

This opening is usually about 3 to 4 feet wide.

The opening on a self-propelled machine is in the center. The opening of a pull-type machine is generally offset to the left, to provide clearance for the tractor on the next round. The twin swath platforms have openings on each end (Fig. 9).

## OPERATION OF THE WINDROWER

The bat-type reel of the windrower divides and delivers the crop. The material is then cut off by the cutterbar. The reel then delivers the cut crop to the canvases behind the cutterbar. From here the crop is then conveyed by the canvases to the opening and discharged to the ground (Fig. 8).

The twin-swath platform on self-propelled windrowers has a movable canvas. This canvas is moved by hydraulics to allow delivery to the opening at either end of the platform. With the pull-type windrower, there is an attachment that will cover the opening with a movable canvas which conveys the material to the left-hand side of the machine, beside the previously laid swath. This twin swath attachment is also moved by tractor hydraulics to open and close the platform opening for proper swath position.

## HOW TO WINDROW GRAIN

The grain is ready to be windrowed when it reaches 35 percent moisture level. This is normally referred to as the late dough stage. This is when the line between the green and gold on the straw stem is about 8 inches above the ground. After the grain reaches this maturity, the crop will not be damaged by cutting. This guideline suggests

Fig. 8 — *The Reel Divides and Delivers the Crop*

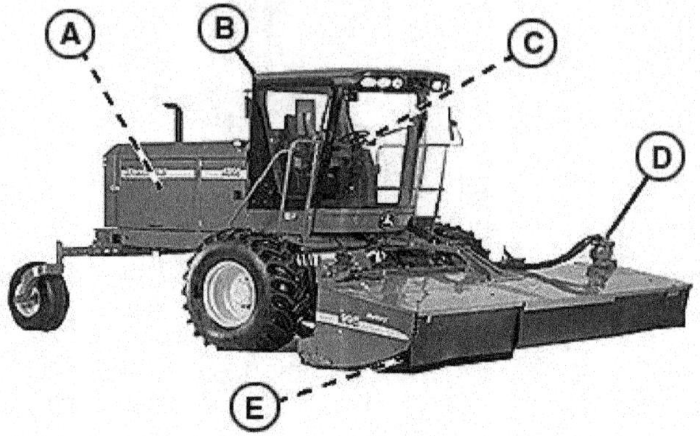

Fig. 9 — *Typical Windrower Construction*

A—Engine
B—Cab
C—Platform Lift Lever
D—Cutterbar Drive Case
E—Cutterbar

windrowing about one week earlier than many consider grain is ripe enough to cut. University tests show, however, that windrowing does not damage the crop at this stage.

DK75838,0000206 -19-26AUG13-4/17

## CHECKING THE MACHINE

Prepare the windrower by thoroughly checking the machine for loose bolts, lubricating properly (refer to the operator's manual), and checking the condition of the cutting parts.

Adjust the windrower before going to the field. Check the condition of the canvas and adjust the tension if necessary. The canvas is usually tightened by a spring-loaded mechanism. Be sure the tighteners are capable of moving freely (Fig. 10). The drive belts and chains must be inspected for wear and then tightened properly if that adjustment is needed.

DXP04247 —UN—23JAN13

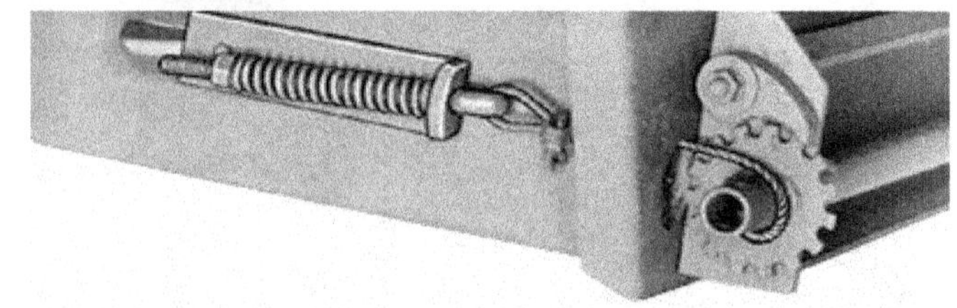

Fig. 10 — *Canvas Adjusting Mechanism*

Continued on next page

DK75838,0000206 -19-26AUG13-5/17



050217
PN=130

## OPENING THE FIELD

To open a field with a self-propelled center delivery platform, go around the field in a clockwise direction and continue until the field is completely cut (Fig. 11).

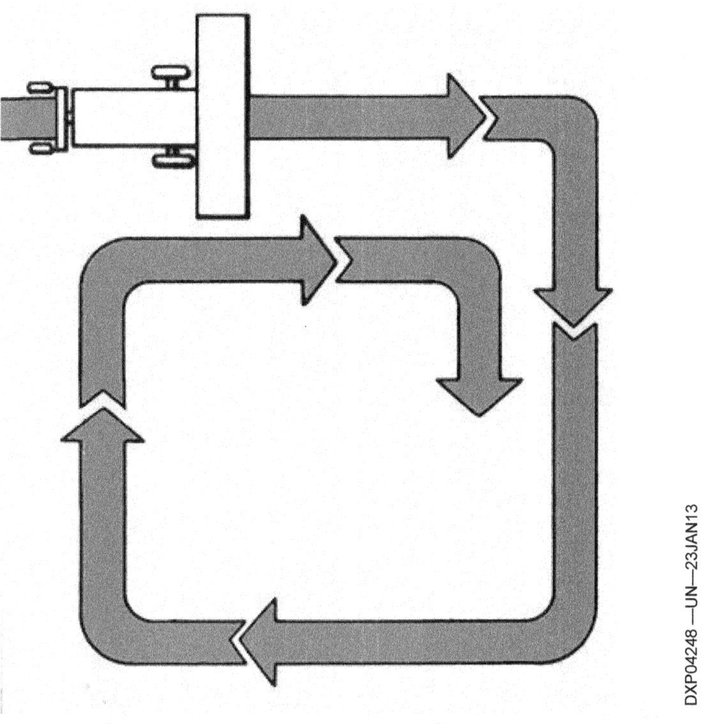

*Fig. 11 — Opening a Field with Self-Propelled Center Delivery Windrower*

DK75838,0000206 -19-26AUG13-6/17

To open a field with a pull-type single swath platform, start in a clockwise direction for a couple of rounds (Fig. 12). Next, reverse direction to pick up the portion knocked down by the tractor on the opening swath. Then go back to cutting the rest of the field in a clockwise direction.

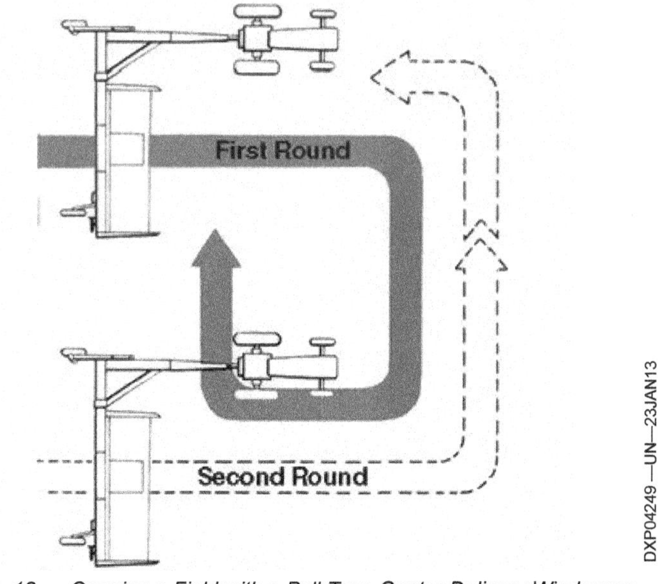


*Fig. 12 — Opening a Field with a Pull-Type Center Delivery Windrower*

**Continued on next page**

DK75838,0000206 -19-26AUG13-7/17

To open a field for a self-propelled twin swath platform, shift the canvas to the left. This delivers the grain to the opening on the right (Fig. 13). Go in a clockwise direction for one round.

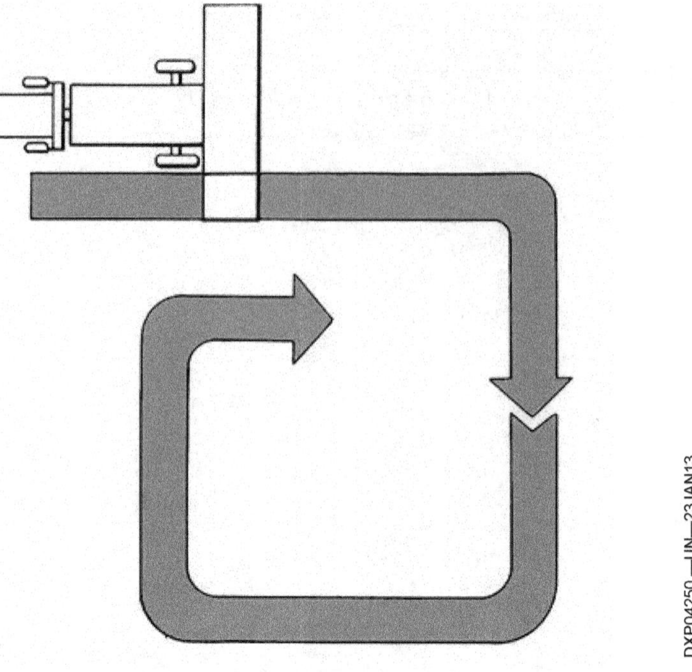

*Fig. 13 — Opening a Field with a Self-Propelled Twin Swath Windrower*

DK75838,0000206 -19-26AUG13-8/17

On the second round, shift the canvas to the right to allow left-hand delivery and continue in a clockwise direction (Fig. 14). This places the two swaths beside each other. Continue in a clockwise direction, alternately placing the windrow to the right and to the left.

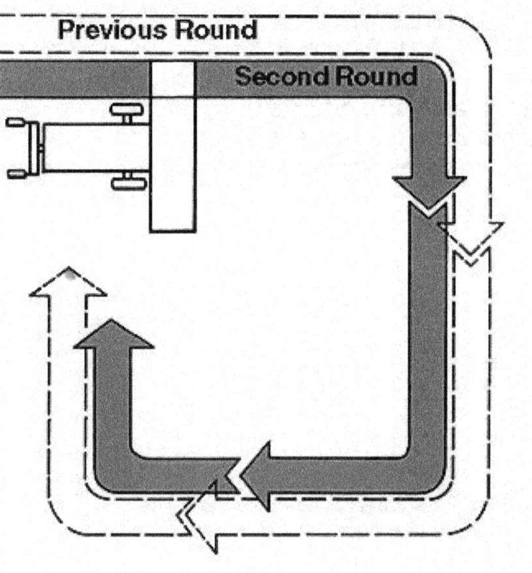


*Fig. 14 — Second Row with a Self-Propelled Twin Swath Windrower*

**Continued on next page**                    DK75838,0000206 -19-26AUG13-9/17

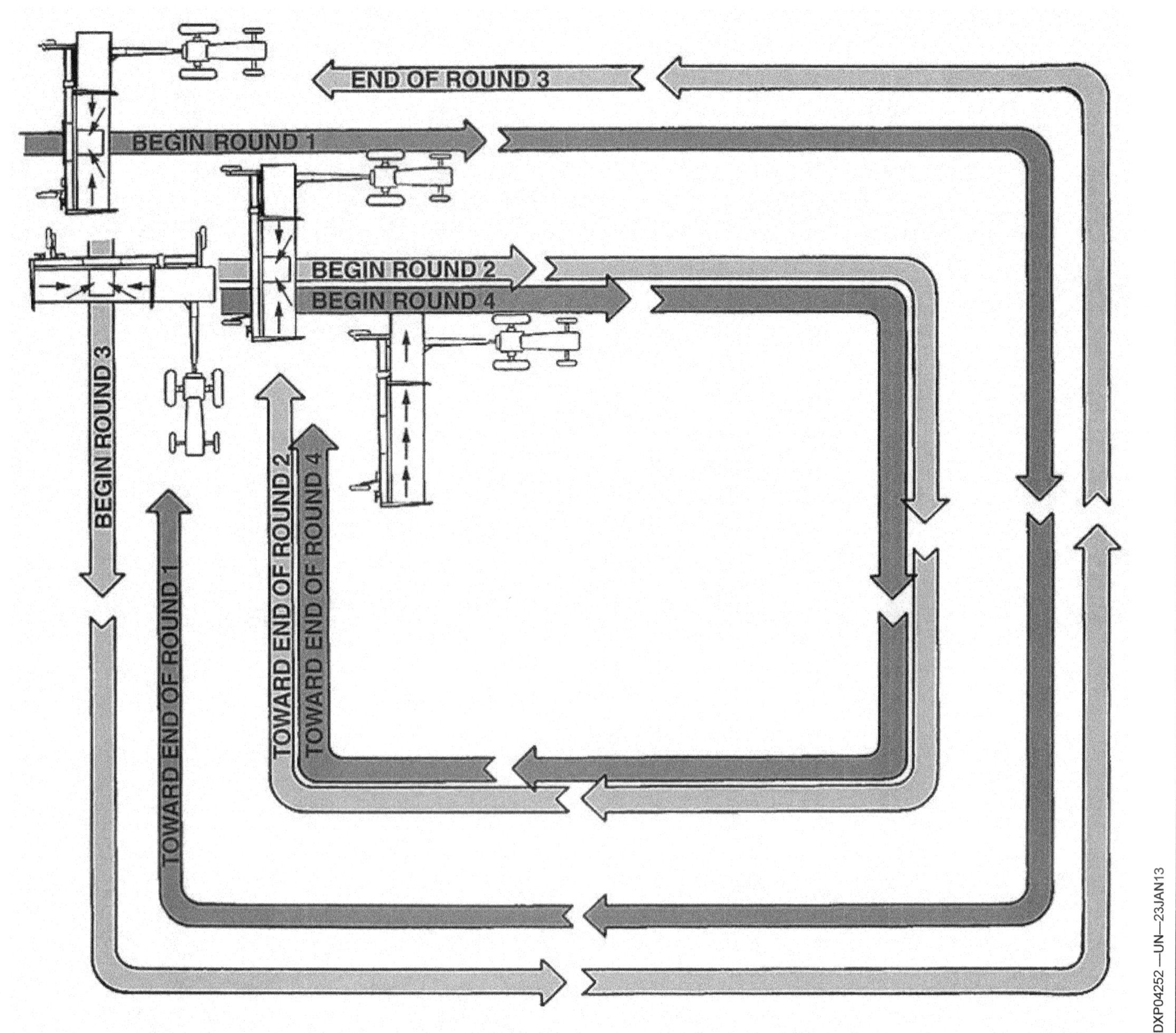


*Fig. 15 — Opening a Field with a Pull-Type Twin Swath Windrower*

To open a field for a pull-type twin swath platform, go in a clockwise direction for two rounds with the canvas in the single swath position (Fig. 15, Rounds 1 and 2). Reverse direction and cut the material knocked down by the tractor on the opening swath (Round 3).

Shift the canvas and start to twin swath beside the second swath laid (Round 4). Then continue to alternately lay the windrow beside and then behind the tractor. (This method may be altered because of tillage practices or because a self-propelled machine may be used to open fields.)

**Continued on next page**

DK75838,0000206 -19-26AUG13-10/17



DXP04252 —UN—23JAN13

050217
PN=133

## FIELD ADJUSTMENTS

The field adjustments necessary are those required to lay a proper type of swath. The most desirable type of swath is a windrow with most of the grain heads in line, with a slight herringbone pattern.

The drawbacks to the herringbone windrow are:

1. With all the heads in the center of the windrow (Fig. 16) during wet weather, the heavier heads will go down and touch the ground.
2. The combine cylinder will only be used at the center since that is the only position of the heads.

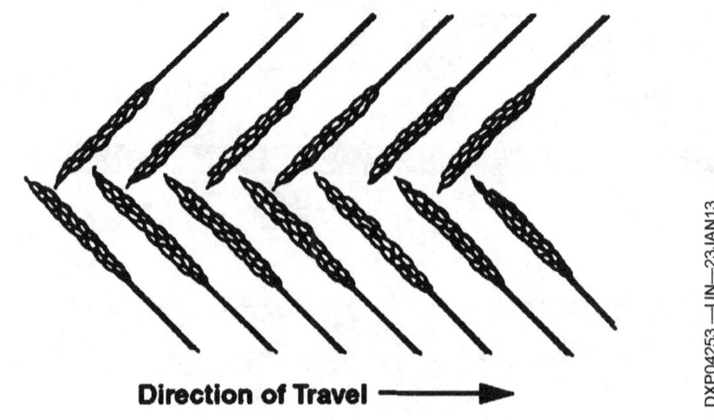


Fig. 16 — Herringbone Swath with Heads in Center (Poor Windrow)

DK75838,0000206 -19-26AUG13-11/17

The drawback to a strict heads-in-line windrow is that some straw will go down between the grain drill rows and never be picked up (Fig. 17).

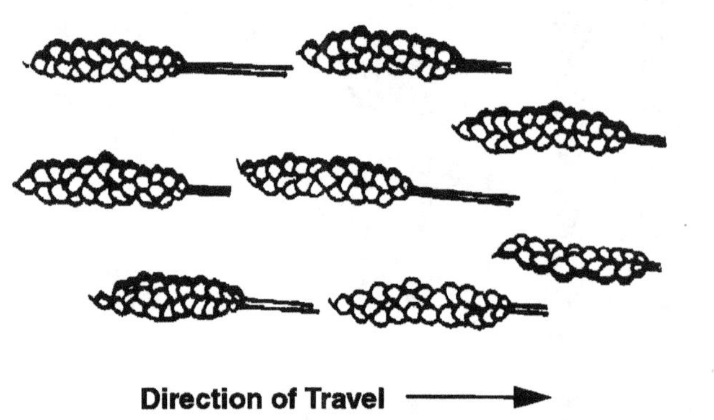


Fig. 17 — Heads-in-Line Windrow (Poor Windrow)

DK75838,0000206 -19-26AUG13-12/17

To obtain the desired heads-in-line with a slight herringbone pattern requires proper speed relationships between the reel, the ground, and the canvas (Fig. 18). It also requires proper height from the ground to the chaffer and from the cutterbar to the reel. These speed and height relationships must be maintained to give the desired windrow.

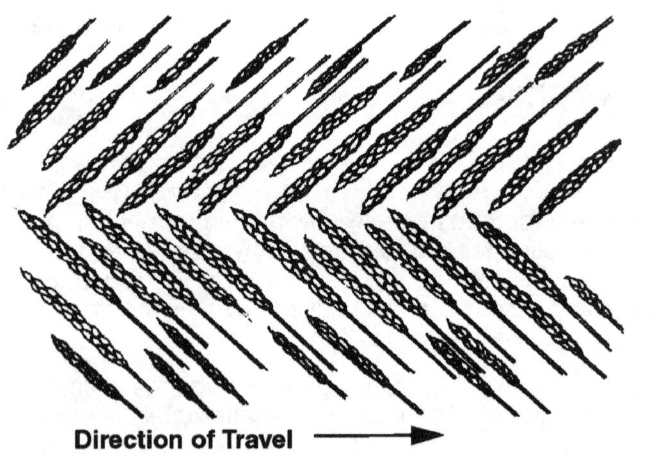


Fig. 18 — Heads-in-Line with Herringbone Pattern (Best Windrow)

Continued on next page

DK75838,0000206 -19-26AUG13-13/17

Start the adjustments by obtaining proper cutterbar height. The stubble should be 4 to 8 inches high (10 to 20 cm) (Fig. 19). This height is enough to allow adequate air circulation and yet not tall enough to allow bending so the heads may touch the ground. The height of the reel from the cutterbar to the reel bat is determined by the height of the grain. Position the reel to let it touch only the top 6 inches (15 cm) of grain. Continual adjustment of the reel height may be necessary due to uneven height of the standing grain.

Once the height relationships have been set, the reel speed and ground speed must be set. Determine the proper ground speed. This will normally be 4 to 6 miles per hour (6.4 to 9.7 km/h) depending on the ground conditions. Next, set the reel speed. The reel speed should be 25 percent faster than ground speed. Finally, set the canvas speed, if adjustable, to lay the desired windrow at the selected ground speed and reel speed settings.

Other aids to forming good windrows are a short piece of canvas at the openings for the butts to fall on and the use of windrow forming rods.

*Fig. 19 — Stubble Should Be 4 to 8 Inches (10 to 20 cm) High*

**Continued on next page**

DK75838,0000206 -19-26AUG13-14/17

## TRANSPORTING THE WINDROWER

The self-propelled windrower should be equipped with flashing warning lights and an SMV emblem (check local regulations). Some pull-type windrowers may be placed in a transport position by repositioning the wheels to the end to allow endways transport (Fig. 20). With either type, safety must be stressed due to the width of the machine.

## MAINTENANCE

The components requiring maintenance for the grain windrower are the belts, chains, canvases, and cutting parts. Check these items periodically while using the machine. Follow the preventive maintenance procedures as outlined in John Deere's FMO Preventive Maintenance manual.

Storage recommendations for a grain windrower are:

1. If possible, shelter the windrower in a dry place, or cover it with a tarpaulin.
2. Clean the windrower thoroughly. Chaff and dirt will draw moisture and rust the steel.
3. Remove and clean the canvases. Hang canvases in a dry place where they will not be subject to damage from inclement weather or rodents.
4. Remove the tension from all belts and clean them thoroughly.
5. Clean the chains thoroughly. Brush heavy oil on the chains to prevent corrosion.
6. Lubricate the windrower completely. Grease the threads on adjusting bolts and the sliding surfaces of the variable-sheave assemblies.
7. Paint all parts from which paint has worn.
8. Support the platform with blocks to level it.
9. Block up the windrower to take the load off the tires. Do not deflate the tires. If the windrower is stored outside, remove the wheels and tires and store them in a cool, dry place away from direct sunlight.

Fig. 20 — Transporting Pull-Type Windrower

### STOCK OF PARTS

It is advisable to have on hand a number of parts that are subject to wear and breakage. The parts list for grain windrowers includes:

- Knife guards
- Knife wear plates
- Knife sections
- Knife hold-downs
- Rivets
- Chain couplers
- Drive belts
- Engine air filters
- Engine oil filters

## FIELD PROBLEMS WITH WINDROWERS

The following chart provides a general guide to solving field problems. Each operator's manual has such a chart to help the operator make proper adjustments or to correct improper operation of windrower.

### CROP LOSS AT CUTTERBAR TROUBLESHOOTING CHART

| PROBLEM | POSSIBLE CAUSE | REMEDY |
|---|---|---|
| Shattering of grain ahead of cutterbar. | Excessive agitation of grain heads due to incorrect entry of reel slats into crop. | Set reel so reel slats feed material smoothly to cutterbar and canvases. |
| | Reel speed not coordinated with ground speed, causing excessive agitation before crop is cut. | Change reel drive to coordinate reel speed with ground speed so reel will move material smoothly and evenly. |
| | Ground speed too fast for condition of crop. | Reduce ground speed so reel will not bat crop, causing shattering of grain heads. |
| Cut crop building up and falling from front of cutterbar or loss of grain heads at cutterbar. | Reel not adjusted low enough for proper delivery of cut crop to canvases. | Set reel low enough to sweep material from cutterbar to canvases. |
| | Cutting platform too high, cutting stalks too short for proper delivery. | Lower cutting platform so stalks of crop will be long enough for smooth, even feeding and to support windrow. |

### POOR CUTTING ACTION TROUBLESHOOTING CHART

| PROBLEM | POSSIBLE CAUSE | REMEDY |
|---|---|---|
| Ragged and uneven cutting action. | Various parts of cutterbar, such as knife sections, guards, and wearing plates are worn, damaged, or broken. | Check and replace all worn and broken parts on cutterbar to obtain an even cutting of crop. |
| | Bent knife, causing binding of cutting parts. | Straighten the bent knife. Check guard alignment and align, if necessary, for a smooth cut. |

Continued on next page

DK75838,0000206 -19-26AUG13-15/17

| | Cutterbar out of register. | Adjust register. (See combine operator's manual.) |
| | Knife clips not adjusted to permit knife to work freely. | Adjust knife clips so knife will work freely, but still keep knife sections from lifting guards. |
| | Looseness between knife back and guard. | Adjust knife clips so knife back is snug to guard. |
| | Lip of guard out of adjustment or bent, causing poor shearing action. | Adjust lip of guard so it is parallel to shear edge of guard. |
| | Loose cutterbar drive belt. | Adjust drive belt. |
| Excessive vibration of cutting parts. | Excessive looseness of cutting parts and knife drive. | Remove all excess play from cutterbar and knife drive to eliminate vibration. After removing excess play, make certain cutterbar and knife drive are properly adjusted and move freely. |

### IMPROPER REEL DELIVERY TROUBLESHOOTING CHART

| PROBLEM | POSSIBLE CAUSE | REMEDY |
|---|---|---|
| Reel wrapping in tangled and weedy crops. | Incorrect location of reel and improper setting of reel slats. | Place reel well ahead and down. |
| | Reel speed too fast. | Reduce speed of reel to allow weedy crops to fall onto platform. |
| Reel carrying crop around. | Tall grain or nodding varieties of crops catch on reels slats and arms. | Increase width of reel slats with wire screen or canvas for nodding varieties of crops. |
| | Reel speed too fast. | Reduce speed of reel. Reel should turn faster than ground travel just enough so that crop heads are laid back on cutting platform. |
| | Reel height too low. | Raise reel height to reduce amount of crop gathered by reel. |
| Crop falling in front of cutterbar after it is cut. | Reel speed too slow. | Adjust reel to turn 25 percent faster than forward travel of combine. |

DK75838,0000206 -19-26AUG13-16/17

## SUMMARY: Windrowers

The grain windrower makes a windrow to be picked up by the grain combine (Fig. 21). It comes in two styles, self-propelled and pull-type. Either type can lay two types of windrows — single swath and twin or double swaths.

*Fig. 21 — The Grain Windrower Makes a Windrow to Be Picked Up by the Combine*

DK75838,0000206 -19-26AUG13-17/17

## WHEN TO COMBINE

Some combine operators may argue about the best time to combine — one operator may like to combine before the crop reaches full maturity because the whole crop can be harvested early; another may like to harvest at the optimum maturity because less crackage and losses will result.

In any case, if the crop is harvested too early or too late, the crop may suffer damage or losses, which could cut profits considerably.

The usual rule of thumb is to harvest as early as possible when the moisture content has dropped to acceptable levels.

Premature harvesting results in yield losses and a reduction in quality. Immature grain produces smaller yields and less weight. The grain is also harder to thresh, which tends to cause threshing damage and incomplete threshing.

Delayed harvesting results in losses from shattered grain, lodged plants, down plants, and reduced weight. Shatter losses occur because the grain is too easily threshed. Losses from lodged and down plants occur because the header cannot gather the grain properly. Reduced weight means weight loss, which may result in penalties of 1/2 to 1 cent per pound under minimum weight.

*Fig. 22 — Combine When the Crop Is Ready*

### BEST HARVESTING TIMES

Best harvesting times vary with the crop and weather. The proper stage for harvesting is when the grain will give the biggest yield at the highest quality.

Here are general suggested harvesting times for corn, soybeans, and wheat:

DK75838,0000207 -19-25JUL13-1/4

### CORN

Usually corn gives its highest yields when the moisture content is between 20 and 30 percent. At higher moisture (35 percent), losses result from cracking and small pieces of kernels (fines). Also, profit is lost from the cost of drying the corn to a moisture content that is safe for storage.

*Fig. 23 — Corn with 20 to 30 Percent Moisture May Be Harvested*

**Continued on next page**

DK75838,0000207 -19-25JUL13-2/4

## SOYBEANS

Usually soybeans are harvested when the moisture content is between 12 and 14 percent. When the plants are very dry, gathering losses may be high. This is because the stems and pods become brittle and shatter when they come in contact with the cutterbar and reel. If the bean crop is too dry, it may help to harvest after a light rain or dew. The extra moisture will make the pods and stems more durable and will cut down on shatter losses.

*Fig. 24 — Soybeans Are Usually Harvested at 12 to 14 Percent Moisture Content*

DK75838,0000207 -19-25JUL13-3/4

## WHEAT

Direct combining of wheat is usually done when the moisture content is below 14 percent. At this moisture content, however, shattering and cutterbar losses are greater than at higher moisture levels.

At 14 to 20 percent moisture, heads will be harder to thresh from the straw. However, wheat can be harvested at any of the above moisture levels because grain can be dried to safe storage levels. It is a matter of figuring the advantages and disadvantages. At the hard dough stage, wheat can be cut, windrowed, and allowed to dry to safe levels.

## OTHER COMMON CROPS

Best moisture levels for harvesting oats, barley, rye, and sorghums is usually 13 to 14 percent. If harvested at higher moisture, drying must be used to bring the grain down to a moisture content safe for storage.

*Fig. 25 — Wheat Is Usually Harvested When Moisture Content Is Below 14 Percent*

DK75838,0000207 -19-25JUL13-4/4

## PRODUCTION CAPACITIES OF COMBINES

The production capacity of a combine is important to the operator. Sometimes, smaller capacity means less grain harvested per day and so smaller profits. By estimating the number of acres per hour a combine can harvest, the operator will know how long it will take to completely combine the fields. This will give the operator a better idea of when to start, and whether more equipment or help will be needed.

Table 1 shows effective harvesting capacity of combines operating at harvesting efficiency of 75 percent, which is considered average. This means that 25 percent of the time is lost due to the following reasons:

- Refueling
- Lubricating
- Adjusting machine
- Unloading
- Turning at ends of fields
- Unclogging machine
- Breakdowns

The percentage of time lost can be more or less than 25 percent according to layout of the field, adjustment of the machine, preventive maintenance practices, and operator's skill.

### HOW TO USE THE CAPACITY CHART

To use Table 1, first determine the width of the header. Our example is a combine with a 16-foot (4.9 m) header. If the combine is equipped with a corn head, figure the width of the header by the row spacing and number of rows the header can harvest (for example, a corn head that is set to combine three 30-inch rows is equivalent to 3 x 30 inches or a header width of 7.5 feet).

Next, draw a vertical line from the appropriate header width to the line representing miles per hour ground speed (example is 3 mph [5 km/h]). Now draw a horizontal line over to the line representing the time lost. From this point draw another vertical line to the top of the chart to

Fig. 26 — Production Capacity Is Important

determine the approximate acres harvested per hour (example is 4.25 acres per hour).

For metric calculation, assume the header width is 5 m (16 feet) and the combine has a field speed of 6 km/h (4 mph). Draw a vertical line from the header width to the line representing ground speed of 6 km/h (4 mph). Next, draw a horizontal line to the line representing time lost. From this point draw another vertical line to the top of the chart and read the approximate hectares harvested per hour (example is 2.25 hectares per hour).

Now you know what the combine is capable of in one hour of time. Divide this figure into the total acres to be combined and you will find the number of hours it will take to harvest the crop.

Keep in mind that you may not be able to travel at the speed you selected because the capacity of the combine is affected by the crop and field conditions.

The average speed most combines travel is between 2.5 and 4.5 mph (4 and 7.25 km/h). This can also vary with the field conditions and the skill of the operator.

**Continued on next page**

DK75838,0000208 -19-25JUL13-1/2

## HARVESTING CAPACITY CHART

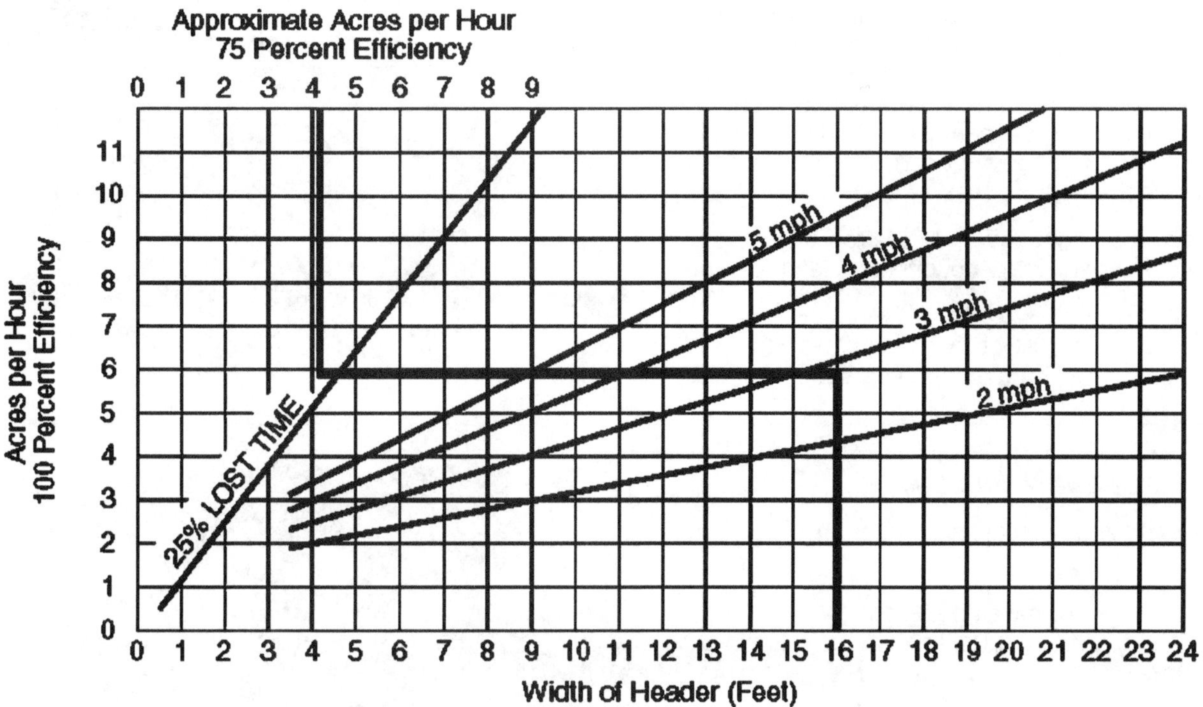



## HARVESTING CAPACITY CHART

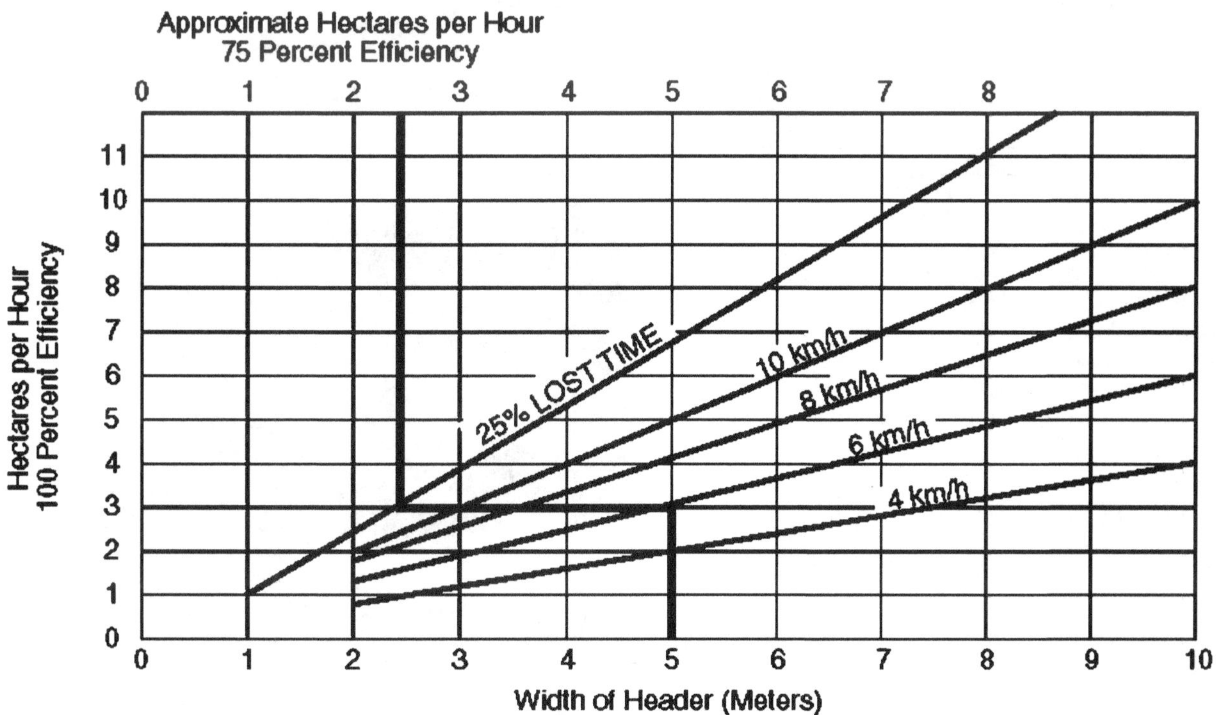



*Table 1 — Harvesting Capacity Charts*

DK75838,0000208 -19-25JUL13-2/2

## MECHANICAL CONDITION OF COMBINE

One of the costliest factors of combine harvesting is a machine that is in poor mechanical condition. A combine that does not receive good preventive maintenance will cost more in repairs, in addition to time lost due to breakdowns in the field. Poor mechanical condition may also cause poor harvest performance. Harvesting components that are damaged or worn cannot be adjusted to perform satisfactorily.

A combine should be checked and repaired between seasons to avoid unnecessary downtime when it can be most costly.

Here are some of the problems that should be corrected before attempting to begin the harvesting season:

Fig. 27 — Check the Condition of the Combine Regularly

DK75838,0000209 -19-25JUL13-1/8

---

### HEADER COMPONENT PROBLEMS

CUTTING PLATFORM PROBLEMS (FIG. 28)

- Bent knife guards
- Badly worn, broken, or missing knives
- Worn or broken cutterbar parts, such as hold-down clips or knife guards
- Broken reel slats
- Broken, bent, or missing reel fingers
- Worn or loose chains and belts
- Worn or bent feeder conveyor chain
- Bent or missing auger fingers
- Worn bearings
- Loose or missing bolts

A— Worn Belts and Chains    C— Bent Reel Slats
B— Damaged Cutterbar

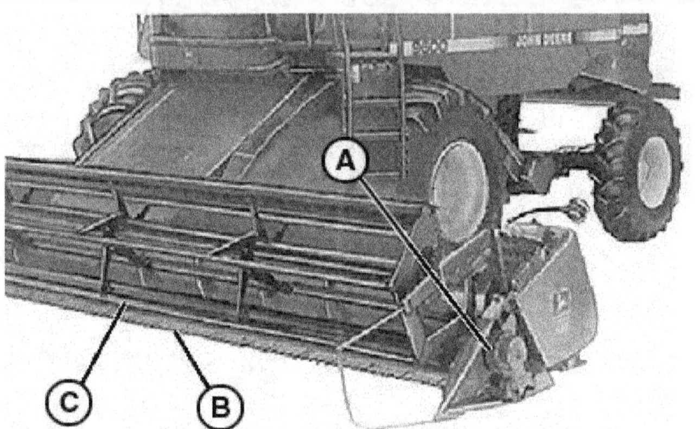

Fig. 28 — Check the Cutting Platform

DK75838,0000209 -19-25JUL13-2/8

---

CORN HEAD PROBLEMS (FIG. 29)

- Faulty gatherer points
- Worn or loose gatherer chains
- Worn snapping plates
- Worn or broken stalk plates
- Worn or bent trash knives
- Worn or bent feeder conveyor chain
- Worn or loose drive chains
- Worn bearings
- Loose or missing bolts

A— Worn Snapping Plates    C— Poorly Operating Gatherer
B— Worn or Loose Drive       Points
     Chains           D— Loose Gatherer Chains

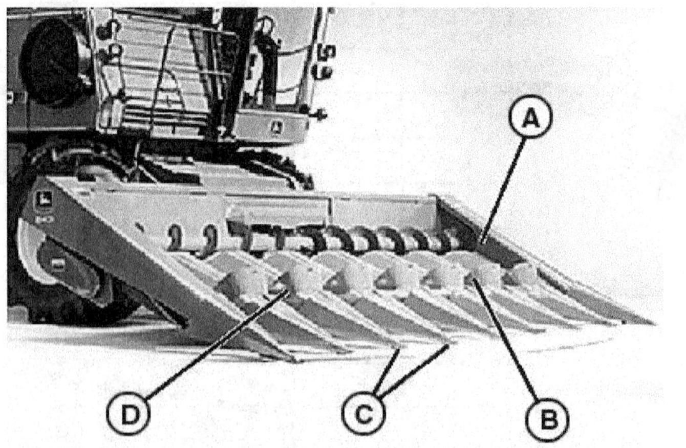

Fig. 29 — Check the Corn Head

Continued on next page

DK75838,0000209 -19-25JUL13-3/8



PN=142

## SEPARATOR COMPONENT PROBLEMS

### THRESHING SECTION PROBLEMS (FIG. 30)

- Worn or bent cylinder or rotor bars
- Worn or bent concave bars
- Dirt packed in cylinder, concave, or stone trap
- Misalignment of cylinder concave
- Worn or loose drive belts or chains

A— Bent Cylinder Bars    B— Bent Concave Bars

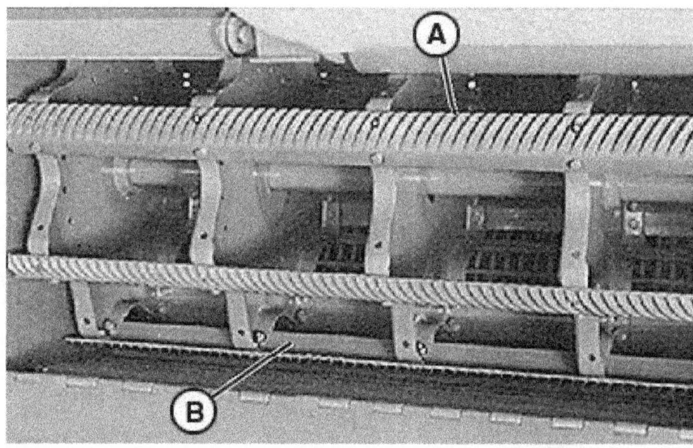

Fig. 30 — Check the Threshing Section

DK75838,0000209 -19-25JUL13-4/8

### SEPARATING SECTION PROBLEMS (FIG. 31)

- Damaged straw walkers
- Torn or missing straw walker curtain
- Worn or loose drive belts

A— Hinged Door    B— Straw Walkers

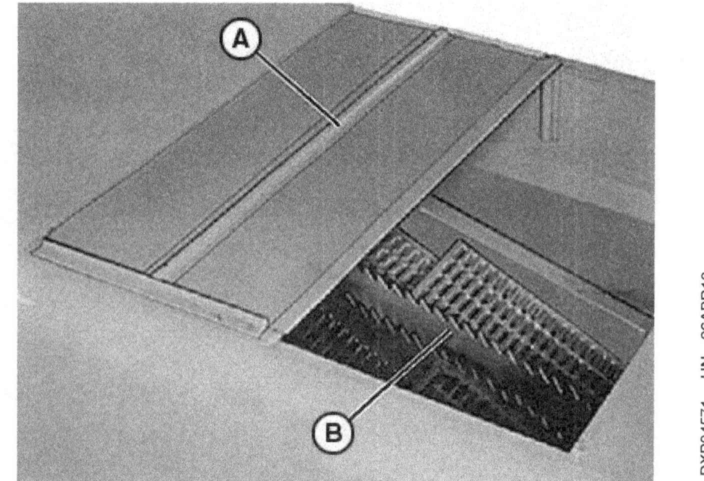

Fig. 31 — Check the Straw Walkers

DK75838,0000209 -19-25JUL13-5/8

### CLEANING SECTION PROBLEMS (FIG. 32)

- Damaged fan blades or fan shutters (if used)
- Damaged chaffer sieve
- Loose or bent shoe hangers

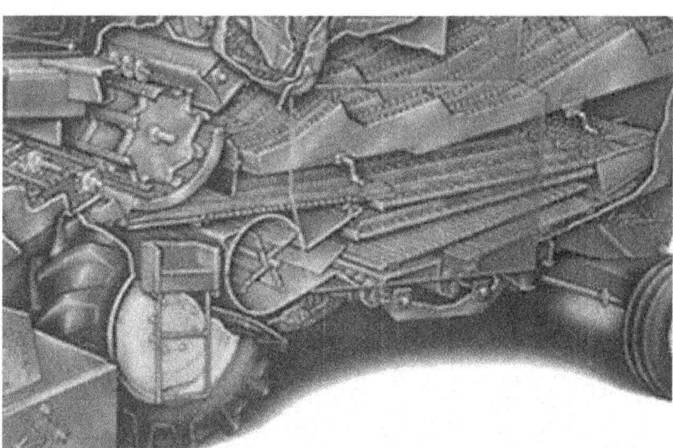

Fig. 32 — Check the Cleaning Section

**Continued on next page**

DK75838,0000209 -19-25JUL13-6/8

## GRAIN HANDLING PROBLEMS (FIG. 33)

- Worn elevator chains with missing paddles
- Badly worn, bent, or broken augers
- Dirt packed in grain tank or elevators

*Fig. 33 — Check the Elevators and Augers*

DK75838,0000209 -19-25JUL13-7/8

## OTHER PROBLEM AREAS (FIG. 34)

- Worn or loose drive chains and belts
- Damaged controls
- Worn bearings
- Loose or missing bolts
- Worn or broken sheaves and sprockets

## ENGINE PROBLEMS

- Engine not properly tuned
- Engine speed too fast or too slow
- Leaks in air, fuel, or oil lines
- Dirty air and oil filters
- Leaking or dirty cooling system

*Fig. 34 — Check All the Drive Chains and Belts*

DK75838,0000209 -19-25JUL13-8/8

## PRELIMINARY SETTINGS OF COMBINE

All combine operator's manuals have charts that recommend preliminary settings of the machine (Fig. 35). These settings are starting points only; each crop and field condition will require further adjustments.

*NOTE: Very few combines will do a good job when adjusted at these starting points unless field conditions are perfect.*

Here we will discuss the following important items:

- Proper operating speed
- Suggested combine settings

### PROPER OPERATING SPEED

Every combine separator is designed by the manufacturer to operate at a particular speed regardless of the crop harvested. Never alter the recommended speed, or the combine will not perform satisfactorily.

If the speed of the separator is slower than normal, the entire combine will run slower. This will cause sluggishness, which will result in clogging and grain losses.

If the speed of the separator is faster than normal, material will pass through the combine too rapidly, which will result in grain losses and excessive strain and wear on all components.

Nearly all new combines are equipped with speed sensors on most important separator functions. The speed of the rotor, cleaning fan, and header drive shaft can be viewed on the tachometer in some machines. Other functions such as chopper, beater, and elevator speeds are monitored also. If the speed of a specific function slows below a set point, a warning light for the particular function will illuminate and an alarm will sound to alert the operator. The operator should stop the engine and investigate the cause of problem.

### Factory-Programmed Separator Settings

Some combines come equipped with factory-programmed separator settings for each crop (corn, soybeans, wheat,

Fig. 35 — Read the Operator's Manual for Recommended Settings

etc.). The operator can change from one crop to another just by pressing a button. If the operator changes from harvesting soybeans to corn, all that is required is to change the crop shown on the display. All separator settings (concave clearance, cylinder speed, fan speed, and chaffer/sieve settings) are automatically changed for that crop without having to the change each of these settings individually. These settings are used as a baseline, and crop losses should be checked to see if any adjustments must be made. Adjustments can be made and saved to each crop so the operator can go right back to them without having to re-enter the changes. The operator can change just one setting, such as cylinder speed, or can create customized settings for each separator function.

**Continued on next page**

DK75838,000020A -19-26JUL13-1/26



050217
PN=145

## SUGGESTED COMBINE SETTINGS

Suggested machine settings for starting points in each crop to be harvested are just that — recommended starting points. They must be adjusted carefully to the crop and field conditions. Usually a range is given that tells the approximate settings for average conditions. When setting a particular combine, consult the operator's manual to determine where in the range to begin.

After combining the crop for a while at these settings, check for grain losses and adjust the combine as necessary; field adjustments will be covered later in this chapter.

On the following pages we will list typical settings for the most popular crops harvested by combines. Crops are listed in alphabetical order by popular name along with an illustration of the harvested crop. While many of these crops may be harvested in other ways for different uses, we will confine this discussion to combine harvesting. Keep in mind that some of these crops are harvested for seed or grain while many varieties are also harvested for hay, silage, and other purposes that require different methods of harvesting. For example, oats may be harvested for seed or grain with a combine, but when it is used for hay or silage a combine isn't used.

Let's look at some of these crops on the following pages. Notice that typical cylinder, concave, chaffer, and sieve settings for a conventional combine are given along with a brief description of the harvesting characteristics. Because cleaning fan settings vary considerably between makes of combines, they are not given here. These settings may vary a great deal with each manufacturer because of the various designs of combines, so always consult the operator's manual for specific recommendations.

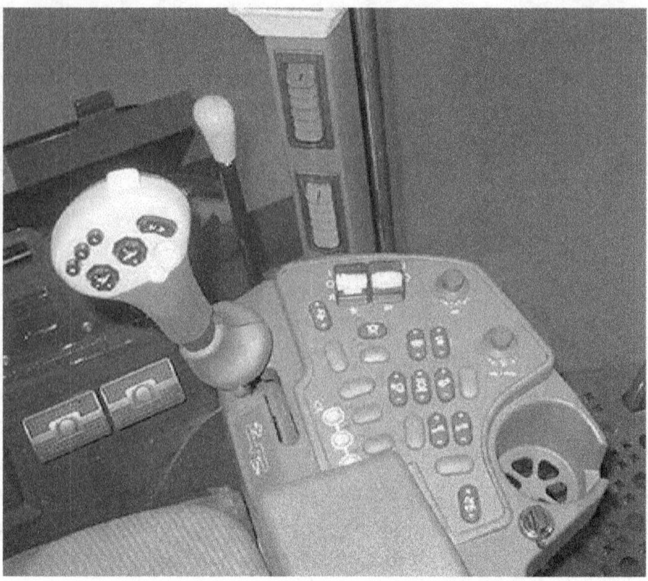

Fig. 36 — Adjust Combine to Suggested Starting Points

*NOTE: The concave settings shown indicate the range for the front of the concave.*

Concave spacing on rotary combines is generally set somewhat wider than on conventional combines, but the range may overlap for some crops. And, although the recommended range of rotor speeds for a particular crop may overlap the speeds suggested here, the best rotor speed for any crop will depend on rotor diameter, combine design, and crop conditions. So, always consult the combine operator's manual for specific recommendations.

DK75838,000020A -19-26JUL13-2/26

---

### ALFALFA

Alfalfa may be combined directly for seeds if the crop is in proper condition. However, when weeds or uneven ripening is a problem, it is usually cut with a windrower and cured before combining.

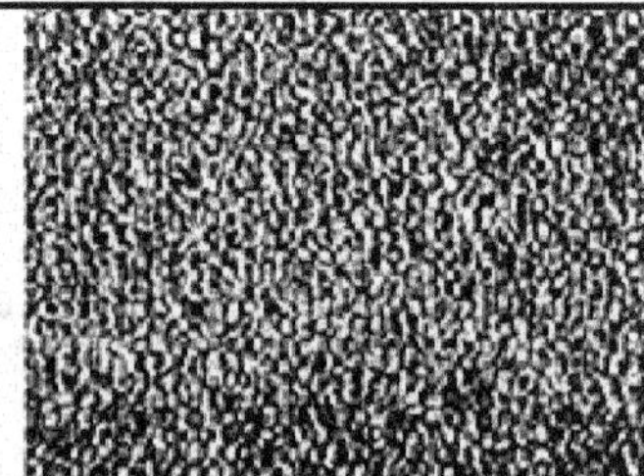

**ALFALFA**

| | |
|---|---|
| Cyl. rpm | 500–1300 |
| Concave | 1/8–3/8 in. (3.2–9.5 mm) |
| Chaffer | 3/8–1/2 in. (9.5 mm–12.7 mm) |
| Sieve | 1/16–1/8 in. (1.6–3.2 mm) |

DK75838,000020A -19-26JUL13-3/26

---

### BARLEY

Barley tends to shatter from the stalk when ripe and under direct combining methods. Shatter losses may be reduced by windrowing the crop when the heads have turned golden yellow and the straw is slightly green. After drying three or four days, it can be combined with a windrow-pickup unit.

**BARLEY**

| | |
|---|---|
| Cyl. rpm | 650–1300 |
| Concave | 1/8–5/8 in. (3.2–15.9 mm) |
| Chaffer | 1/2–3/4 in. (12.7 mm–19 mm) |
| Sieve | 1/4–1/2 in. (6.4–12.7 mm) |

Continued on next page

DK75838,000020A -19-26JUL13-4/26



PN=146

## BEANS

Beans (edible beans) are usually harvested when most of the pods have turned yellow, but before they shatter due to excessive dryness. They may be combined directly or windrowed first. Usually a low cylinder speed and a wide concave clearance are recommended to prevent cracking or splitting the beans.

DK75838,000020A -19-26JUL13-5/26

## BUCKWHEAT

Buckwheat usually ripens after the first frost. It is threshed easily by direct combining or it can be windrowed to aid curing and later combined with a windrow-pickup unit.

DK75838,000020A -19-26JUL13-6/26

## CLOVER

Clover seed is usually harvested with the windrow method because of weed and moisture problems. After the windrow is dry and cured, it is combined with a windrow-pickup unit. If the seed is combined directly from a standing crop (when 70 to 80 percent of the heads are ripe), the combine header must cut just under the heads because the stems are still green.

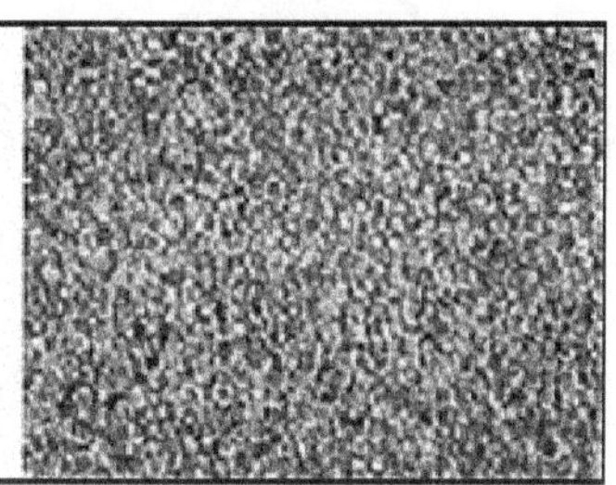


DK75838,000020A -19-26JUL13-7/26

## CORN

Corn (field shelled) is harvested with a combine equipped with a corn head. Cylinder filler plates are used to hold the ears close to the concave for complete shelling. These plates also keep cobs from passing through the cylinder unthreshed. Early harvesting is popular, but because the moisture content may be 20 to 30 percent, artificial drying is used to bring it to a safe storage level of 13 to 14 percent.

DK75838,000020A -19-26JUL13-8/26

## CORN-COB MIX (CRACKED KERNEL)

Corn-Cob Mix (Cracked Kernel) is used for feeding purposes. The threshing cylinder is operated at a fast speed, which cracks the kernels. The sieve must be removed to allow the pieces of cob to fall through and be carried to the grain tank.

Continued on next page

DK75838,000020A -19-26JUL13-9/26

## CORN-COB MIX (WHOLE KERNEL)

Corn-Cob Mix (Whole Kernel) is also used for feeding purposes. The threshing cylinder is operated at slower speeds to prevent cracking the kernels. The sieve must be removed in this case also.

DK75838,000020A -19-26JUL13-10/26

## FLAX

Flax is often windrowed before combining, but it can be harvested as a standing crop when thoroughly dry and free of weeds. The best harvesting conditions, however, are produced with the windrower method. Harvesting is usually started after the majority of the bolls are ripe.

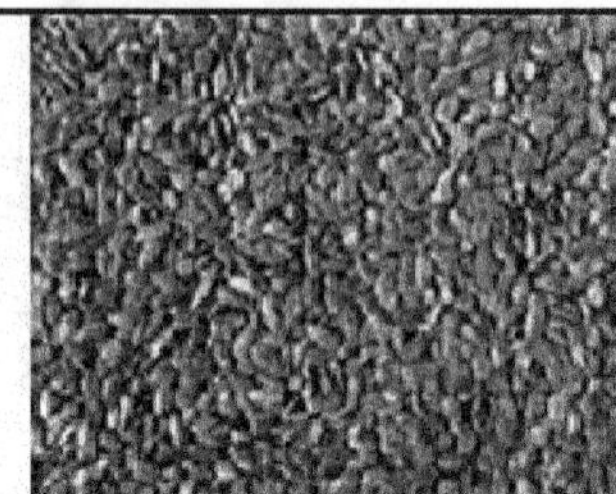


DK75838,000020A -19-26JUL13-11/26

## CANOLA

Canola (Rape) seeds are usually combined directly after the leaves drop off the stalk. The seeds are easily threshed from the pods even when the stems are green. Care must be used to avoid taking too much of the stalk into the combine because the high moisture content will make separation of the seed difficult.

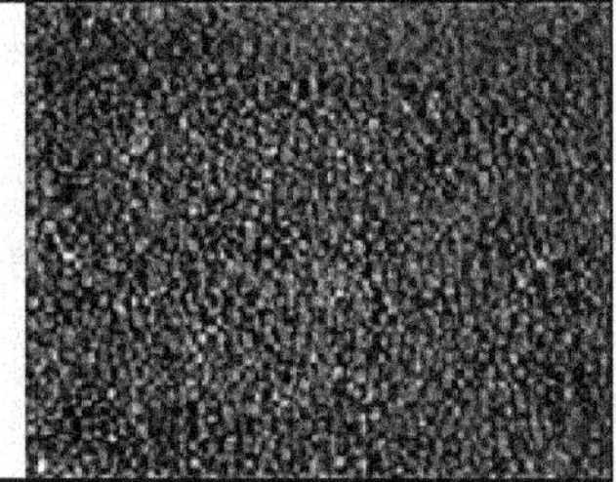


DK75838,000020A -19-26JUL13-12/26

## GRASS

Grass seed (most varieties) may be combined directly, but usually it is better to combine from a windrowed crop. Each method depends on the height of the grass, moisture content, and ripeness of the head. When ripe, the seed is easily shattered from the stems. Because of this, the crop is usually cut when the stems are still green.

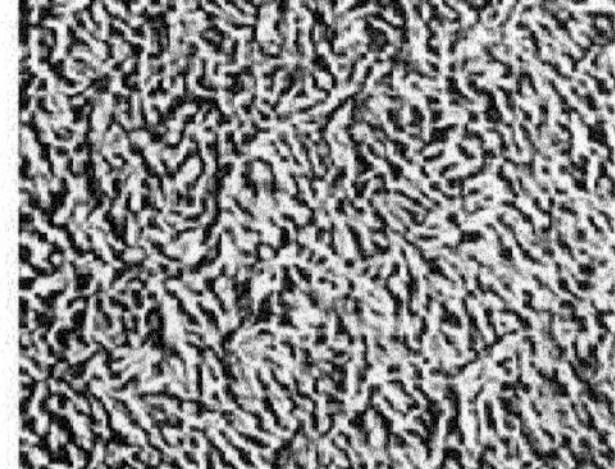


DK75838,000020A -19-26JUL13-13/26

## LESPEDEZA

Lespedeza is often harvested as a standing crop, but it can be windrowed. It should be cut when slightly green to prevent shattering. Otherwise, the straw may break easily and add more material to the cleaning shoe. The fan speed must be low to prevent blowing the light seed out of the combine. This can result in increased tailings, which may be unavoidable.



Continued on next page

DK75838,000020A -19-26JUL13-14/26

## MILLET

Millet seeds are difficult to combine directly because the seeds shatter from the stem before all the seeds are ripe. Therefore, the crop is windrowed before combining with a windrow-pickup unit.

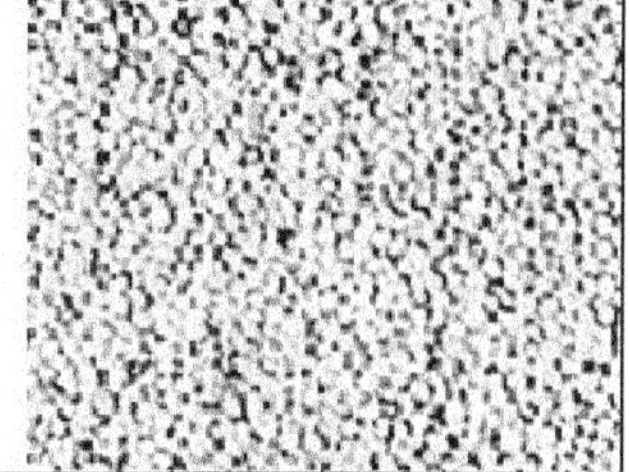


DK75838,000020A -19-26JUL13-15/26

## MUSTARD

Mustard seeds may be combined directly if the pods have matured properly while standing. If not, then the crop is windrowed to cure and combined with a windrow-pickup unit.

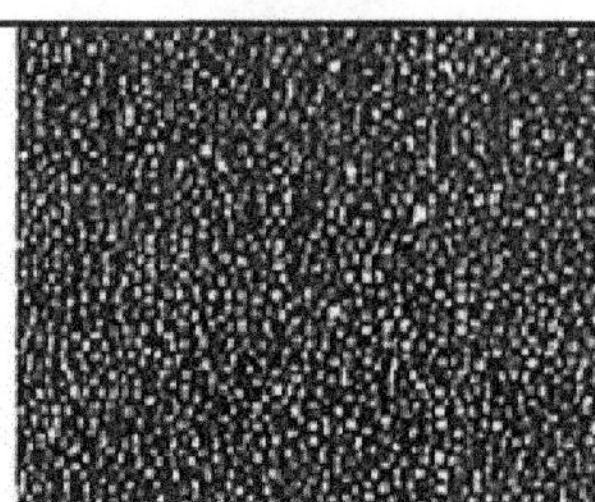


DK75838,000020A -19-26JUL13-16/26

## OATS

Oats are usually combined directly; however, when weeds or uneven ripening is a problem, windrowing is recommended. Windrows allow the green weeds to dry and reduce the moisture of the oats. Then the crop can be combined with a windrow-pickup unit.

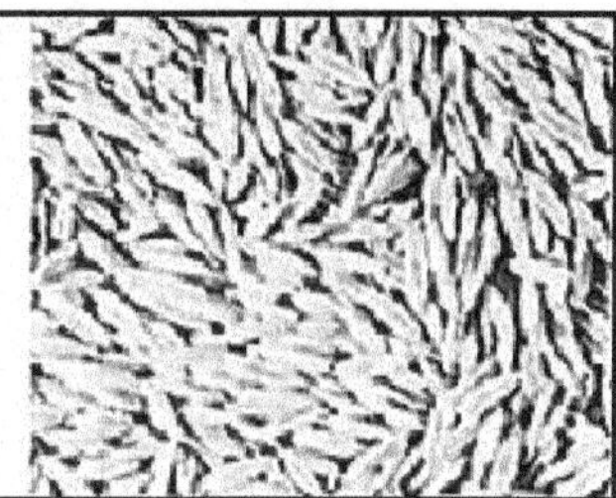


DK75838,000020A -19-26JUL13-17/26

## PEAS

Peas (edible) may be either windrowed and combined after curing or combined directly when ripe and dry. If peas are too dry, they will crack easily.

DK75838,000020A -19-26JUL13-18/26

## RICE

Rice is harvested by direct combining. It is a difficult grain to thresh because it is hard to strip from the straw. A spike-tooth threshing cylinder is usually used because of its aggressive threshing action. Rice should be threshed as soon as it ripens to avoid crackage by the sun if allowed to stand in the field too long. Rice may often be down or lodged following storms, which makes harvesting more difficult.

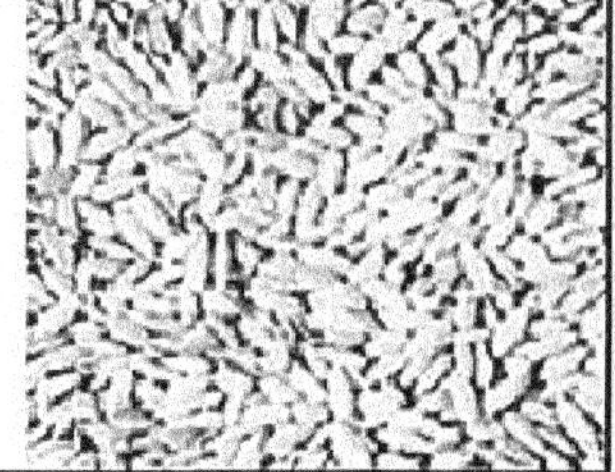


Continued on next page          DK75838,000020A -19-26JUL13-19/26

## RYE

Rye can be combined directly or harvested by the windrow-pickup method, depending on conditions. It tends to shatter easier than wheat and may ripen unevenly.

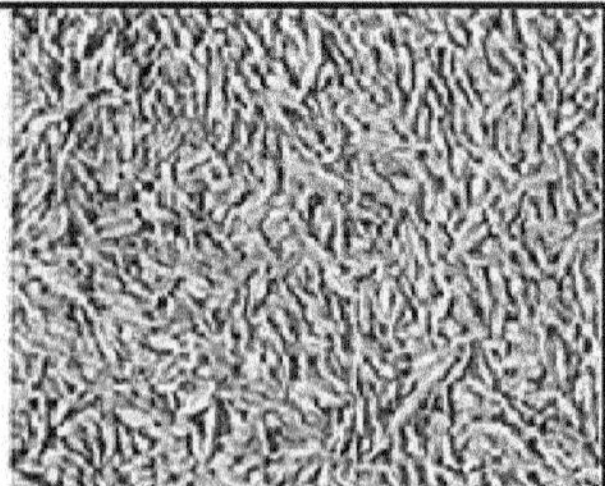


DK75838,000020A -19-26JUL13-20/26

## SORGHUM

Sorghum (grain) or milo-maize may be harvested directly by a combine. When combining sorghum, the cutting platform must be carried as high as possible to avoid taking in too much of the stalk. The stalk is pithy and damp; if too many stalks are threshed, they will carry seeds out of the combine. The dampness will also add excessive moisture to the seeds.

DK75838,000020A -19-26JUL13-21/26

## SOYBEANS

Soybeans are combined directly when the moisture content drops to 12 to 14 percent. Soybeans tend to shatter and crack when ripe; therefore, header and threshing operations must be carefully adjusted to match conditions. If beans are extremely dry, it may be advisable to harvest at night when pods become tough from dew to reduce shatter losses.

DK75838,000020A -19-26JUL13-22/26

## SUNFLOWER

Sunflower seeds may be combined directly at 8 or 9 percent moisture content. When combining directly, the cutting platform must be carried as high as possible to avoid taking in too much stalk, which will make separating more difficult. Special attachments may be used on the platform.

DK75838,000020A -19-26JUL13-23/26

## TREFOIL

Trefoil (birdsfoot) seeds may be combined directly or the windrow-pickup method may be used. When combining directly, a desiccant (drying agent) is applied to the crop prior to threshing. The seed shatters easily after the pods are ripe.

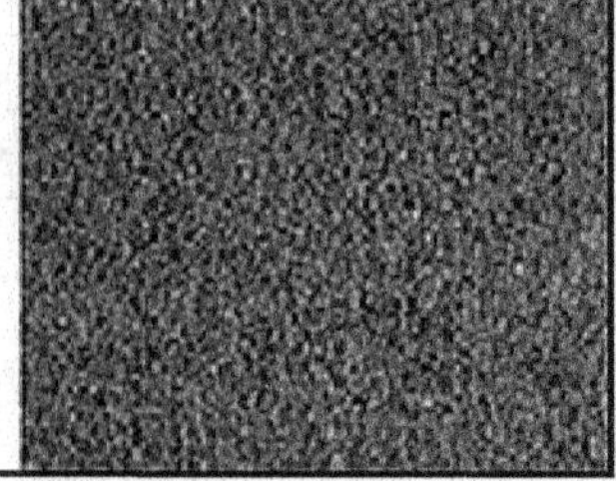


Continued on next page

DK75838,000020A -19-26JUL13-24/26

## VETCH

Vetch seeds are harvested by direct combining or by the windrow-pickup method, depending on the locality and variety. To avoid shattering, it is usually windrowed when the lower pods are ripe.

DK75838,000020A -19-26JUL13-25/26

## WHEAT

Wheat is primarily combined directly as a standing crop. However, if the crop has excessive weeds and does not ripen evenly, or may be damaged by bad weather before it ripens completely, the windrow-pickup method is used. It is usually harvested when moisture content is below 14 percent.

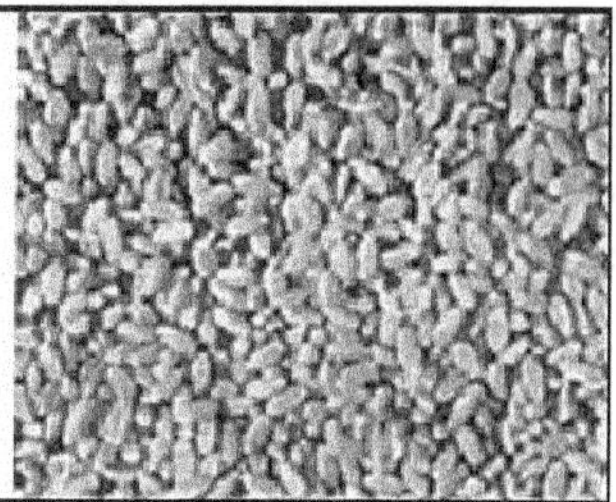


DK75838,000020A -19-26JUL13-26/26

# FIELD OPERATION AND ADJUSTMENTS

Operating the combine in the field and making the necessary adjustments requires the skills of a good operator — not just a combine driver (Fig. 37). A good combine operator can save several dollars per acre over a combine "driver." In this section we will discuss what a good combine operator must know to do quality harvesting.

## OPERATING PROPERLY IN THE FIELD

One important thing that the operator must keep in mind when operating in the field is the relationship between combining speed and crop losses. Even when the combine is properly adjusted, losses can occur because of excessive speed. The operator must judge what acceptable losses are. Some operators will tolerate more losses and crackage when they want to get the crop harvested faster.

Fig. 37 — Operating in Conditions Such As These Requires a Good Operator

The combine operator's manual is an excellent guide to proper operation and adjustment. Refer to it frequently when problems are encountered because it has the answers to many troubles.

Continued on next page

DK75838,0000211 -19-26JUL13-1/3

## GENERAL RULES FOR GOOD COMBINING

Here are some general practices for good combining:

1. When beginning to combine a field, get the feel of the combine's ability to handle the crop by operating at a slow ground speed. Use a lower gear than normal, but don't reduce engine speed; if you do, the combine will not perform efficiently. Gradually increase ground speed and check the results until problems are encountered, such as unacceptable losses or threshing damage. Under normal conditions you should be able to operate at ground speeds between 2-1/2 and 3-1/2 miles per hour (4 and 5.6 km/h).
2. Don't hesitate to make adjustments to the combine if they are necessary. Know why you must make the adjustment before doing it.

   NOTE: Make only one adjustment at a time and check the results before making any others.

3. Check frequently for proper threshing action and adjust the cylinder speed and concave spacing as necessary.
4. Check for grain losses when checking threshing action. Make adjustments as necessary, such as reducing ground speed, changing cylinder-concave, fan, chaffer, and sieve settings.
5. When using a cutting platform in a standing crop, cut as high as possible without missing too many low grain heads. In down and tangled crops, use lifting guards and keep the header low and the ground speed slow.
6. Keep the reel height and speed adjusted for changes in crop height and ground speed.
7. When using a corn head, keep the header low to get low ears. Keep the corn head centered in the rows to prevent stalks from bending and losing ears.

Fig. 38 — When Operating in Weedy Conditions, Slow Down

8. Adjust the cleaning units when losses over the shoe or excessive tailings occur.
9. Don't overload the combine by operating at faster ground speeds; losses will increase tremendously.
10. When operating in poor conditions, such as a weedy crop or a hard-to-thresh crop, slow down and make frequent checks for combine performance (Fig. 38).

Now let's look at some of these operations and adjustments more closely. Here we will discuss the following:

- Operating and adjusting cutting and feeding units
- Operating and adjusting threshing units
- Operating and adjusting cleaning units

Later in this section we will tell how to calculate crop losses and how to reduce them.

DK75838,0000211 -19-26JUL13-2/3

## HEADER AND FEEDER HOUSE PLUGGING

Difficult field conditions or incorrect operation settings such as excess ground speed may cause plugging in the header and feeder house. Follow your operator's manual safety procedures when you manually remove plugs.

A special reverser has been developed that will remove plugs with machine power. The reverser is a planetary gearbox assembly that is controlled in the cab for operator convenience and safety. The planetary gearbox is driven by the primary countershaft. The gearbox in turn drives the header and feeder house.

When the control is actuated, the planetary gears reverse the drive to the header and feeder. The gears create slower speeds but more torque than normal forward operating conditions. The plugs are backed out of the machine.

The reverser assembly allows the operator to work at faster ground speeds without worrying about plugging. It

Fig. 39 — Planetary Gear Case Location for Reverser Feature

also is a safety feature that keeps the operator in the cab during the unplugging process (Fig. 39).

DK75838,0000211 -19-26JUL13-3/3

# OPERATING AND ADJUSTING CUTTING AND FEEDING UNITS

The crop must be cut, gathered, and fed properly to the threshing unit, or shattering and uneven feeding will result. This in turn will affect the overall efficiency of the combine.

Here we will discuss operation and adjustment of:

- Cutting platform
- Windrow pickup
- Corn head
- Feeder conveyor

## CUTTING PLATFORM ADJUSTMENTS

To operate and adjust the cutting platform properly, consider the following:

- Ground speed
- Cutting height
- Reel adjustments
- Auger adjustments

### GROUND SPEED

Ground speed is governed by the yield of the crop, the capacity of the combine, and the skill of the operator. When the crop yields are high, ground speed should be reduced. Maximum effective ground speed, however, is affected by the capacity of the combine to thresh, separate, and clean the crop. If the ground speed is too fast and overloads the combine, losses will be excessive. Even with a properly adjusted combine, the results of excessive speed are unthreshed grain or overthreshed straw. In each case, grain is lost over the straw walkers and shoe.

Usually, most operators combine at a ground speed that is easy for them to handle, and this depends a great deal on the operator's skill and condition of the crop.

### CUTTING HEIGHT

Cutting height is determined by the condition of the crop. In standing grain, the cutterbar should usually be adjusted to cut low enough to get most of the grain without leaving too many low grain heads (Fig. 40). This reduces the amount of straw that would otherwise add to the load of

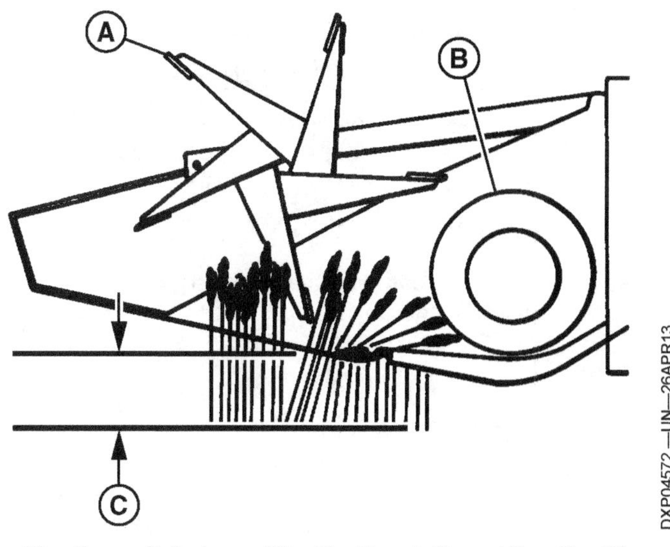

Fig. 40 — *Proper Cutterbar and Reel Positions in Normal Crop Conditions*

A— **Reel Position**        C— **Cutterbar Height**
B— **Auger**

the straw walkers and shoe. Excessive straw can cause separating and cleaning problems.

Continued on next page

DK75838,000020B -19-31JUL13-1/20

In crops that are down and tangled, lifting guards should be used to lift the crop onto the cutterbar (Fig. 41). An automatic header height control or platform float springs help maintain a close relationship of the cutterbar to the ground. Adjust the header height during operation for variations in the height of the crop.

### REEL ADJUSTMENT

The reel must be carefully adjusted to push the crop against the cutterbar and into the path of the platform auger. This requires adjustment of the reel position and reel speed.

### REEL POSITION IN STANDING CROPS

The reel height is usually adjusted so that the slats of the reel touch the crop about midway between the cutoff point and the top of the crop (Fig. 40). If the reel speed is correct, the position will cause the crop to immediately fall onto the platform when cut.

The forward position of the reel must be adjusted so that the reel shaft is slightly ahead of the cutterbar (Fig. 40). If the reel is too far forward, the crop isn't pushed against

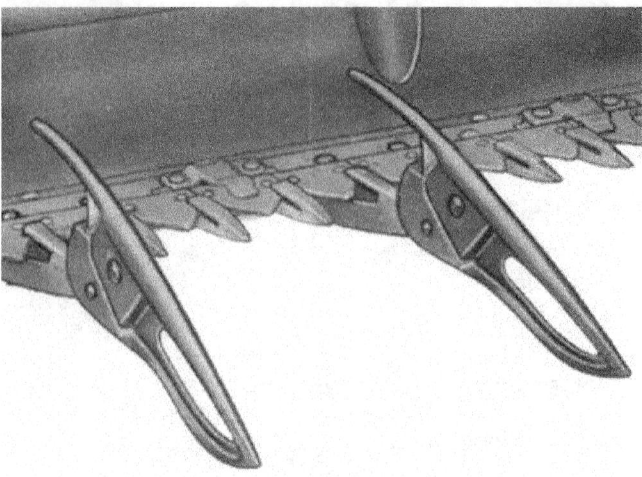

*Fig. 41 — Lifting Guards Help in Down and Tangled Crops*

the cutterbar and cut grain will fall on the ground. If the reel is too far back, the crop is pushed down too low; when it is cut, some of the heads are missed.

DK75838,000020B -19-31JUL13-2/20

### REEL POSITION IN DOWN CROPS

A pickup reel is required. The reel height in down crops is low so that the fingers can lift the crop up and onto the cutterbar (Fig. 42). The fingers must be adjusted so that they pick up the crop and lift it onto the cutterbar without carrying the crop around the reel again.

The forward position of the reel must be about a foot (30 cm) in front of the cutterbar so that the crop is lifted before it gets to the cutterbar. Care must be used with a pickup reel so that the fingers don't come in contact with the cutterbar, which may cause the knives to break. The fingers should just clear the cutterbar. A hydraulically powered reel fore/aft control is available. The reel position is controlled by a switch on the control panel in the operator station.

Adjust the reel so it looks as though the reel is pulling the combine through the field. If the reel speed is too slow, the crop isn't pushed against the cutterbar; then the crop will fall on the ground because the reel doesn't push it onto the platform.

*Fig. 42 — Proper Reel Position in Down Crops*

If the reel speed is too fast, the crop may be shattered by the impact of the reel. Also, the crop may be pushed down before it can be cut, and the uncut grain will be left in the field.

**Continued on next page**

DK75838,000020B -19-31JUL13-3/20



PN=154

## AUGER ADJUSTMENTS

The auger must be positioned correctly and operated at the proper speed to feed the crop evenly to the feeder conveyor without bunching.

### AUGER POSITION

The auger must be adjusted so that it clears the bottom of the platform by 1/4 to 1/2 inch (6 to 13 mm), depending on the crop (Fig. 43). If there is too much clearance, uneven feeding can result. If there is not enough clearance, the grain will be shattered and eventually fall on the ground, and the straw will be chewed excessively.

### AUGER STRIPPER

The auger stripper, located behind the auger, must be adjusted to prevent the crop from going around the auger (Fig. 43). Poor feeding can result if too much clearance exists between the auger and stripper.

### AUGER FINGERS

The auger fingers must be adjusted to provide positive feeding. In light crops, adjust the fingers to reach out and pull in the crop. In heavy crops, adjust the fingers inward. If the fingers extend too far in heavy crops, material will be carried around the auger, resulting in poor feeding. To adjust the auger on a typical combine, turn the slotted bracket down (Fig. 43).

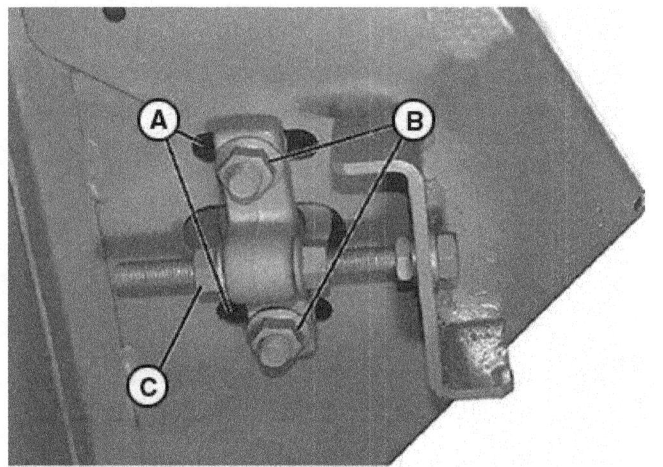

Fig. 43 — Typical Auger Adjustment

A— Auger Finger Adjustment  
B— Auger Fore-and-Aft Adjustment  
C— Auger Height Adjustment

On some platforms, auger height must be adjusted according to finger-to-floor clearance to avoid damaging fingers or floor. See the operator's manual for each platform for specific adjustment procedures.

DK75838,000020B -19-31JUL13-4/20

### AUGER SPEED

If the auger speed is adjustable, use the lowest speed that will give good feeding action. Fast speeds can throw the crop forward or force-feed the feeder conveyor. If combining in short crops, auger flight extensions may be required to get the auger to feed evenly (Fig. 44). These extensions bunch the crop together before delivery to the feeder conveyor. In taller crops, the extensions should not be used. They may cause the taller crop to bunch, which will result in uneven feeding to the cylinder.

A— Platform Auger Flight Extensions

Fig. 44 — Auger Flight Extensions Help in Short Crops

**Continued on next page**

DK75838,000020B -19-31JUL13-5/20



PN=155

## WINDROW-PICKUP ADJUSTMENTS

When using a windrow pickup, keep the windrow centered so that the material is fed evenly to the threshing cylinder. If the heads of grain are lying in one general direction, operate the combine so that the heads are picked up first (Fig. 45). This ensures better threshing and separation of the grain. Losses will also increase if the windrows are picked up butt end first.

Two adjustments must be considered when combining a windrow:

• Speed of the pickup
• Position of the pickup

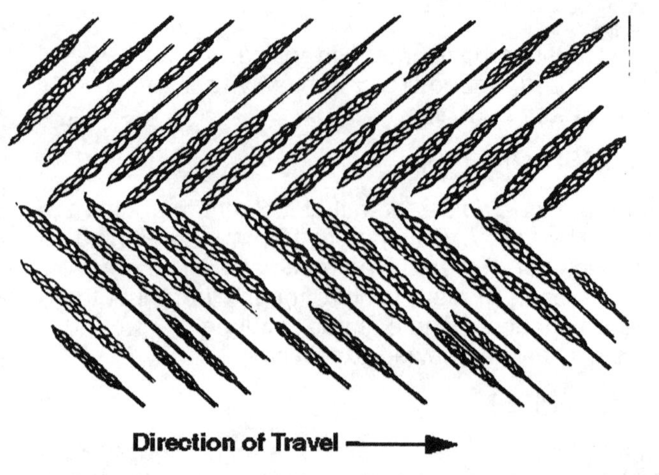


Fig. 45 — Combine in the Direction That Will Pick Up Heads First

DK75838,000020B -19-31JUL13-6/20

## PICKUP SPEED

Just as the reel on a cutting platform must be adjusted to the ground speed, the pickup must be adjusted to forward travel speed also. If the pickup is operated at excessive speeds, shattering will result in losses due to the impact of the pickup fingers. Fast speeds will also tear the windrow apart, which will cause uneven feeding to the threshing cylinder and will result in poor threshing and separating action. Stones may also be picked up and damage the threshing cylinder.

The speed of the pickup must be adjusted to operate at a speed which makes it appear that the windrow is simply lifted up as the pickup goes underneath it. The windrow should be gathered in an unbroken, even flow (Fig. 46). Don't use ground speeds that are faster than the pickup can handle properly; bunching will occur, shatter losses will increase, and uneven feeding will result.

Fig. 46 — Gather Windrow in an Unbroken, Even Flow

**Continued on next page**

DK75838,000020B -19-31JUL13-7/20

## PICKUP POSITION

The pickup must be positioned so that the teeth clear the ground yet are low enough to get all the material (Fig. 47). If the teeth are allowed to contact the ground, stones may be picked up or teeth may be bent and damaged. Also, dirt will be picked up, which will make cleaning more difficult. A windrow hold-down may be used to keep the windrow down on the pickup for proper feeding.

## *CORN HEAD ADJUSTMENTS*

Proper operation of the corn head requires careful operation and adjustment of these things:

- Ground speed
- Header height
- Gatherer points
- Gatherer chains
- Stalk rolls
- Trash knives and shields
- Snapping plates
- Auger
- Row spacing

## GROUND SPEED

The proper ground speed depends on the capacity of the combine, crop condition, and yield. Ground speed must not exceed the ability of the combine to harvest the crop with very little shattering and losses.

A— Pitch Line        B— Parallel

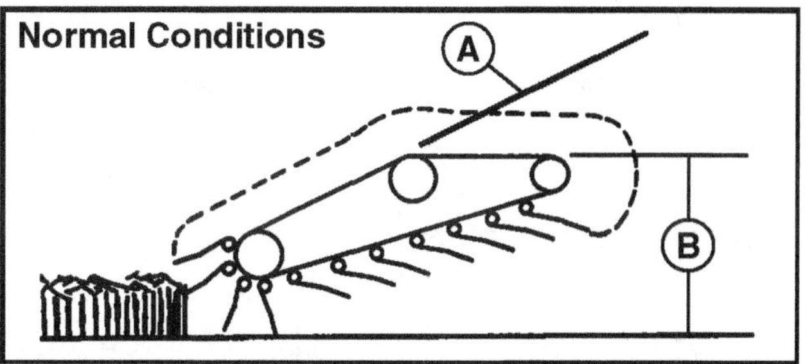


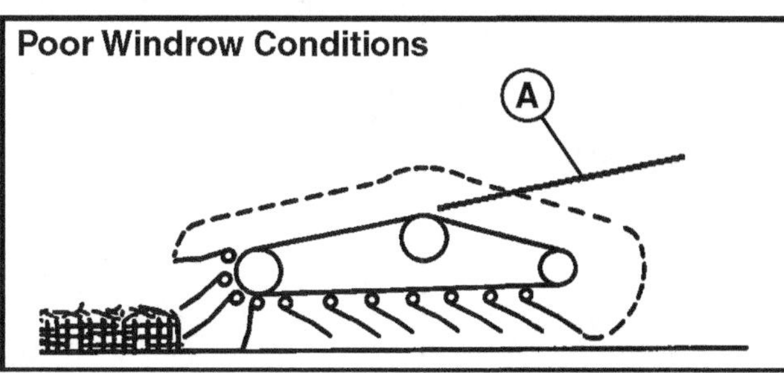


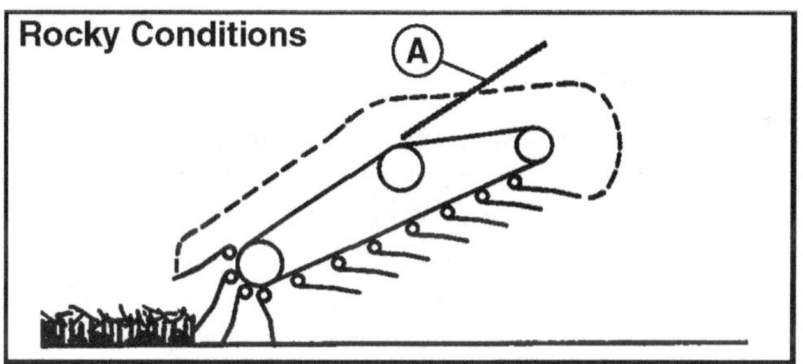


Fig. 47 — Windrow-Pickup Positions

DK75838,000020B -19-31JUL13-8/20

## HEADER HEIGHT

The header must be operated low enough to gather the lowest hanging ears without knocking them to the ground or shattering corn from the cob on impact of the header (Fig. 48). In down and tangled corn, the header must skim the ground without picking up dirt or stones.

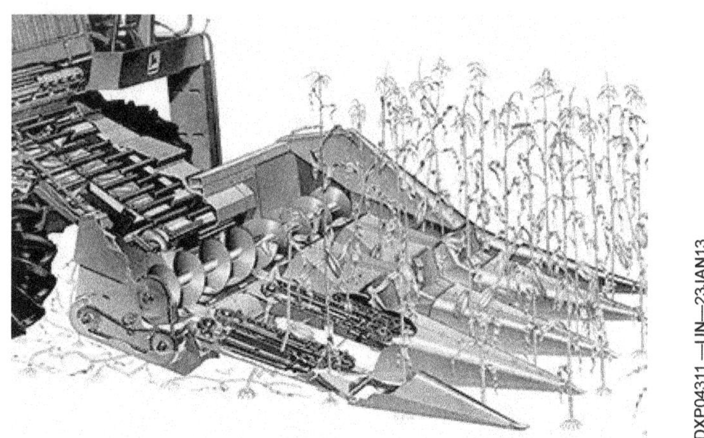

Fig. 48 — Operate Header to Gather Lowest Ears of Corn

**Continued on next page**                         DK75838,000020B -19-31JUL13-9/20

## GATHERER POINTS

Under normal conditions, operate the gatherer points so that they follow the contour of the ground when operating in standing corn. In down corn, adjust the gatherer points in a more downward position to pick up down stalks (Fig. 49). When the corn is down and tangled, operate at a slower ground speed to reduce losses.

A— Down Position

B— Gatherer Points in Normal Position

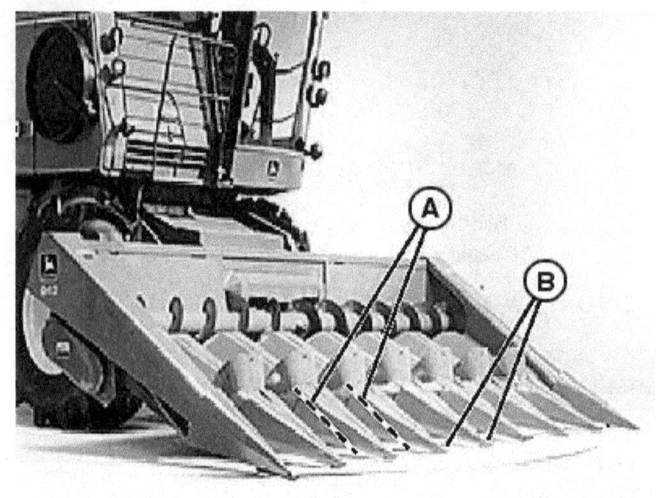

*Fig. 49 — In Down Corn, Adjust Gatherers in Down Position*

DK75838,000020B -19-31JUL13-10/20

## GATHERER CHAINS

Gatherer chains require adjustment only when worn (Fig. 50). Some combines have spring-loaded tighteners that maintain tension. The speed of the gatherer chains is very important in relation to ground speed. The speed of the gatherer chains is controlled by the overall speed of the corn head. The gatherer chains should move rearward at about the same speed the combine travels forward — just fast enough to guide corn stalks into the snapping rolls. Excessive gatherer chain speed causes stalks to break and throws ears to the ground. Also, the entire stalk may be pulled up and fed into the threshing unit if the chain speed is too fast. If ground speed is too fast (chain speed too slow), gatherer chains will push stalks forward and knock off ears.

## STALK ROLLS

If the spacing of the stalk rolls is adjustable, adjust this distance according to the diameter of the corn stalk. The speed of the header drive controls the speed of the stalk rolls, and this speed must match the forward speed of the combine to keep from losing ears.

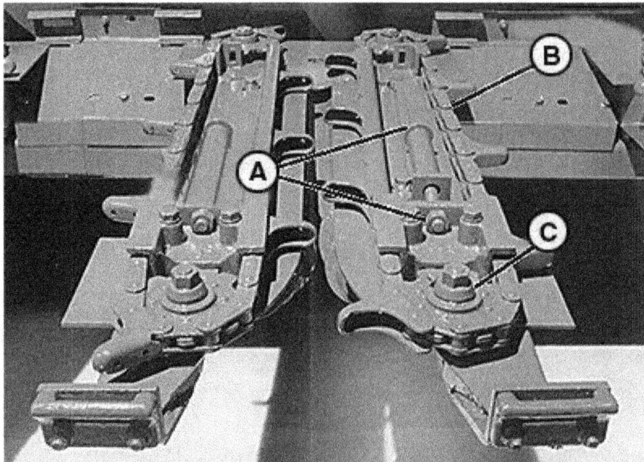

*Fig. 50 — Typical Gatherer Chain Adjustment for Corn Head*

A— Adjusting Nut
B— Gatherer Chain

C— Idler Sprocket

Continued on next page

DK75838,000020B -19-31JUL13-11/20

## TRASH KNIVES AND SHIELDS

Trash knives and shields must be adjusted to prevent wrapping of weeds and stalks (Fig. 51). Usually, these parts are adjusted as close as possible to the stalk rolls without touching them.

A— Right-Side Adjustment Bolts
B— Trash Knives
C— Left-Side Adjustment Bolts

Fig. 51 — Typical Adjustment of Trash Knives for Corn Head

DK75838,000020B -19-31JUL13-12/20

## SNAPPING PLATES

The snapping plates snap the ears from the stalks as the stalks are pulled down by the stalk rolls. Two snapping plates are provided for each row unit. These are adjustable to different clearances to accommodate the various sizes of stalks and ears of corn. The snapping plates are usually adjusted at about 1-3/16 inches (30 mm) apart at the front and 1-1/4 inches (32 mm) apart at the rear (Fig. 52). The wider clearance at the rear provides adequate stalk clearance so that pieces of stalk or leaves are not torn from the stalk and allowed to enter the combine. Excessive trash such as this will hinder the separating and cleaning action of the combine.

**IMPORTANT: The snapping plates must be adjusted open as far as possible without causing shelling to minimize the amount of trash taken into the combine. The front spacing should always be approximately 1/16 to 5/64 inch (1.5 to 2.0 mm) less than the rear to maintain the proper snapping plate action.**

A— 1-1/4 Inches at Rear
B— Gatherer Chain Guide
C— Adjust Here: Be Sure Plates Are Centered over Stalk Rolls
D— 1-3/16 Inches at Front
E— Snapping Plates
F— Stalk Rolls

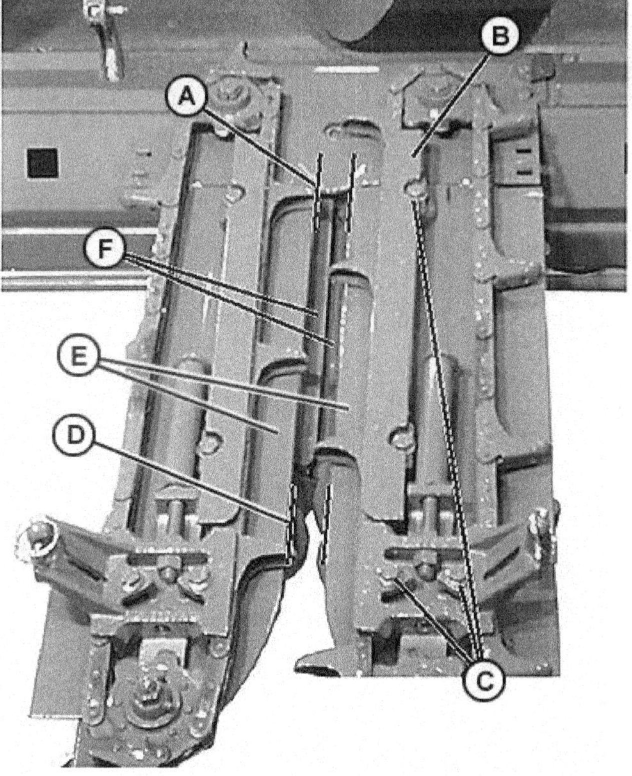

Fig. 52 — Typical Adjustment of Snapping Plates for Corn Head

**Continued on next page**　　　DK75838,000020B -19-31JUL13-13/20

Some combines are equipped with hydraulic snapping plates that are adjustable from the operator seat. A hydraulic cylinder is mounted to the bottom of the corn head (Fig. 53). The cylinder is connected to one snapping plate on each row unit by a linkage. A sensor mounted at the end of the linkage monitors snapping plate position. The operator presses a button on the control handle to open or close the spacing of the snapping plates and views their position on a display in the cab.

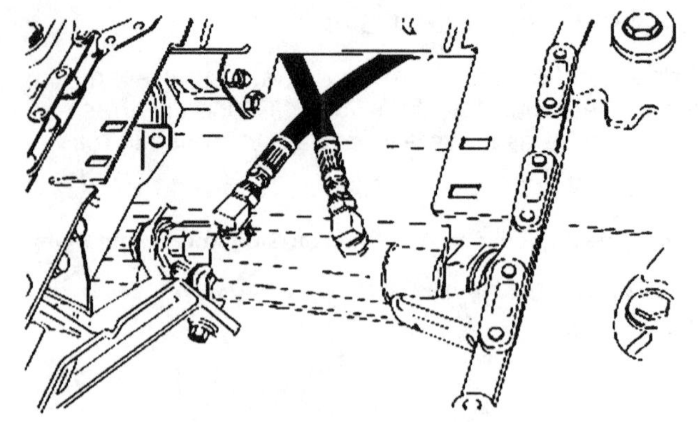

*Fig. 53 — Deck Plate Cylinder*

DK75838,000020B -19-31JUL13-14/20

## AUGER

The auger must be adjusted for the condition of the field and the size of the ears (Fig. 54). Too much clearance can cause shelling if ears are small enough to be ground under the auger flights. Usually, the auger must be adjusted down and to the rear in normal conditions. In damp, sticky, or heavy trash conditions, the auger should be adjusted slightly up and forward to move material away from the row units. The auger stripper must always be within 1/4 inch from the auger to prevent carrying over material, which can cause poor feeding.

**A— Auger Adjustment**

*Fig. 54 — Typical Auger Adjustment for Corn Head*

**Continued on next page**

DK75838,000020B -19-31JUL13-15/20

## ROW SPACING

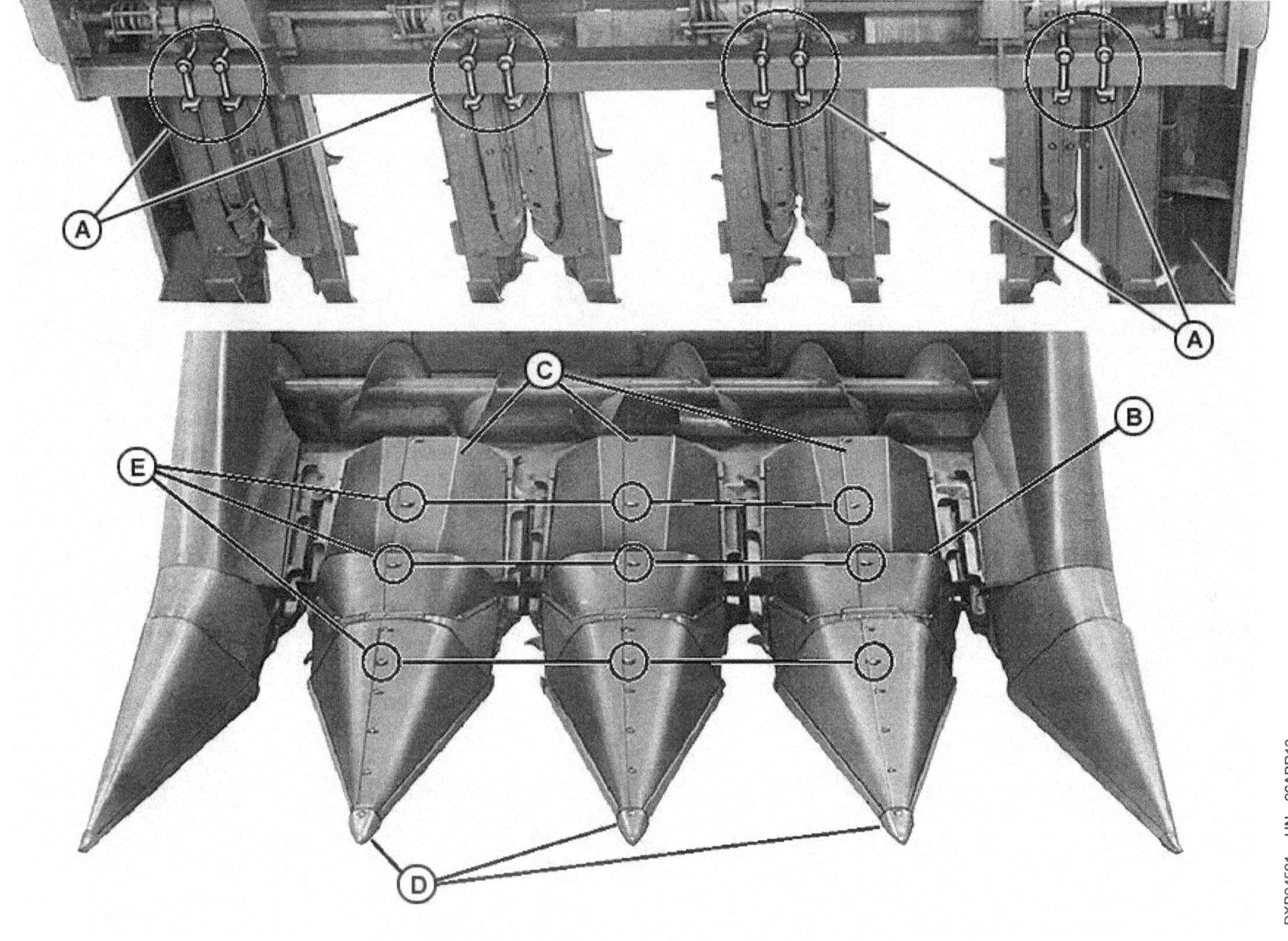

Fig. 55 — Typical Adjustments for Corn Head Row Spacing

A— **Adjust Row Unit Spacing Here**
B— **Center Shields Extension**
C— **Center Shields**
D— **Center Gatherer Points**
E— **Adjustment Points**

The row units must be adjusted for the proper row spacing (Fig. 55) so that the stalks are not pushed aside by the gatherer points. If the stalk is deflected too much by improper spacing, ears can be dislodged or missed and will fall on the ground. The difference between spacing and row width is multiplied by the number of rows, so improper spacing becomes very serious with wider corn heads.

**Continued on next page**                    DK75838,000020B -19-31JUL13-16/20

## DRUM HEIGHT ADJUSTMENT

No tools are required for this adjustment (Fig. 56). Placing the handle on each side of the feeder house in the vertical position raises the front feed drum to increase conveyor-to-floor clearance for large grain. Placing the handle in the horizontal position lowers the drum for small grain. See the operator's manual for each machine for specific adjustment recommendations.

### ROW-CROP HEAD ADJUSTMENTS

The row-crop head performs best when all components are properly adjusted and function together. This requires proper adjustment or control of the following factors:

- Ground speed
- Gathering belt speed
- Gatherer point clearance and spacing
- Operating height

Fig. 56 — Drum Height Adjustment

DK75838,000020B -19-31JUL13-17/20

## GROUND SPEED

Always adjust ground speed to combine capacity and ground and crop conditions (Fig. 57). Excessive speed increases shatter losses and chances of plugging the row-crop head. If speed is too slow, combine capacity and time are wasted. Also, gatherer chains may pull plants up by the roots, causing excessive shattering and poor cutting action.

## GATHERING BELT SPEED

Gathering belts should operate at approximately the same speed as the combine to prevent excessive shattering of seeds, plugging of row units, or pulling plants out of the ground before they are cut off. Gathering belt speed may be adjusted by changing drive chain sprockets or adjusting the variable-speed feeder house drive, if used.

## GATHERING POINT CLEARANCE AND SPACING

Gatherer points must be set high enough to prevent them from scooping dirt into the gathering belts, but low enough to slide under low-growing or lodged plants. If soybeans are standing well, set points to clear the ground by about 2 inches (50 mm). However, if beans are lodged, lower gatherer points to about 1/2 to 1 inch (13 to 25 mm) from the ground. On wide row-crop heads (six or eight rows), outer gatherer points may be set slightly higher than the other points for smoother operation over uneven terrain.

Always operate the row-crop head on the same rows as they were planted to avoid cutting odd or guess rows,

Fig. 57 — Adjust Ground Speed to Crop and Ground Conditions

which may not properly align with row unit spacing. Also be sure that row unit spacing on the row-crop head matches the spacing between crop rows to avoid excessive shattering and crop losses. Row units can be shifted on the head to match row spacings within certain limitations, but cannot be set to match all possible row widths.

## OPERATING HEIGHT

Each row unit on the row-crop head is designed to float independently for close cutting of low-growing crops. The units also may be locked rigidly in place for operation in taller crops such as sorghum or sunflowers. To permit free flotation of row units, be sure there is adequate clearance between the gatherer shields over each row unit.

**Continued on next page**                    DK75838,000020B -19-31JUL13-18/20

Float springs on each row unit must also be adjusted so that units float freely over uneven terrain, but drop quickly to lower levels when necessary (Fig. 58). Skid shoes on each row unit may be adjusted up for closer cutting, or down to increase cutting height in low-growing crops such as soybeans. Automatic header height control automatically raises or lowers the entire row-crop head whenever changes in ground elevation cause any row unit to move up or down more than the float range provided between units. For instance, if the float range is 6 inches (15.25 cm) and one row unit passes over an object that is 7 inches (18 cm) high, the entire row-crop head will automatically be raised 1 inch (2.5 cm) until the obstruction is passed, and then lowered to the proper height above the ground.

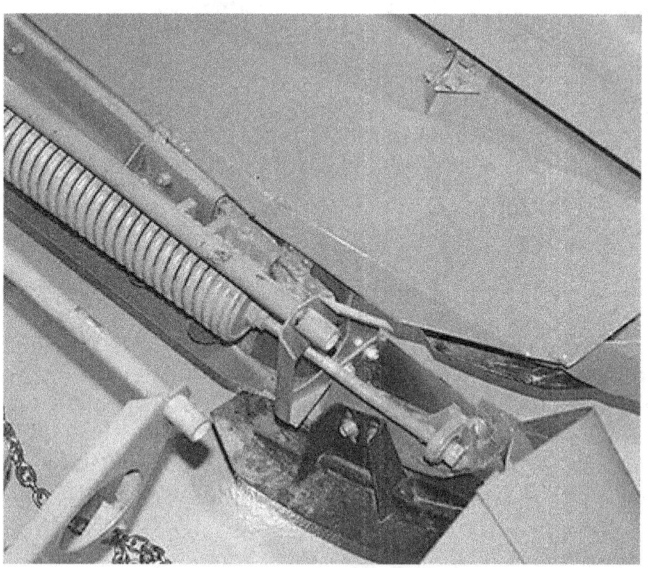

*Fig. 58 — Float Spring Tension Adjustment*

DK75838,000020B -19-31JUL13-19/20

## FEEDER CONVEYOR ADJUSTMENTS

The feeder conveyor chain must operate properly to feed the material smoothly and evenly to the threshing cylinder. The chain must be tight enough so the slats don't slap the bottom of the feeder house. On combines that have adjustable conveyors, adjust the chain clearance for the size of crop. For grain crops, the space beneath the slats should be about 1/8 inch (3 mm), and for corn about 1 inch (25 mm).

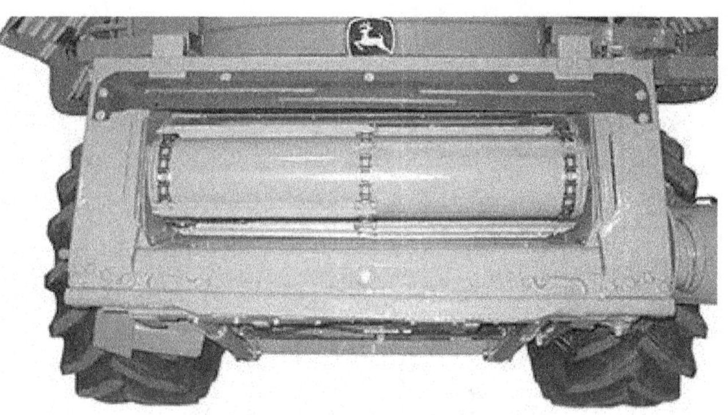

*Fig. 59 — Typical Feeder Conveyer*

DK75838,000020B -19-31JUL13-20/20

## OPERATING AND ADJUSTING THRESHING UNITS

The threshing section of the combine is the "heart" of the harvesting function. Many combine problems can occur if the operator does not know why or how to make the proper adjustments.

### GENERAL RULES

The cylinder or rotor and concave must be adjusted as follows:

- In small grain and seeds, all grain must be removed from the stalk without damaging the grain, breaking the straw, or causing excess chaff.
- In corn, all kernels must be removed from the cob without damaging the kernels or breaking many of the cobs.

The general rule of thumb is:

- Small grain, high threshing speeds and narrow concave spacing
- Large grain or seeds, low threshing speeds and wide concave spacing

### THRESHING ACTION

Remember: Usually, the higher the cylinder or rotor speed, the narrower the concave spacing. This provides the best threshing action; the opposite will decrease threshing action.

Here are the results of overthreshing, underthreshing, and proper threshing:

- Overthreshing

*Fig. 60 — Check the Grain Tank for Cracked Kernels*

  - Cracked grain
  - Broken and chewed straw, which overloads the shoe
  - Grain losses over the shoe or rotor

- Underthreshing
  - Unthreshed heads and excessive tailings
  - Overloaded straw walkers or rotor
  - Grain losses over the straw walkers or discharge area of the rotor

- Proper threshing
  - No cracked grain
  - No unthreshed heads or excessive tailings
  - No broken or chewed straw
  - No grain losses over the straw walkers, discharge area of the rotor, or shoe

**Continued on next page**

DK75838,000020D -19-26JUL13-1/4

## HOW TO DETERMINE THRESHING ACTION

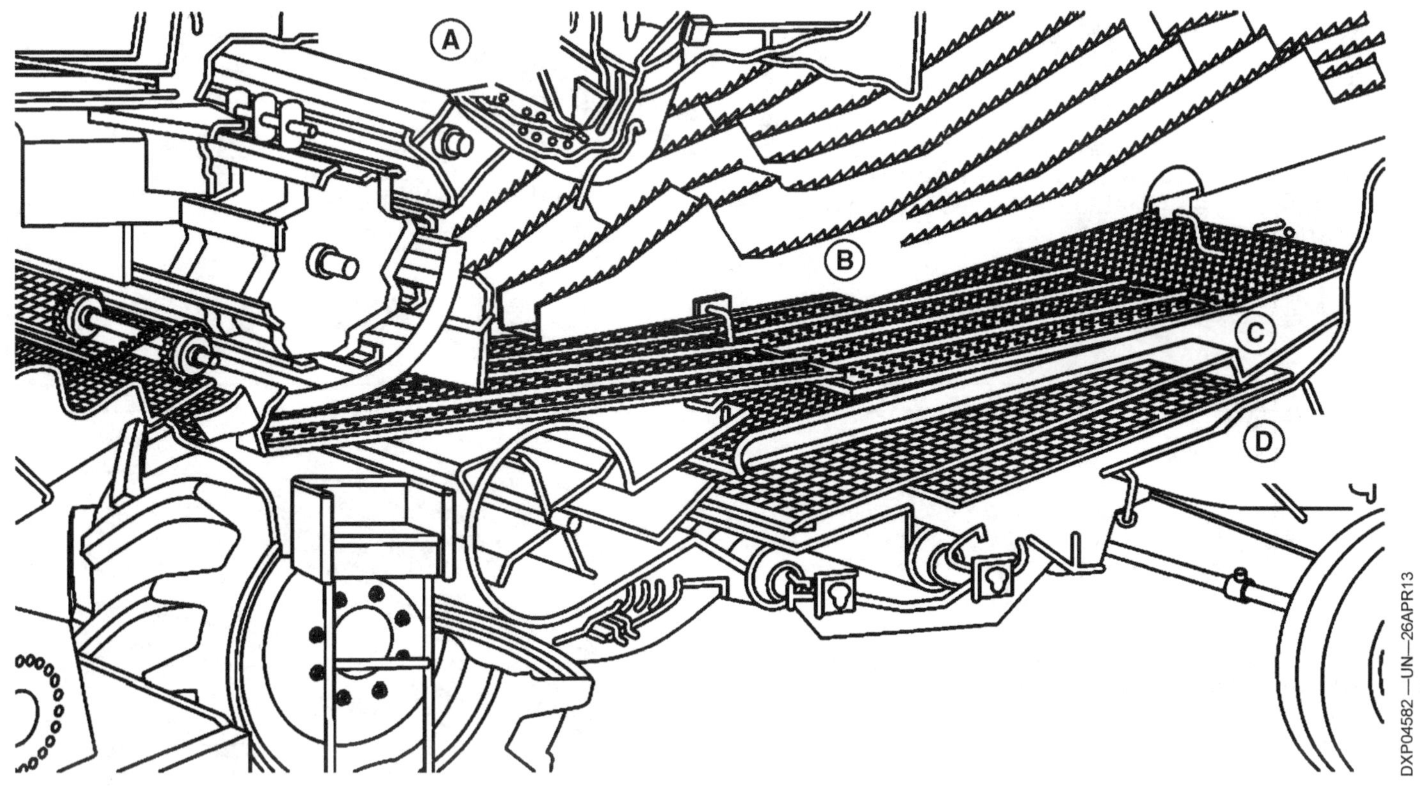

Fig. 61 — Check Threshing Results in These Areas

A— Check Tailings Return
B— Check Straw Walkers
    (Underthreshing)
C— Check Shoe (Overthreshing)
D— Check Straw at Rear of
    Combine

In order to correct improper threshing action, the operator must be able to recognize it and determine what is causing it. This involves examining the straw and grain discharged from the rear of the combine and checking the grain tank sample, the tailings returns, straw walkers or separator, and shoe (Fig. 61).

Straw discharged from the rear of the combine should not be broken or chewed, and very few kernels of grain should still be attached to the stalk. In corn, the cobs must not be broken excessively and few if any kernels should remain on the cob.

The grain tank sample should have very few cracked kernels (Fig. 60). If many cracked kernels are found, the problem may be excessive tailings returned to the cylinder for rethreshing rather than fast cylinder speeds or narrow concave spacing.

Tailings returns should be very few. Each paddle of the elevator should contain no more than a tablespoon of tailings or less, and the sample must contain mostly unthreshed heads rather than loose kernels or straw and chaff.

Straw walkers or rotor will be overloaded if underthreshing is occurring at high speeds. Grain cannot be separated from the thick masses of straw on the walkers. Thus, grain is carried out the rear of the machine with the straw.

The cleaning shoe will be overloaded with excessively chewed and broken straw when overthreshing is occurring. Kernels cannot be separated from the mass of straw and chaff on the chaffer. Thus, grain is carried out the rear of the machine with the straw.

### HOW TO CORRECT THRESHING PROBLEMS

After determining that the problems may be improper threshing rather than excessive tailings or a poorly adjusted shoe, use one of the following procedures:

Combine operating speed may be too fast or too slow. If engine or primary countershaft speed is more than ±10 rpm off the recommended speed, have the engine adjusted to the proper speed. Adjust governor on gasoline engine.

Continued on next page

DK75838,000020D -19-26JUL13-2/4

## OVERTHRESHING

Overthreshing may be reduced by slowing cylinder or rotor speed by 5 percent at a time (Fig. 62). Check the results of these changes before making additional changes. Try opening the concave spacing slightly if reducing threshing speed 10 percent doesn't help. If overthreshing cannot be corrected by these measures, try slowing the ground speed; too much material at too fast a feed rate may cause overthreshing.

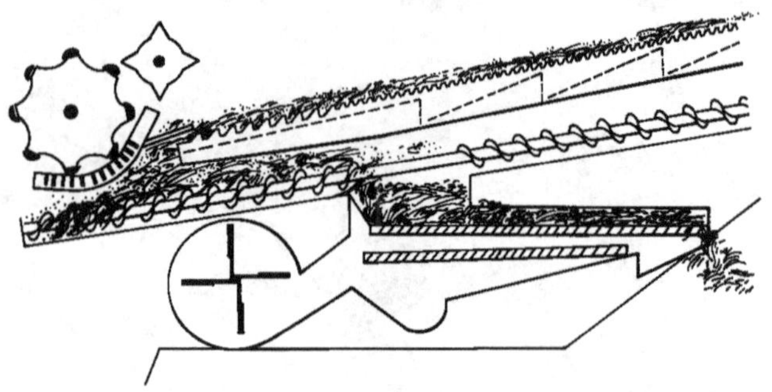

Fig. 62 — Overthreshing Chews and Breaks the Straw, Overloading the Shoe

DK75838,000020D -19-26JUL13-3/4

## UNDERTHRESHING

Underthreshing may be caused by too slow a cylinder or rotor speed and wide concave spacing (Fig. 63). Try increasing the threshing speed by 5 percent. If this doesn't correct the problem, narrow the concave spacing slightly. Check the results of these changes frequently. If after trying a few of these adjustments threshing action still hasn't improved, increase ground speed by about half a gear range. Check the results and make additional adjustments as necessary.

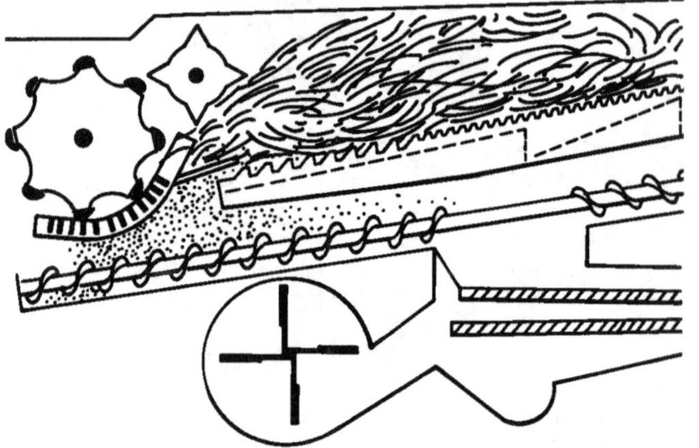

Fig. 63 — Underthreshing Overloads the Walkers

DK75838,000020D -19-26JUL13-4/4

# OPERATING AND ADJUSTING CLEANING UNITS

The cleaning fan, chaffer, and sieve must be operated and adjusted in conjunction with each other to get the best cleaning action. Fine chaff and straw must be removed with as much air as possible without blowing kernels out of the combine.

## ADJUSTING THE FAN SPEED

Before adjusting the fan speed, open the chaffer and sieve to the maximum recommended openings for the crop being harvested. Then start with the lowest fan speed suggested and gradually increase the speed without blowing kernels out of the combine or into the tailings return. Check the results carefully. After reaching the maximum acceptable fan speed, continue to make minor adjustments to the fan speed and to chaffer and sieve openings until best results are achieved.

## ADJUSTING THE CHAFFER

Open or close the chaffer just enough so that the grain falls through before passing the length of the chaffer. If the chaffer is opened too wide, it may overload the sieve with chaff and straw and increase the tailings returns. If the chaffer is not opened wide enough, excess grain will be moved to the tailings return and some will be lost out the rear of the combine. Most combines have a chaffer extension that should be opened slightly more than the chaffer and raised a little to allow tailings to drop through easily.

Grain losses in the cleaning unit may be caused by:

1. Too little fan blast or too narrow chaffer openings, which cause a layer of straw and chaff to carry over the grain (Fig. 64).
2. Too much fan blast, which will blow the grain over the shoe (Fig. 65).

It is important to know which of the reasons is causing the shoe losses so that you can make the proper adjustment to the area that is causing the problem. In item 1 above, the problem may be caused by overthreshing as well as too little air from the fan. Check the amount and condition of the straw; if the straw appears excessively broken and chewed, the cylinder and concave must be adjusted to reduce overthreshing. If the straw is whole and unbroken, then more air is needed to suspend the straw and chaff so that the grain can drop through to the sieve, or the chaffer needs to be opened slightly. Be sure to check the results before going too far.

If too much fan blast is the problem, then there will be very little chaff and straw on the shoe, which indicates

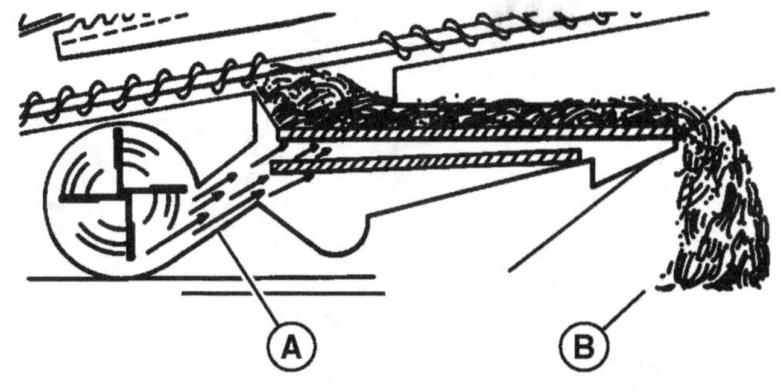

Fig. 64 — Too Little Fan Blast

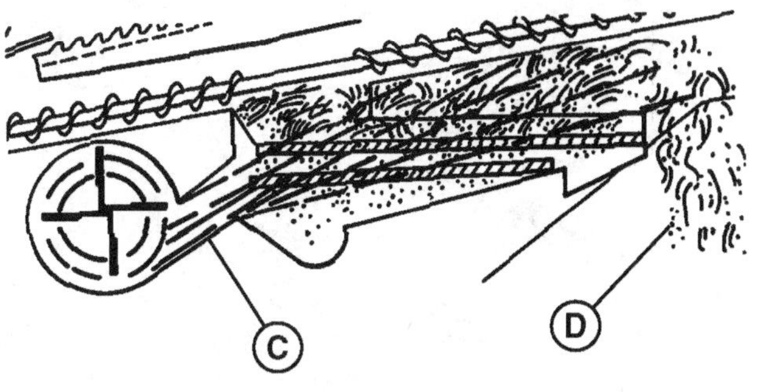

Fig. 65 — Too Much Fan Blast

A— Too Little Fan Blast
B— Grain Carried out of Combine
C— Too Much Fan Blast
D— Grain Blows over Shoe

too much air is being used. Reduce the fan speed and check the results.

## ADJUSTING THE SIEVE

The sieve does the final job of cleaning. It must be opened far enough to allow the kernels to fall through easily, but not so far that chaff and straw are allowed to drop through.

If the sieve is closed too far, the grain will move to the tailings return and overthreshing will occur because of excessive tailings returns. This may also cause excessive kernel cracking.

To adjust the sieve, open it until too much foreign material appears in the grain tank and then close the sieve slightly until the grain tank sample becomes acceptable. If the combine has a lower position for the front of the sieve, use this position because it will retard the movement of the material and allow better cleaning.

DK75838,000020E -19-26AUG13-1/1

# DETERMINING GRAIN LOSSES

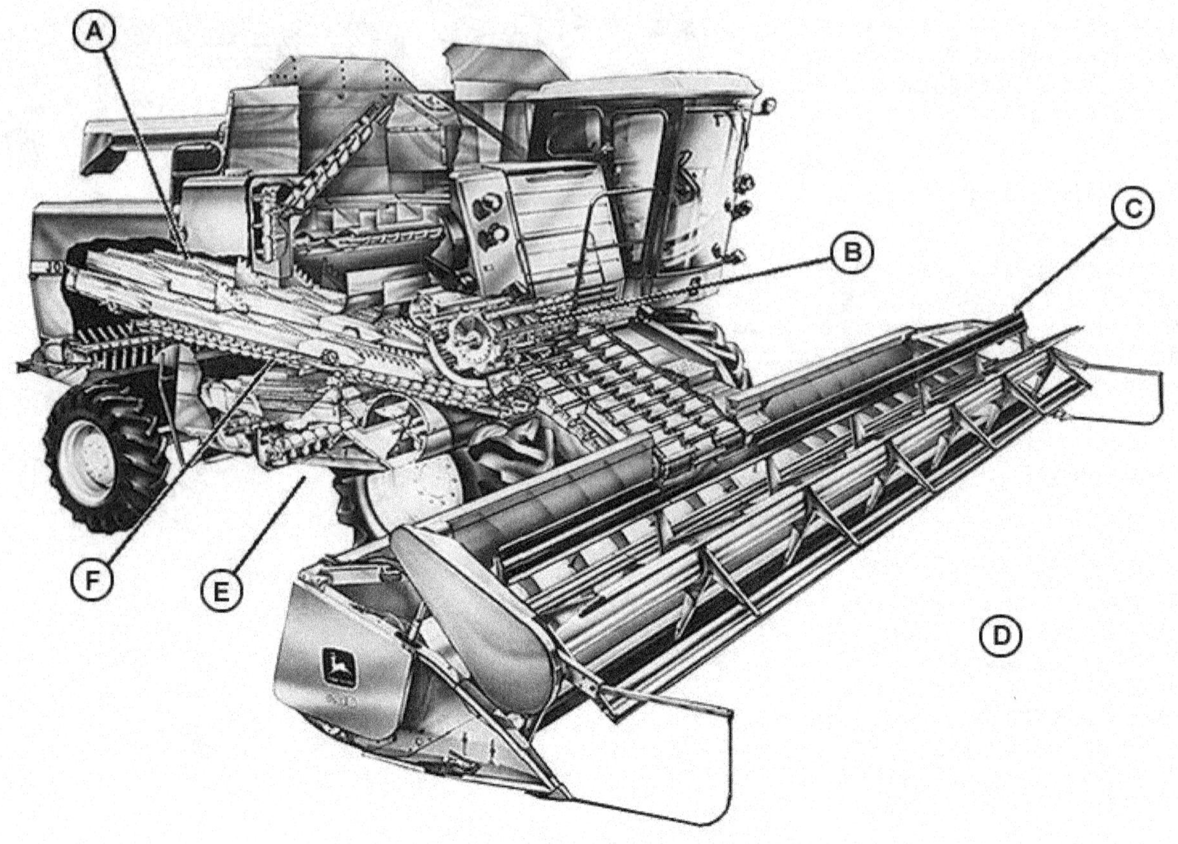

Fig. 66 — Typical Sources of Losses

A— Straw Walker Losses    C— Header Losses    E— Leakage Losses
B— Threshing Losses    D— Preharvest Losses    F— Shoe Losses

After adjusting the harvesting units to get the best job, the operator must also determine what grain losses are occurring. The few minutes spent in determining machine losses and making the proper adjustments are well worth the time.

Suppose you are harvesting a 20-acre (8.1 ha) corn field with an average yield of 100 bushels per acre (5 t/ha). After checking the machine losses, you find you are losing 6 bushels an acre (300 kg/ha) in the harvesting process. However, if it is properly adjusted, the combine should lose not more than 3 bushels per acre (150 kg/ha) in this field. By properly adjusting the machine, you can save 3 bushels an acre (150 kg/ha). Figuring corn at $2.50 per bushel ($98/t/ha), in the 15 or 20 minutes it takes to measure the loss and make the corrective adjustments, you can save $150.00 for the 20-acre (8.1 ha) field!

Grain losses result from improper adjustment of the combine. Even if the machine has been adjusted to cut, thresh, separate, and clean the grain acceptably, the operator should check the combine occasionally to make sure it is continuing to operate properly.

If the grain losses are not acceptable, the operator must reduce them by adjusting the components that are causing the costly losses.

To reduce losses, the operator must know:

- The source of losses
- How to measure losses

## SOURCES OF GRAIN LOSSES

Unless the operator knows the source of grain losses, they cannot be reduced. Some losses are due to improper operation, and others are caused by improper adjustment.

Here are the sources of losses the operator must identify:

- Preharvest losses
- Header losses
- Threshing losses
- Separator losses
- Shoe losses
- Leakage losses

Continued on next page

DK75838,0000212 -19-26AUG13-1/14



DXP04585 —UN—26APR13

050217
PN=168

## PREHARVEST LOSSES

Preharvest losses are those that occurred in the field before combining. Such losses show up as grain on the ground as a result of wind shatter, lodging, down crop, or weather conditions (Fig. 67). Corn and soybeans are two common crops that may have large preharvest losses.

## HEADER LOSSES

These losses occur when the header is operated improperly or when the crop tends to shatter easily. Each type of header has operating characteristics that can cause losses. Losses caused by faulty adjustment and operation of cutting platforms, corn heads, and windrow-pickup units are discussed below.

## CUTTING PLATFORM LOSSES

Usual causes of cutting platform losses are:

- Grain heads missed by cutterbar
- Grain shattered by action of knife
- Grain missed because of improper reel speed
- Grain shattered by too fast a reel speed
- Grain thrown over in front of the reel by too low a reel height
- Grain shattered by too fast a ground speed
- Grain uncut by damaged knife or guard

## CORN HEAD LOSSES

Usual causes of corn head losses are:

- Ears missed by gatherers
- Kernels shattered from cobs by impact with header
- Ears missed because of improper gatherer chain speed
- Ears deflected to ground by too fast a ground speed

## WINDROW-PICKUP LOSSES

Usual causes of windrow-pickup losses are:

- Grain shattered by too fast a pickup speed
- Grain missed because of improper pickup height
- Grain shattered by too fast a ground speed
- Grain shattered by too slow a pickup speed

## THRESHING LOSSES

Threshing losses are caused by:

- Unthreshed grain carried over straw walkers or out the rear of the rotor

Fig. 67 — Weather Conditions May Cause Great Preharvest Losses

- Cracked grain due to overthreshing
- Cracked grain due to excessive tailings

## STRAW WALKER OR SEPARATOR LOSSES

Straw walker or separator losses are usually caused by feeding too much material over them at slow cylinder speeds and by wide concave spacing when the combine is operating at excessive ground speeds. Too much material prevents the grain from falling through the straw walkers or separator grates and onto the cleaning shoe.

## SHOE LOSSES

Shoe losses may be caused by:

- Too much air from fan
- Too much material on chaffer
- Improperly adjusted chaffer and sieve

Too much air from excessive fan speeds will blow chaff and grain over the shoe and out the rear of the combine.

Too much material deposited on the chaffer from overthreshing prevents the grain from falling through to the sieve, and the fan blast cannot blow the straw and chaff free of the chaffer. Thus the grain rides out the rear of the combine with the straw and chaff.

An improperly adjusted chaffer and sieve will not allow grain to fall through if openings are not wide enough. Grain either rides out the rear of the combine on the chaffer, or it is delivered to the tailings return system, which delivers the grain to the cylinder for rethreshing. Too many tailings can cause excessive kernel crackage.

**Continued on next page**          DK75838,0000212 -19-26AUG13-2/14

## LEAKAGE LOSSES

Leakage losses can occur almost anywhere on the combine. To guard against leakage, inspect the combine to see that all inspection doors, cleaning doors, and drainage doors are in the proper position and closed securely (Fig. 68). Also check for torn seals, damaged sheet metal, or holes.

All leaks must be repaired before accurate measurement of losses can be made. Otherwise, it is difficult to determine where losses are occurring.

### MEASURING LOSSES

We will now discuss these losses as they pertain to harvesting small grain and then corn.

### DETERMINING LOSSES IN SMALL GRAINS AND SOYBEANS

The following losses can be measured to determine where losses are occurring:

- Preharvest losses
- Header losses
- Threshing losses
- Straw walker (separator) and shoe losses

When checking small-grain and soybean losses, always use a typical harvest area, well in from the edges of the field. It is best to follow sequential steps to avoid mistakes while determining losses.

After each explanation of the checking areas, a typical example of a loss problem is given. This example is only for the explanation and should not be used as a guide in field operation.

This procedure may be used to determine losses in wheat, oats, barley, soybeans, etc. Follow the procedures outlined below to find where losses must be reduced.

First determine preharvest losses so that you can compare this with the machine losses. The difference in these two figures will indicate whether or not you are doing a good job of combining and whether or not adjustments are necessary.

## DETERMINING PREHARVEST LOSSES

To determine preharvest losses, select a typical unharvested area of the field, well in from the edges. Place a frame 12 inches square (1 square foot), or make it 31.6 cm square (equal to 0.1 m$^2$), in the standing crop. Count all the kernels lying on the ground within the frame.

Fig. 68 — Keep Doors and Covers Tightly Closed

Refer to Table 2 to determine how much grain is already on the ground. Make several random samples and average them to find average grain lost per acre (hectare).

For example, if you are harvesting wheat and find 20 kernels within the frame, the preharvest loss is one bushel per acre (67 kg/ha).

| Preharvest and Header Loss Chart Small Grains and Soybeans | | |
|---|---|---|
| CROP | One Bushel/Acre Loss (Kernels/ft$^2$) | 50 kg/ha Loss (Kernels/0.1 m$^2$) |
| Barley | 13—15 | 13—15 |
| Beans-red kidney | 1—2 | 1—2 |
| Beans-white pea | 3—4 | 3—4 |
| Oats | 10—12 | 15—18 |
| Rice | 29—31 | 31—33 |
| Rye | 21—24 | 18—21 |
| Sorghum | 19—22 | 16—19 |
| Soybeans | 4—5 | 3—4 |
| Wheat | 18—20 | 14—16 |

Table 2 — Preharvest and Header Loss Chart

## DETERMINE MACHINE LOSSES

When checking machine losses, do not use any straw spreading device, such as a straw chopper or straw spreader, because the loss count will be inaccurate. Harvest a typical area. Allow the machine to clear itself of material and then back the combine a distance equal to the length of the machine and stop the combine. This will allow you to check all the loss points without starting and stopping the combine several times.

Continued on next page

DK75838,0000212 -19-26AUG13-3/14

## DETERMINING HEADER LOSSES

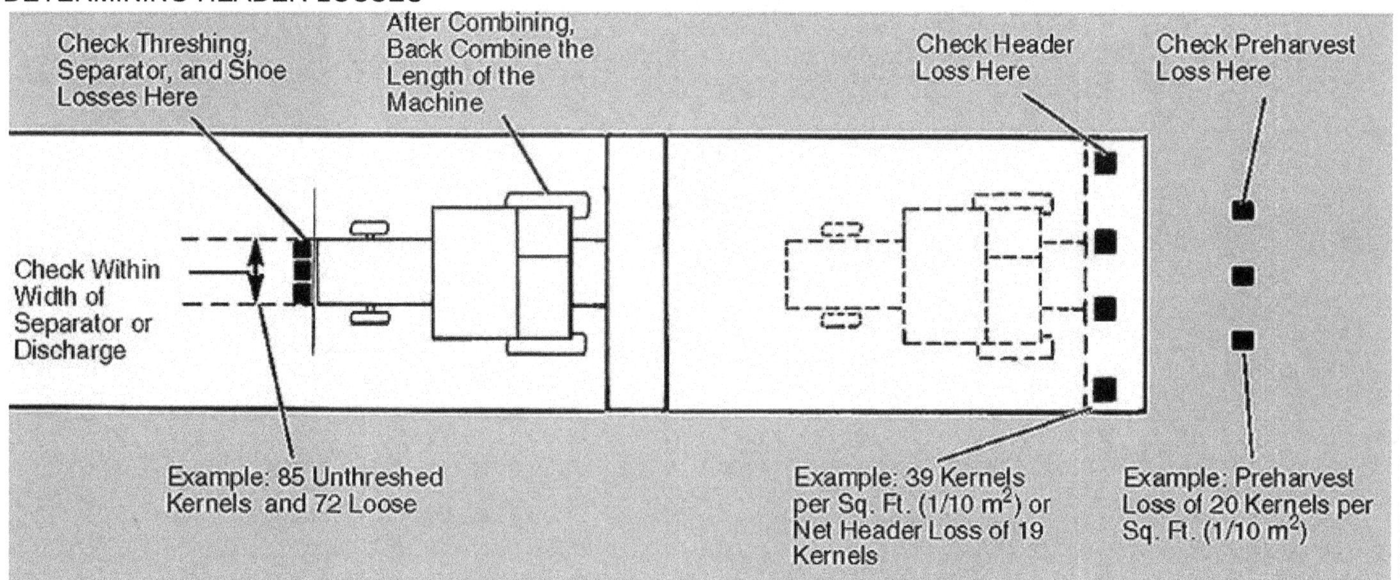


*Fig. 69 — Checking Small Grain Losses*

To determine header losses, after backing the length of the machine, place the one-square-foot (0.1 m²) measuring frame on the ground in front of the combine within the harvested area (Fig. 69). Count the number of kernels found in the frame. Check several other sample areas and average the kernel count. Finally, subtract the number of kernels found in the preharvest loss check. Use Table 2 to determine the loss in bushels per acre (kg/ha).

For example, if the combine has a 14-foot (4.3 m) cutting platform and is harvesting wheat, suppose you find 39 kernels of wheat within the frame. Preharvest losses were 20 kernels. Subtract the preharvest loss and you are left with 19 kernels.

By checking Table 2, we find the header loss is one bushel per acre (67 kg/ha). Typical header losses for small grain crops, when the combine is adjusted and operated correctly, may vary from 1/2 to 2 percent of the average yield. For soybeans, the range is from 10 to 12 percent of the average yield.

If losses are not acceptable, consult the troubleshooting charts on field problems (page 59) to determine possible remedies for header losses.

| Crop | Crop Separator Width (in.) | 10 | 13 | 14 | 15 | 16 | 18 | 20 | 22 | 24 | 30 |
|---|---|---|---|---|---|---|---|---|---|---|---|
| Barley | 38 | 55 | 71 | 76 | 82 | 87 | 98 | 109 | - | - | - |
|  | 44 | 47 | 61 | 66 | 71 | 76 | 85 | 95 | 104 | - | - |
|  | 55 | - | 49 | 53 | 57 | 61 | 68 | 76 | 83 | 91 | 113 |
|  | 60 | - | 45 | 48 | 52 | 55 | 62 | 69 | 76 | 83 | 104 |
|  | 65 | - | - | - | 48 | 51 | 57 | 64 | 70 | 77 | 96 |
| Beans (Red Kidney) | 38 | 4 | 5 | 6 | 7 | 8 | 9 | - | - | - | - |
|  | 44 | 4 | 5 | 5 | 6 | 6 | 7 | 7 | 8 | - | - |
|  | 55 | - | 4 | 4 | 4 | 5 | 5 | 6 | 6 | 7 | 9 |
|  | 60 | - | 4 | 4 | 4 | 4 | 5 | 5 | 6 | 6 | 8 |
|  | 65 | - | - | - | 4 | 4 | 4 | 5 | 5 | 6 | 7 |
| Beans (White Pea) | 38 | 7 | 9 | 10 | 10 | 11 | 12 | 14 | - | - | - |
|  | 44 | 6 | 8 | 8 | 9 | 10 | 11 | 12 | 13 | - | - |
|  | 55 | - | 6 | 7 | 7 | 8 | 9 | 10 | 11 | 12 | 14 |
|  | 60 | - | 6 | 6 | 7 | 7 | 8 | 9 | 10 | 11 | 13 |
|  | 65 | - | - | - | 6 | 6 | 7 | 8 | 9 | 10 | 12 |
| Oats | 38 | 7 | 9 | 10 | 10 | 11 | 12 | 14 | - | - | - |
|  | 44 | 6 | 8 | 8 | 9 | 10 | 11 | 12 | 13 | - | - |
|  | 55 | - | 6 | 7 | 7 | 8 | 9 | 10 | 1 | 12 | 14 |
|  | 60 | - | 6 | 6 | 7 | 7 | 8 | 9 | 10 | 11 | 13 |
|  | 65 | - | - | - | 6 | 6 | 7 | 8 | 9 | 10 | 12 |

**Continued on next page**

DK75838,0000212 -19-26AUG13-4/14

| | | 10 | 13 | 14 | 15 | 16 | 18 | 20 | 22 | 24 | 30 |
|---|---|---|---|---|---|---|---|---|---|---|---|
| Rice | 38 | 53 | 69 | 75 | 80 | 85 | 96 | 107 | - | - | - |
| | 44 | 46 | 60 | 65 | 69 | 74 | 83 | 92 | 101 | - | - |
| | 55 | - | 48 | 52 | 55 | 59 | 66 | 74 | 81 | 88 | 111 |
| | 60 | - | 44 | 47 | 51 | 54 | 61 | 68 | 74 | 81 | 101 |
| | 65 | - | - | - | 47 | 50 | 56 | 62 | 69 | 75 | 94 |

**Machine Loss Chart for Small Grain (Number of Kernels per Square Foot Equivalent to 1 Bushel per Acre)**

| Crop | Crop Separator Width (in.) | 10 | 13 | 14 | 15 | 16 | 18 | 20 | 22 | 24 | 30 |
|---|---|---|---|---|---|---|---|---|---|---|---|
| Rye | 38 | 116 | 151 | 163 | 174 | 186 | 209 | 232 | - | - | - |
| | 44 | 100 | 130 | 141 | 151 | 161 | 181 | 201 | 221 | - | - |
| | 55 | - | 104 | 112 | 120 | 128 | 145 | 161 | 177 | 193 | 241 |
| | 60 | - | 96 | 103 | 110 | 118 | 132 | 147 | 162 | 177 | 221 |
| | 65 | - | - | - | 102 | 109 | 122 | 136 | 149 | 163 | 204 |
| Sorghum | 38 | 61 | 80 | 86 | 92 | 98 | 110 | 123 | - | - | - |
| | 44 | 53 | 69 | 74 | 79 | 85 | 95 | 106 | 116 | - | - |
| | 55 | - | 55 | 59 | 63 | 68 | 76 | 85 | 93 | 102 | 127 |
| | 60 | - | 50 | 54 | 58 | 62 | 70 | 78 | 85 | 93 | 116 |
| | 65 | - | - | - | 54 | 57 | 64 | 72 | 79 | 86 | 107 |
| Soybeans | 38 | 11 | 15 | 16 | 17 | 18 | 20 | 23 | - | - | - |
| | 44 | 10 | 13 | 14 | 15 | 16 | 18 | 20 | 22 | - | - |
| | 55 | - | 10 | 11 | 12 | 13 | 14 | 16 | 17 | 19 | 24 |
| | 60 | - | 9 | 10 | 11 | 12 | 13 | 14 | 16 | 17 | 22 |
| | 65 | - | - | - | 10 | 11 | 12 | 13 | 15 | 16 | 20 |
| Wheat | 38 | 61 | 79 | 85 | 91 | 97 | 109 | 121 | - | - | - |
| | 44 | 52 | 68 | 73 | 79 | 84 | 94 | 105 | 115 | - | - |
| | 55 | - | 54 | 59 | 63 | 67 | 75 | 84 | 92 | 101 | 126 |
| | 60 | - | 50 | 54 | 58 | 61 | 69 | 77 | 84 | 92 | 115 |
| | 65 | - | - | - | 53 | 57 | 64 | 71 | 78 | 85 | 106 |

**Machine Loss Chart for Small Grain (Number of Kernels per Square Foot Equivalent to 1 Bushel per Acre)**

| Crop | Separator Width (cm) | 3 | 4 | 4.3 | 4.6 | 4.9 | 5.5 | 6.1 | 6.7 | 7.3 | 9.1 |
|---|---|---|---|---|---|---|---|---|---|---|---|
| Barley | 97 | 54 | 71 | 77 | 82 | 87 | 98 | 109 | - | - | - |
| | 112 | 46 | 62 | 66 | 71 | 76 | 85 | 94 | 103 | - | - |
| | 140 | - | 49 | 53 | 57 | 61 | 68 | 75 | 83 | 90 | 112 |
| | 152 | - | 46 | 49 | 52 | 56 | 63 | 69 | 76 | 83 | 104 |
| | 165 | - | - | - | 48 | 51 | 58 | 64 | 70 | 77 | 95 |
| Beans (Red Kidney) | 97 | 4 | 5 | 6 | 6 | 7 | 7 | 8 | - | - | - |
| | 112 | 4 | 5 | 5 | 6 | 6 | 7 | 7 | 8 | - | - |
| | 140 | - | 4 | | 4 | 45 | 5 | 6 | 6 | 7 | 8 |
| | 152 | - | 3 | 4 | 4 | 4 | 5 | 5 | 6 | 6 | 8 |
| | 165 | - | - | - | 4 | 4 | 4 | 5 | 5 | 6 | 7 |
| Beans (White Pea) | 97 | 6 | 9 | 9 | 10 | 11 | 12 | 13 | - | - | - |
| | 112 | 6 | 8 | 8 | 9 | 9 | 10 | 11 | 13 | - | - |
| | 140 | - | 6 | 6 | 7 | 7 | 8 | 9 | 10 | 11 | 14 |
| | 152 | - | 6 | 6 | 6 | 7 | 8 | 8 | 9 | 10 | 13 |
| | 165 | - | - | - | 39 | 42 | 47 | 52 | 57 | 62 | 77 |
| Oats | 97 | 48 | 58 | 62 | 66 | 71 | 79 | 88 | - | - | - |
| | 112 | 38 | 50 | 54 | 58 | 61 | 69 | 76 | 84 | - | - |
| | 140 | - | 40 | 43 | 46 | 49 | 55 | 61 | 67 | 73 | 91 |
| | 152 | - | 37 | 40 | 42 | 45 | 51 | 56 | 62 | 67 | 84 |
| | 165 | - | - | - | 39 | 42 | 47 | 52 | 57 | 62 | 77 |

**Continued on next page**

DK75838,0000212 -19-26AUG13-5/14

| Crop | Separator Width | 3 | 4 | 4.3 | 4.6 | 4.9 | 5.5 | 6.1 | 6.7 | 7.3 | 9.1 |
|---|---|---|---|---|---|---|---|---|---|---|---|
| Rice | 97 | 56 | 74 | 80 | 85 | 91 | 102 | 113 | - | - | - |
| | 112 | 48 | 64 | 69 | 74 | 79 | 88 | 98 | 108 | - | - |
| | 140 | - | 51 | 55 | 59 | 63 | 71 | 78 | 86 | 94 | 117 |
| | 152 | - | 47 | 51 | 54 | 58 | 68 | 72 | 79 | 86 | 108 |
| | 165 | - | - | - | 50 | 53 | 60 | 67 | 73 | 80 | 99 |

**Machine Loss Chart for Small Grain (Number of Kernels per 1/10 Square Meter to Equal 50 Kilograms per Hectare)**

| Crop | Separator Width | 3 | 4 | 4.3 | 4.6 | 4.9 | 5.5 | 6.1 | 6.7 | 7.3 | 9.1 |
|---|---|---|---|---|---|---|---|---|---|---|---|
| Rye | 97 | 97 | 130 | 140 | 149 | 159 | 179 | 198 | - | - | - |
| | 112 | 84 | 112 | 121 | 129 | 138 | 155 | 172 | 188 | - | - |
| | 140 | - | 90 | 97 | 104 | 110 | 124 | 137 | 151 | 167 | 205 |
| | 152 | - | 83 | 89 | 95 | 102 | 114 | 126 | 139 | 151 | 289 |
| | 165 | - | - | - | 88 | 94 | 105 | 116 | 128 | 139 | 174 |
| Sorghum | 97 | 51 | 68 | 74 | 79 | 84 | 94 | 104 | - | - | - |
| | 112 | 44 | 59 | 64 | 68 | 73 | 90 | 99 | 108 | - | - |
| | 140 | - | 47 | 51 | 55 | 58 | 65 | 72 | 79 | 87 | 108 |
| | 152 | - | 44 | 47 | 50 | 54 | 60 | 67 | 73 | 80 | 99 |
| | 165 | - | - | - | 46 | 49 | 55 | 61 | 67 | 73 | 91 |
| Soybeans | 97 | 9 | 12 | 13 | 14 | 15 | 16 | 18 | - | - | - |
| | 112 | 8 | 10 | 11 | 12 | 13 | 14 | 16 | 17 | - | - |
| | 140 | - | 8 | 9 | 10 | 10 | 11 | 13 | 14 | 15 | 19 |
| | 152 | - | 8 | 8 | 9 | 9 | 10 | 12 | 13 | 14 | 17 |
| | 165 | - | - | - | 8 | 9 | 10 | 11 | 12 | 13 | 16 |
| Wheat | 97 | 48 | 64 | 68 | 73 | 78 | 87 | 97 | - | - | - |
| | 112 | 41 | 55 | 59 | 63 | 67 | 76 | 84 | 92 | - | - |
| | 140 | - | 44 | 47 | 51 | 54 | 60 | 67 | 74 | 80 | 100 |
| | 152 | - | 41 | 44 | 47 | 50 | 56 | 62 | 68 | 74 | 92 |
| | 165 | - | - | - | 43 | 46 | 51 | 57 | 63 | 68 | 85 |

**Machine Loss Chart for Small Grain (Number of Kernels per 1/10 Square Meter to Equal 50 Kilograms per Hectare)**

## DETERMINING THRESHER LOSS

To determine threshing unit loss, after backing the length of the machine, check the ground in a few places directly behind the separator or machine discharge, using the one-square-foot (1/10 m$^2$) frame (Fig. 69). Count all the kernels remaining on partially threshed heads or pods. Do not include kernels lying loose on the ground. Then check "Machine Loss Chart for Small Grain" on page 48 to determine the loss per acre (ha).

For example, if a combine with a 14-foot (4.3 m) cutting platform and 38-inch (100 cm) separator were being used to harvest wheat, and 85 kernels were found on partially threshed heads, the loss would be one bushel per acre (59 kg/ha).

## LOSSES ON COMBINES WITH DIFFERENT SEPARATOR WIDTH

The chart does not list all possible separator or discharge widths. For combines with discharge widths different from those listed, the procedure for determining threshing losses is basically the same.

Again, place the frame on the ground in a few places directly behind the machine discharge and count the number of kernels remaining on partially threshed heads or pods within the frame. Then find the percentage by which the discharge width varies from one of the discharge widths listed in the chart. Multiply the number of kernels counted by this percentage.

For example, suppose the discharge or separator width on a combine is 24 inches (60 cm). The combine is using an 18-foot (5.5 m) header. Place the frame within this 24-inch (60 cm) wide path and count the number of kernels on partially threshed heads or pods.

The 24-inch (60 cm) discharge width is 63 percent as large as the 38-inch (100-cm) separator listed in the chart. If 155 kernels were counted within the frame: 155 x 0.63 = 98 kernals. Compare this to the chart (38-inch separator, 18 ft. head) and find that this figure equals one bushel per acre loss. Check "Machine Loss Chart for Small Grain" on page 48.

Typical threshing unit loss ranges from 1/2 to 1 percent of the average yield. Acceptable losses are largely a matter of operator preference. If losses are not acceptable, consult the troubleshooting charts on field problems (page 59) to determine possible remedies for threshing unit losses.

Continued on next page

DK75838,0000212 -19-26AUG13-6/14

## DETERMINING SEPARATOR AND CLEANING SHOE LOSSES

To determine separator and shoe losses, after backing the length of the machine, place the one-square-foot measuring frame on the ground directly behind the separator or machine discharge (Fig. 69). Then count the kernels lying loose within the frame. Do not include kernels on partially threshed heads. On combines with separator widths different from those listed in the chart, multiply the number of kernels found lying loose within the frame by the percentage the width varies.

Then subtract the number of kernels found in the header-loss check and the preharvest-loss check. The remaining figure will be the number of kernels lost over the separator and shoe. Check the "Machine Loss Chart for Small Grain" on page 48 to find the loss per acre.

For example, if a combine with a 14-foot (4.3 m) cutting platform and a 38-inch (100 cm) separator is being used to combine wheat, suppose you find 72 loose kernels within the frame. On a machine with a 24-inch (60 cm) separator or discharge, 115 kernels were counted. Multiply 115 x 0.63 = 72 kernels. Remember, 20 of the kernels are from preharvest loss and 19 of the kernels are from header loss. By subtracting these counts (total 39) from 72, the example shows 33 kernels are being lost over the straw walkers (or separator) and shoe. Checking the "Machine Loss Chart for Small Grain" on page 48 we find the loss is about 1/3 bushel per acre (18 kg/ha).

Typical separator and shoe losses should be less than 1 percent of the average yield. Acceptable losses are largely a matter of operator preference.

If losses are not acceptable, consult the field problems chart on (page 59) to determine possible remedies for separator and shoe loss.

## ACCEPTABLE LOSSES IN SMALL GRAINS

What are acceptable losses? They may vary with the operator. Some operators aren't concerned with machine losses of 5 percent, and others don't want any. Generally, acceptable crop losses range from 3 to 5 percent of the yield.

## *DETERMINING LOSSES IN CORN*

While procedures used to check losses in corn are similar to those used in small grain, some items are different. The following points should be checked to determine corn losses:

- Preharvest ear losses
- Corn head ear losses
- Corn head kernel losses
- Threshing unit losses
- Separator and shoe losses
- Leakage losses

When checking corn losses, always use a typical harvest area, well in from the edges of the field. It is best to follow steps in sequence to avoid mistakes while determining losses.

## DETERMINING PREHARVEST LOSSES

To determine preharvest ear losses, select a typical unharvested area of the field, well in from the edges (Fig. 71). Place a marker on one of the rows and measure the necessary distance for 1/100 acre (1/200 ha) according to "Row Length In Feet per 1/100 Acre (1/200 ha)" on page 53. Use the appropriate figures for the row spacing of the corn and the number of row units on the corn head.

Gather all ears found on the ground in this area. Count each 3/4-pound (1/3 kg) ear or equivalent smaller ears as one bushel lost per acre (50 kg/ha of shelled corn). For example, if two 3/4-pound (1/3 kg) ears are found in the area, preharvest loss is two bushels per acre (100 kg/ha).

**Continued on next page**    DK75838,0000212 -19-26AUG13-7/14

## DETERMINING CORN HEAD EAR LOSSES

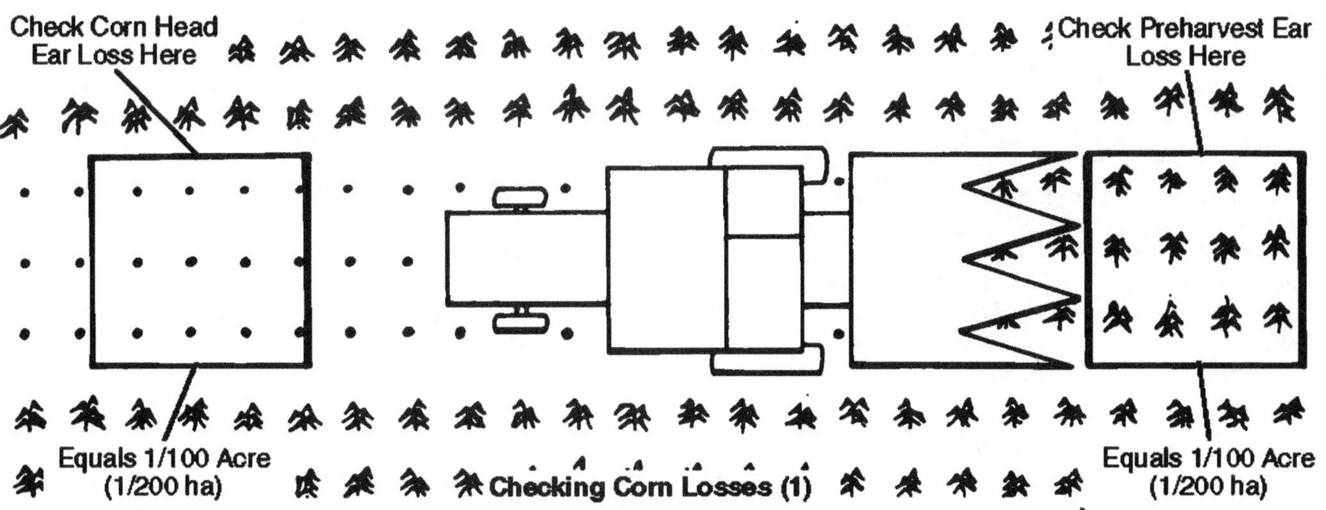


Fig. 70 — Checking Preharvest Ear Loss and Corn Head Ear Loss

After operating the combine for some distance, check the rows harvested behind the combine (Fig. 71). Then count the number of 3/4-pound (1/3 kg) ears picked up in 1/100 acre (1/200 ha). Subtract the preharvest loss figure from this amount to determine corn head ear loss.

For example, if four 3/4-pound (340 g) ears are found in 1/100 acre (1/200 ha) and preharvest loss was two 3/4-pound (340 g) ears (two bushels per acre or 100 kg/ha), corn head ear loss equals two bushels per acre (100 kg/ha).

Less than 1 percent loss can be expected when the machine is adjusted and operated correctly. A 1-to-2 bushel (60 to 120 kg/ha) loss may be considered acceptable, but this depends on operator judgment. If losses are not acceptable, consult the field problems troubleshooting chart on (page 59) to determine possible remedies for corn head ear losses.

## DETERMINING MACHINE KERNEL LOSSES

When checking kernel losses, first harvest a typical area. Allow the machine to clear itself of material and then back the combine a distance equal to the length of the machine and stop the combine. This will allow you to check all loss points without starting and stopping the combine several times.

If machine losses are not acceptable, check the individual units of the combine as described in the following paragraphs to determine where the losses are occurring.

## DETERMINING CORN HEAD KERNEL LOSSES

To determine corn head kernel losses, after backing the length of the machine, use a measuring frame to enclose 10 square feet (1 m$^2$) of the sample area directly in front of the corn head (Fig. 73). Consult the Loss-Measuring Frame Dimension Chart for Corn (Fig. 71) to determine the correct inner dimensions of the frame. Then count the number of kernels within the frame. Twenty kernels per 10 square feet equals one bushel per acre (16 kernels per square meter equals 50 kg/ha). For example, if 10 kernels are found in the 10-square-foot (1 m$^2$) area, the loss is 1/2 bushel per acre (25 kg/ha). Expected normal loss is less than one percent of the average yield when the combine is adjusted and operated properly. If losses are not acceptable, consult the field problems troubleshooting chart on page 59 to determine possible remedies for corn head kernel loss.

| Loss-Measuring Frame Dimension Chart for Corn | | | |
|---|---|---|---|
| 10 ft$^2$ | | 1 m$^2$ | |
| Width x Length in. | | Width x Length cm | |
| 20 | 72 | 50 | 200 |
| 28 | 51.5 | 70 | 143 |
| 30 | 48 | 75 | 133 |
| 36 | 40 | 90 | 111 |
| 38 | 38 | 95 | 105 |
| 40 | 36 | 100 | 100 |

Fig. 71 — Loss-Measuring Frame Dimension Chart for Corn

## DETERMINING THRESHING UNIT LOSSES

Continued on next page

DK75838,0000212 -19-26AUG13-8/14



050217
PN=175

To determine threshing unit losses, check the ground in a few places directly behind the separator or machine discharge (Fig. 72). Use the 10-square-foot (1 m$^2$) frame and count all kernels remaining on the cobs. Do not include kernels lying loose on the ground. Combines may have separator widths different from those listed in "Machine-Kernel Loss Chart for Corn" on page 53 in column 2. For these combines, use a 10-square-foot (1 m$^2$) frame with the width dimension no greater than the width of the separator or machine discharge. Again count the kernels remaining on the cobs within the frame. Multiply this number by the percentage the separator or discharge width varies from one of the widths listed. Again check the chart on page 53 in column 2 to determine the loss in bushels per acre. For example, if a combine with a three-row corn head and 38-inch separator were being used on 40-inch rows and 29 kernels were found on threshed cobs in a 10-square-foot (1 m$^2$) area, threshing unit loss would be one-half bushel per acre. If this combine had only a 24-inch (60 cm) separator, 46 kernels would have been counted on the threshed cobs. This separator is 63 percent as wide as the 38-inch (100 cm) separator. Therefore, multiply 46 x 0.63 = 29 kernels to obtain the equivalent threshing unit loss.

Less than one percent of the total yield may be lost when the combine is operating properly. If losses are not acceptable to the operator, consult the field problems troubleshooting chart on page 59 to determine possible remedies for threshing unit losses.

| Row Length in Feet per 1/100 Acre (1/200 ha) | | | | | | | |
|---|---|---|---|---|---|---|---|
| Row Width (Spacing) Inches (cm) | Row Width (Spacing) Inches (cm) | | | | | | |
| | One-Row Head | Two-Row Head | Three-Row Head | Four-Row Head | Six-Row Head | Eight-Row Head | Twelve-Row Head |
| 20 (50) | 262 (100) | 131 (50) | 87.3 (33.3) | 65.5 (25) | 43.6 (16.7) | 32.7 (12.5) | 21.8 (8.3) |
| 28 (70) | 187 (71.4) | 93.5 (35.7) | 61.3 (23.8) | 46.7 (17.8) | 31.1 (11.9) | 23.4 (8.9) | 15.6 (5.9) |
| 30 (75) | 174 (66.7) | 87 (33.3) | 58 (22.2) | 43.6 (16.7) | 29 (11.1) | 21.8 (8.3) | 14.5 (5.6) |
| 36 (90) | 145 (55.6) | 72.5 ( 27.7) | 48.3 (18.5) | 36.2 (13.9) | 24.2 (9.3) | 18.1 (7) | 12.1 (4.6) |
| 38 (95) | 138 (52.6) | 69 (26.3) | 46 (17.5) | 34.5 (13.2) | 23 (8.8) | 17.2 (6.6) | 11.5 (4.4) |
| 40 (100) | 131 (50) | 65.5 (25) | 43.6 (16.7) | 32.7 (12.5) | 21.8 (8.3) | 16.4 (6.2) | 10.9 (4.2) |

## DETERMINING SEPARATOR AND SHOE LOSSES

To determine separator and shoe losses, after backing the machine, use a 10-square-foot (1 m$^2$) frame to check several areas directly behind the separator (Fig. 72). Consult the "Loss Measuring Frame Dimension Chart for Corn" on page 52 for proper inside dimensions of the frame (one dimension should be no wider than the width of the separator or machine discharge). Then count the number of loose kernels lying on the ground. Do not include kernels remaining on threshed cobs (threshing unit losses). For combines with separator widths different from those listed in the chart, multiply the number of kernels counted by the percentage the width varies.

Then subtract the number of kernels found under corn head kernel losses. Use the "Machine-Kernel Loss Chart for Corn" on page 53 to determine total losses per acre from the remaining kernel count.

By checking the chart it is determined that one bushel per acre is being lost on the separator and shoe.

Less than one percent of the total yield may be lost on the straw walkers and shoe when the combine is adjusted and operated properly. If the losses are unacceptable, consult the field problems troubleshooting chart on page 59 to determine possible remedies for straw walker and shoe losses.

## ACCEPTABLE LOSSES FOR CORN

What are acceptable losses? They may vary with the operator. Some operators aren't concerned with total machine losses of 5 percent, and others don't want any. Generally, total acceptable corn losses range from 3 to 5 percent of the yield.

| Machine-Kernel Loss Chart for Corn | | | | | | |
|---|---|---|---|---|---|---|
| Approximate Kernels per 10 ft$^2$ (1 m²) to Equal 1 Bushel per Acre (50 kg/ha) | | | | | | |
| Corn Head Size | Row Spacing (Inches) | Separator Width (Inches) | | | | |
| | | 38 (97) | 44 (112) | 55 (140) | 60 (152) | 65 (165) |
| 3-row | 36 (90) | 52 (44) | 45 (38) | - | - | - |
| | 38 (95) | 55 (46) | 48 (40) | - | - | - |
| | 40 (100) | 58 (49) | 50 (42) | - | - | - |

Continued on next page

DK75838,0000212 -19-26AUG13-9/14

| Rows | Spacing | | | | | |
|---|---|---|---|---|---|---|
| 4-row | 28 (70) | 54 | 47 | - | - | - |
|  |  | (49) | (39) |  |  |  |
|  | 30 (75) | (45) | (39) | - | - | - |
|  |  | 58 | 50 |  |  |  |
|  | 32 (80) | 62 | 54 | - | - | - |
|  |  | (52) | (45) |  |  |  |
|  | 36 (90) | 70 | 60 | 48 | 44 | - |
|  |  | (58) | (51) | (40) | (37) |  |
|  | 38 (95) | 74 | 64 | 51 | 47 | - |
|  |  | (62) | (53) | (43) | (39) |  |
|  | 40 (100) | 77 | 67 | 54 | 49 | - |
| 5-row | 28 (70) | - | 59 | 47 | - | - |
|  |  |  | (49) | (39) |  |  |
|  | 30 (75) | - | 63 | 50 | - | - |
|  |  |  | (53) | (42) |  |  |
|  | 36 (90) | - | 75 | 60 | 55 | 51 |
|  |  |  | (63) | (51) | (47) | (43) |
|  | 38 (95) | - | 79 | 64 | 58 | 54 |
|  |  |  | (67) | (53) | (49) | (45) |
|  | 40 (100) | - | 100 | 80 | 74 | 68 |
|  |  |  | (70) | (56) | (52) | (48) |
| 6-row | 28 (70) | - | 70 | 56 | 52 | 48 |
|  |  |  | (59) | (47) | (44) | (40) |
|  | 30 (75) | - | 75 | 60 | 55 | 51 |
|  |  |  | (63) | (51) | (47) | (43) |
|  | 36 (90) | - | 90 | 72 | 66 | 61 |
|  |  |  | (76) | (61) | (56) | (52) |
|  | 38 (95) | - | 95 | 76 | 70 | 65 |
|  |  |  | (80) | (64) | (59) | (54) |
|  | 40 (100) | - | 100 | 80 | 74 | 68 |
|  |  |  | (84) | (68) | (62) | (57) |
| 8-row | 28 (70) | - | 94 | 75 | 67 | 63 |
|  |  |  | (79) | (63) | (58) | (53) |
|  | 30 (75) | - | 100 | 80 | 74 | 68 |
|  |  |  | (84) | (68) | (62) | (57) |
|  | 36 (90) | - | - | 96 | 88 | 82 |
|  |  |  |  | (81) | (75) | (69) |
|  | 38 (95) | - | - | 102 | 93 | 86 |
|  |  |  |  | (86) | (79) | (73) |
|  | 40 (100) | - | - | 107 | 98 | 91 |
|  |  |  |  | (90) | (83) | (76) |
| 12-row | 28 (70) | - | - | 112 | 103 | 95 |
|  |  |  |  | (94) | (87) | (80) |
|  | 30 (75) | - | - | 120 | 110 | 102 |
|  |  |  |  | (101) | (93) | (86) |

**Continued on next page**

DK75838,0000212 -19-26AUG13-10/14

## GUIDELINES FOR CORN LOSSES

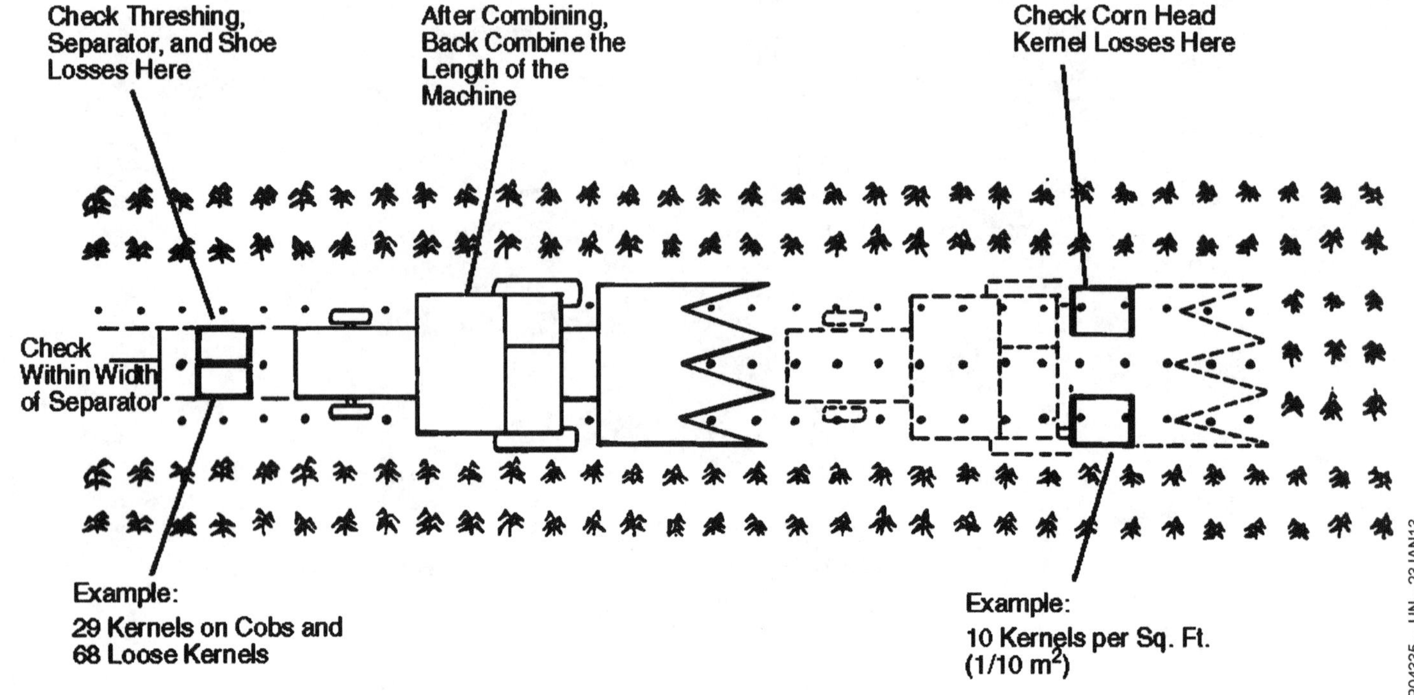


**Checking Corn Losses (2)**

*Fig. 72 — Checking Kernel Losses in Corn*

Remember that a few kernels on the ground do not mean the combine is doing a poor job. No combine is 100 percent efficient.

If total losses are acceptable, no adjustment of the machine or operating procedure is needed.

Check the losses occasionally to make sure that crop or field conditions haven't changed and increased losses. Changes in the moisture level, field conditions, how well the crop is standing, and crop variety all affect the rate of loss.

Make only one adjustment at a time. Then check the losses. This will help you pinpoint the adjustment problem.

### WEEDY FIELD CONDITIONS

Many field conditions affect the combine's efficiency. When special conditions are encountered, always check the grain loss to be sure it is held at a minimum.

Weedy fields increase losses and the moisture content of the harvested grain. Much of the heavy green material accumulates on the straw walkers or rotor, chaffer, and sieves, and does not permit grain to be separated from the straw.

To reduce crop losses due to weeds, reduce the amount of green material taken into the machine by raising the cutter bar if the weeds are shorter than the crop. Go around especially weedy areas. If many weeds must be taken into the combine to harvest the crop, some steps can be taken to reduce the crop loss:

1. First, slow down or take a narrower cut. This decreases the amount of material entering the machine and allows the walkers to clear the trash and separate the grain.
2. It may help to increase the cylinder clearance so the green material is broken up as little as possible. However, this may result in overloading the straw walkers or rotor and can increase grain losses. If the cylinder spacing is increased, keep a careful check to see that the walkers are not overloaded.
3. On machines with fan air deflectors, change the direction of the fan blast so that a strong gust of air is present where the grain leaves the grain pan and passes on to the chaffer; this will aid in removing green material that is present with the grain.

Continued on next page

DK75838,0000212 -19-26AUG13-11/14

4. Windrowing the crop may also be helpful (Fig. 73). The crop should be cut and windrowed about a week or ten days before a clean crop would be harvested. Using this procedure, the weeds will dry and less wet material will be taken into the machine. To ensure proper drying, the windrow should be kept light and fluffy and resting on top of the crop stubble.

## *WEATHER CONDITIONS*

Weather greatly affects the harvestability of the crop. If the crop moisture level is too dry at harvest time, harvesting in early morning or late afternoon when dew is present may help. The opposite, of course, is true for wet crop conditions.

Excessive moisture should be avoided. If snow or ice is present on the crop, the heat of the machine will melt the snow or ice and cause high grain moisture content. The moisture may also cause dust to turn to mud, which can clog the straw walkers, sieves, and chaffers. The same is true for rain or heavy dew.

Fig. 73 — *Windrowing May Help Reduce Losses in Weedy Conditions*

DXP04336 —UN—23JAN13

**Continued on next page**

DK75838,0000212 -19-26AUG13-12/14



050217
PN=179

## CROP MOISTURE CONTENT

*Fig. 74 — Harvesting Soybeans When Dew Is on Them Helps Reduce Shatter Losses*

The moisture content of the crop affects not only the harvesting process, but also the price received for the crop and storage factors.

### WHEAT

Wheat should be harvested when the moisture level is below 14 percent. This permits safe storage of the wheat and selling without dockage due to excess moisture. Combining the wheat at a higher moisture level may damage the grain. The extra threshing action needed to separate a higher moisture crop, coupled with the softer grains, can crack the wheat.

### SOYBEANS

Soybeans should be harvested with the moisture content below 14 percent, with the optimal level about 12 percent moisture. Most bean losses occur at the gathering unit due to shattering. If the beans can be harvested with some moisture on the pod, the shatter losses will be reduced. Early morning or late afternoon harvesting is best for dry beans, but care should be taken to avoid taking in too much moisture.

### CORN

Corn should be harvested below 15 percent moisture for best storage results. Higher moisture content requires drying the grain and can increase the harvesting cost. Combining high moisture corn increases kernel damage at the cylinder because the crop is difficult to thresh and the kernels are softer. But drier corn has a higher preharvest loss. Both factors should be considered to determine the optimum time to harvest.

Variety also affects the optimum harvesting moisture. The best harvest moisture varies from operation to operation, but corn should not be harvested at moisture higher than 30 percent.

**Continued on next page**

DK75838,0000212 -19-26AUG13-13/14

## LODGED CROPS

When the crop to be harvested has lodged (is bent over or lying on ground), extra care must be taken in harvesting. Slower travel speeds and lower header heights help increase the percentage of the crop harvested. Special attachments are available to aid in harvesting down crops. Be careful to avoid damage from stones, sticks, or other obstructions that may damage the machine when the header is operated at a lower height.

*Fig. 75 — Lodged Crops Require Careful Combining*

DK75838,0000212 -19-26AUG13-14/14

# TROUBLESHOOTING FIELD PROBLEMS

The majority of combine operating problems can be traced to improper adjustment. The troubleshooting charts that follow will help you when problems develop by suggesting a probable cause and a recommended remedy. Always check the operator's manual of the combine you are operating for specific adjusting procedures and correct settings.

These suggested remedies should be applied with caution. Make an effort to understand the combine and know why you are making an adjustment. When you are trying to solve a problem, make certain that its source does not come from somewhere other than the apparent cause. For example, a plugged cylinder may result from improper feeding at the feeder house, rather than an improperly adjusted cylinder.

To overcome harvesting problems, special attachments are available to meet various ground and crop conditions. But when you have difficulties, make every effort to correct the problem before purchasing a special attachment. If the

Fig. 76 — Combining a Poor Windrow May Present Field Problems

combine is not adjusted properly, the special attachment will not eliminate the difficulty.

**CUTTING PLATFORM PROBLEMS TROUBLESHOOTING CHART**

| PROBLEM | CAUSE | REMEDY |
|---|---|---|
| Shattering of grain ahead of cutterbar. | Reel speed not coordinated with ground speed, causing excessive agitation before crop is cut. | Change reel drive sprockets or adjust hydrostatic drive reel to coordinate reel speed with ground speed so reel will move crop smoothly and evenly. Reel should turn 25 percent faster than ground travel speed. |
| | Ground speed too fast for conditions of crops. | Slow down ground speed so reel will not bat crop, causing shattering. |
| Cut crop building up and falling from front of cutterbar or loss of grain heads at cutterbar. | Reel not adjusted low enough for proper delivery of cut crop to auger. | Set reel low enough to sweep material from cutterbar. |
| | Reel set too far forward. | Set reel back closer to knife. |
| | Auger clearance too high from platform bottom. | Adjust outer ends of auger to 1/8–5/8 inch (3.2–15.6 mm) clearance from platform bottom and check finger clearance. |
| | Cutting platform too high; cutting stalks too short for proper delivery. | Lower cutting platform so stalks of cut material will be long enough for smooth, even feeding to auger. |
| | Reel speed too slow. | Increase reel speed. |
| Ragged and uneven cutting of crop. | Cutting mechanism not operating at recommended speed. | Check basic speed of combine. (See combine operator's manual.) |
| | Various parts of cutterbar, such as knife sections, guards, or wearing plates, are worn, damaged, or broken. | Check and replace all defective parts on cutterbar to obtain even cutting of crop. |
| | Bent knife, causing binding of cutting parts. | Straighten the bent knife. Check the guard alignment and align if necessary for a smooth cut. |
| | Knife clips not adjusted to permit knife to work freely. | Adjust knife clips so knife will work freely, but still keep knife sections from lifting off guards. |
| | Cutting edge of guards not close enough or parallel to knife sections. | Adjust guards. |
| | Looseness between knife back and guards. | Adjust wearing plates so knife back is snug to guard. |
| | Lips of guards out of adjustment or bent, causing poor shearing action. | Adjust lips of guards so they are parallel to shear edge of guards. |
| | Improper knife register. | Adjust knife register so knife sections pass an equal distance through adjacent guards at each end of pitman stroke. |
| Crop falling in front of cutterbar after it is cut. | Reel speed too slow. | Adjust reel speed to deliver crop to platform. Reel should turn 25 percent faster than forward travel of combine. |

Continued on next page

DK75838,000020F -19-31JUL13-1/8



DXP04339 —UN—23JAN13

050217
PN=182

| | Reel set too far forward. | Set reel back closer to knife. |
|---|---|---|
| | Cutting mechanism not at recommended speed. | Check basic speed of combine. (See combine operator's manual). |
| Excessive vibration of parts. | Cutting mechanism not at recommended speed. | Check basic speed of combine. (See combine operator's manual). |
| | | Check platform and feeder house drive. (See combine operator's manual.) |
| | Excessive looseness of cutting parts and knife drive. | Remove all excessive play from cutterbar and knife drive to eliminate vibration. After removing excessive play, make certain cutterbar and knife drive are properly adjusted. |
| | Improper knife register. | Adjust knife register. |
| Reel wrapping in tangled and weedy crops. | Slat reel not efficiently delivering crop. | Install pickup reel. |
| | Incorrect location of reel. | Place reel well ahead and down. If wrapping occurs at ends of reel, install reel end shields. |
| | Reel speed too fast. | Reduce speed of reel to allow crops to fall onto platform. |
| Reel carrying straw around. | Reel speed too fast. | Reduce speed of reel so straw will not carry over top of reel. Reel should turn 25 percent faster than ground speed. |
| | Reel height too low. | Raise reel height to reduce amount of straw gathered by reel. |
| | Pickup tines pitched too much. | Reduce pitch of tines. |
| Too much material entering combine. | Cutting too low in order to get all down and tangled crops. | Use lifting guards or pickup reel in down and tangled conditions. |
| Uneven or bunched feeding of crop to cylinder. | Auger clearance too high from platform bottom. | Adjust outer ends of auger to proper clearance from bottom. |
| | Buildup of grain on cutterbar. | Lower height of reel and set fore-and-aft position as close as possible to cutterbar and auger. |
| | Fingers in platform auger not adjusted to convey material properly to feeder conveyor chain. | Adjust finger height of auger so there is even feeding from the platform auger to feeder conveyor chain. |
| | Platform drive belt slipping. | Spring-loaded tightener must be free and tight against belt. (See combine operator's manual.) |
| | Auger too far ahead of stripper. | Adjust auger back closer to stripper. |
| | Platform auger slip clutch set too loose. | Tighten auger slip clutch. |
| | Feeder drum lower stops set too high. | Adjust lower stops. (See combine operator's manual.) |

## ROW-CROP HEAD PROBLEMS TROUBLESHOOTING CHART

| PROBLEM | CAUSE | REMEDY |
|---|---|---|
| Plugging row unit at knife. | Belt speed too high. | Reduce header speed. |
| Loss of crop in the field. | Gatherer points set too high. | Lower gatherer points to pick up lodged or down crop. |
| | Ground speed too fast or too slow. | Adjust ground speed to meet field and ground conditions. |
| | | Adjust speed of gathering belts. |
| | Not harvesting planter rows. | Harvest rows as they were planted. It will be easier to follow rows and reduce crop losses. |
| | Row units not centered on rows. | Adjust row unit spacing to match row spacing of crop. |
| | Speed of gathering belts too fast or too slow. | Adjust speed of gathering belts. |
| | Skid shoes set to cut crop too high. | Adjust skid shoes for lower cut. |
| Soybean pods left on stubble. | Skid shoes set to cut crop too high. | Adjust skid shoes for lower cut. |
| | Excessive row unit float spring tension. | Adjust row unit float springs. |
| | Automatic header height control improperly adjusted. | Reset automatic header height control. (See combine operator's manual.) |
| | Gatherer points not properly positioned between rows. | Carefully operate row-crop head on rows; reduce ground speed if necessary. |
| Poor cutting action (rugged and uneven cutting of crop). | Excessive gathering belt speed. | Reduce gatherer speed. |

Continued on next page

DK75838,000020F -19-31JUL13-2/8

| | Incorrectly adjusted rotary knife. | Adjust rotary knife for proper cutting action. |
|---|---|---|
| | Bent, worn, or broken rotary knife sections. | Straighten or replace rotary knife sections. |
| | Worn or broken stationary knife. | Reverse or replace stationary knife. |
| Excessive wear or damage to rotary knife. | Rocky field conditions. | Adjust skid shoes for higher cut. |
| | | Be alert for obstructions during field operation. |
| | Rotary knife is adjusted too close to stationary knife. | Adjust rotary knife for proper cutting action. |
| | Knife speed too high. | Reduce knife speed. |
| Excessive wear of stationary knife blades. | Speed of gathering belts is too fast. | Reduce speed of gathering belts to match ground speed or increase ground speed. |
| | Excessive gap between rotary and stationary knife blades. | Be sure row unit frame and stationary knife are not bowed. Check mounting surface for stationary knife, to be certain it is free of weld spatter, paint, mud, and crop residue. |
| | | Be sure all knife sections on each rotary knife contact the stationary knife evenly. |
| | | Tap top end of rotary knife drive shaft with a brass or lead hammer to check that bearings are fully seated. |
| | | Clean knives and adjust knife spacing. (See combine operator's manual.) |
| | Dirt is forced into rotary knives by the skid shoes and gathering belts. | Increase row unit float spring tension. |
| | | Adjust skid shoes for a higher cut position. |
| | | Adjust automatic header height control to activate sooner. |
| | | Adjust gatherer points upward so they do not funnel dirt clods onto bean ridge. |
| Worn or damaged gathering belts or gathering belts jumping out of time. | Inadequate gathering belt chain tension. | Adjust gathering belt chain tension. |
| | Gathering belts tear off grass, weeds, or crop before they can be cut. | Reduce speed of gathering belts to match ground speed or increase ground speed. |
| | Rocky field conditions. | Install rock deflectors. |
| | Poor cutting by rotary knives. | Adjust rotary knives for proper cutting action. |
| | | Reduce speed of gathering belts. |
| | | Adjust gatherer points so dirt is not funneled into bean ridge. |
| | Trough under gathering belts fills with dirt and cut stalks. | Increase row unit float spring tension. |
| | | Adjust skid shoes for higher cut. |
| | | Adjust automatic header height control to activate sooner. |
| | Inadequate gathering belt chain lubrication. | Lubricate gathering belt chains. |
| Dirt in grain tank or dirt plugs trough under gathering belts. | Excessive pressure on skid shoes causes them to push dirt. | Increase row unit float spring tension. |
| | | Operate row-crop head as high as possible so skid shoes touch the ground. |
| | | Adjust automatic header height control to activate sooner. |
| | | Adjust gatherer points so dirt clods are not funneled into bean ridge. |
| | Gathering belts pull grass, weeds, or crop out of the ground so dirt is brought in with the roots. | Reduce speed of gathering belts to match ground speed or increase ground speed. |
| | | Adjust rotary knives for proper cutting. |
| | Rows are ridged high enough to allow gathering belts or rotary knives to contact dirt. | Adjust skid shoes for higher cut position. |
| | Dirt not being removed by combine. | Use perforated elevator doors, auger connections, and feeder house bottom insert when combining soybeans, edible beans, and similar crops to help remove the dirt and weed seed. |

Continued on next page

DK75838,000020F -19-31JUL13-3/8

| | | |
|---|---|---|
| Skid shoes dig into ground. | Insufficient float spring tension. | Increase row unit float spring tension. |
| | Automatic header height control improperly adjusted. | Reset automatic header height control. |
| Skid shoes will not float properly. | Excessive skid shoe pressure on ground. | Increase row unit float spring tension. |
| | | Adjust automatic header height control. |
| Row units do not move freely through float range. | Insufficient clearance or binding gatherer sheets. | Adjust gatherer sheet clearance between row units. |
| | Row units do not move freely on pivot. | Lubricate row unit pivots at grease fittings provided. |
| | Trash collects at row unit pivots. | Clean around row unit pivots. |
| Row-crop head is too slow to return to operating level after irregular ground. | Excessive float spring tension. | Adjust row unit float spring tension. |
| | Insufficient clearance or binding between gatherer sheets. | Adjust gatherer sheet clearance between row units. |
| | Automatic header height control improperly adjusted. | Adjust drop rate control. |
| | Drop rate valve slows drop rate more than desired. | Readjust drop rate valve. |
| | Deck plates not adjusted properly. | Adjust deck plates. |
| | Row-crop head set too high. | Lower row-crop head. |

## CORN HEAD PROBLEMS TROUBLESHOOTING CHART

| PROBLEM | CAUSE | REMEDY |
|---|---|---|
| Loss of ears from the head. | Gatherer points set too high. | Adjust gatherer points so they just touch the ground. When picking low hanging ears, raise front tip of gatherer points just enough to run the corn head with skids close to the ground. |
| | Ground speed too fast or too slow. | Operate at a speed to meet field and ground conditions. Excessive ground speed can cause ears to fall off the stalks ahead of the gatherer chains. Too slow a ground speed can cause the ears to slide forward out of the unit. Operate at a speed at which the gatherer chains merely help guide the stalks into the rolls. |
| | Not picking planter rows. | Pick rows as they were planted. It will be easier to follow the rows and eliminate loss of ears. |
| | Row units not centered on rows. | Adjust corn head row spacing to equal row spacing of corn. |
| | Ears sliding out over gatherer chains. | Use ear savers and center shield extensions. |
| | Gatherer chain speed too fast or too slow. | Obtain the correct gatherer chain speed by changing the feeder house powershaft drive sprocket. |
| Ear shelling at stalk rolls. | Deck plate not adjusted properly. | Adjust deck plates. |
| | Corn head set too high. | Lower corn head. |
| Ears not shelled completely. | Moisture content of corn too high. | Wait for moisture content of corn to drop. Corn kernels tend to cling to the cobs when the moisture content is above 30 percent. Best shelling is obtained and crackage is at a minimum when the moisture content is under 27 percent. |
| | Cylinder or rotor speed too slow. | Increase cylinder or rotor speed by 5 percent and check combine primary countershaft speed or combine beater speed. Also, check separator drive belt tension to make sure it is not slipping. (See combine operator's manual.) |
| | Rasp-bars bent. | Straighten or replace rasp-bars. |
| | Concave bent. | Replace if necessary. |
| | Concave not level (not parallel with cylinder rotor). | Adjust concave spacing equally on both ends of cylinder or rotor. |
| | Too wide a space between cylinder or rotor and concave. | Close cylinder-to-concave (or rotor-to-concave) spacing to increase shelling action. |
| | Ears going between rasp-bars of cylinder without being shelled. | Install cylinder filler plates. |

Continued on next page

DK75838,000020F -19-31JUL13-4/8

| | | |
|---|---|---|
| Ears not shelled completely (continued). | Cobs being split without the corn being shelled from them (corn is attached to half or smaller section of cob). | Open concave spacing just enough to get proper shelling action. |
| | Ground speed too fast. | Reduce ground speed. |
| Excessive damage to shelled corn or cracked corn. | Concave too close to cylinder or rotor bars. | Increase concave spacing. |
| | Cylinder or rotor too fast. | Reduce cylinder or rotor speed. Also check combine beater speed or combine primary countershaft speed as shown in combine operator's manual. |
| | Moisture content of corn to high. | Wait until moisture content of corn drops. Corn above 30 percent moisture has a tendency to crack and is easily crushed. It is best to wait until moisture is under 27 percent. |
| | Concave not level. | Level concave. |
| | Damaged rasp-bars or concave. | Replace as necessary. |
| | Dented auger housing. | Straighten or replace as necessary. |
| | Excessive tailings. | Reduce ground speed. |
| | | Open or clean the sieve and increase the fan speed. |
| Shelled corn coming out rear of combine. | Corn carrying over straw walkers. | Extend pans at rear of straw walkers, if straw walkers are so equipped. |
| | Corn carrying over chaffer. | Reduce ground speed. |
| | | Adjust the chaffer if too far closed or too far open and plugged with pieces of cob. Refer to combine operator's manual. |
| | | Clean sieve completely if sieve is closed or clogged with cobs. |
| | | Increase the speed of cleaning fan if volume of air does not appear to be adequate. Check belt tension. |
| | | If combine is equipped with sieve tilt adjustment, lower front of sieve to allow air to the chaffer. |
| | | Too much material in combine. Check corn head for excessive stalk breakage, which could be due to rolls not being properly timed or deck plates closed to much. |
| Ears sliding out through the throat. | Ear savers not properly adjusted. | Adjust ear savers. |
| | Ear savers worn out. | Replace ear savers. |
| Pulling up cornstalks. | Deck plates set too close together. | Spread deck plates, a little at a time, until stalks feed through rolls more freely. |
| | Traveling too fast for gatherer chain speed. | Slow down to meet crop conditions or increase row unit drive speed. |
| | Gatherer chain flights digging into cornstalk roots. | Lower gatherer points. |
| | Corn extremely dry or down. | Remove center shield extensions and ear savers. |
| | Worn stalk rolls. | Replace stalk rolls. |
| Plugging. | Stalks breaking in stalk rolls or deck plates. | Adjust the opening of deck plates. Check the stalk roll timing so stalk roll flutes do not break stalks. Also make sure deck plates are set equidistant. |
| | Loose gatherer chains. | Check gatherer chain mechanism. |
| | Not picking planter rows. | Pick rows as they were planted. It will be easier to follow the rows, reduce plugging, and eliminate loss of ears. |
| | Material catching on sheet metal. | Check for broken or bent sheet metal that may prevent flow of material. |
| | Ground speed too fast, causing too much material to go into corn head too fast. | Slow down. Operate at a speed to meet the yield and ground conditions. Faster speeds will cause plugging. |
| | Material not flowing through cross auger. | Check for obstructions in cross auger housing and for roughness of auger. |

Continued on next page

DK75838,000020F -19-31JUL13-5/8



050217
PN=186

| | Corn stalks plugging in gatherer throat opening. | Remove ear savers and center shield extensions. |
|---|---|---|
| Cobs and foreign material in grain tank. | Moisture content of corn too high. | Check moisture content of corn before harvesting. Do not harvest if over 30 percent moisture. |
| | Insufficient air blast from cleaning fan. | Increase fan speed to obtain sufficient air blast. Lower front of sieve. Keep sieve clear of pieces of cob and obstructions. |

## WINDROW PICKUP PLATFORM PROBLEMS TROUBLESHOOTING CHART

| PROBLEM | CAUSE | REMEDY |
|---|---|---|
| Windrow difficult to pick up or strips of crop not being picked up. | Short crop down in the stubble embedded in furrows. | Raise gauge wheels to allow fingers to run closer to ground. |
| | Gauge wheels not always contacting ground. | Lower platform to allow pickup to operate at less pitch. |
| | Pickup speed to slow. | Increase pickup speed. |
| Picking up dirt. | Improper gauge wheel setting. | Lower gauge wheels to prevent pickup fingers from digging in ground. |
| | Pitch of pickup too steep. | Raise platform so area between center and drive rollers is parallel with ground. |
| Picking up stones. | Improper gauge wheel setting. | Lower gauge wheels to increase finger clearance to ground. |
| | Excessive pickup speed. | Reduce pickup speed. |
| | Pitch of pickup too flat. | Raise platform to increase pitch. |
| Pickup feeds too high on auger. | Pickup feeds too high on auger. | Lower platform so area between center and drive rollers is parallel with ground. |
| | Hold-down fingers not adjusted properly. | Lower hold-down fingers to provide more compression of crop at rear of pickup. |
| | Excessive pickup speed. | Reduce speed of pickup. |
| Windrows rolls ahead of pickup. | Speed of pickup too slow. | Increase speed of pickup. |
| Material builds up at pickup stripper or builds up between pickup stripper and auger. | Speed of pickup too slow. | Increase speed of pickup. |
| | Platform auger not adjusted properly. | Adjust platform auger down. Check auger finger adjustment. (See combine operator's manual.) |

## FEEDER HOUSE PROBLEMS TROUBLESHOOTING CHART

| PROBLEM | CAUSE | REMEDY |
|---|---|---|
| Uneven or bunched feeding of crop to cylinder. | Auger clearance too high from platform bottom. | Adjust outer ends of auger to proper clearance of bottom. (See combine operator's manual.) |
| | Buildup of grain on cutter bar. | Lower height of reel and set fore-and-aft position as close as possible to cutter bar and auger. (See combine operator's manual.) |
| | Front of feeder conveyor chain adjusted too high. | Adjust feeder drum so conveyor clears bottom by 1/8 inch. |
| | Feeder conveyor chain too tight and holds drum up. | Adjust conveyor chain to proper tension. |
| | Fingers in platform auger not adjusted to convey material properly to feeder conveyor chain. | Adjust finger height of auger so there is even feeding from the platform auger to feeder conveyor chain. (See combine operator's manual.) |
| | Platform drive belt slipping. | Spring-loaded idler must be free and tight against belt. |
| | Auger too far ahead of stripper. | Adjust auger back to stripper. (See combine operator's manual.) |
| | Feeder drum swing arms binding and not letting drum down. Spring mechanism on arms not functioning. | Check inside of feeder house and remove any material buildup from around swing arms and springs. |
| | Feeder conveyor slats bowed. | Straighten or replace bent slats |
| | Dirt buildup on feeder house bottom. | Clean bottom. |
| | Too great a distance between platform auger and feeder house front drum. | Add links to feeder house chain. |
| | Conveyer drive chain not on correct sprocket. | Place chain on correct sprocket. |

Continued on next page

DK75838,000020F -19-31JUL13-6/8

## THRESHING PROBLEMS TROUBLESHOOTING CHART

| PROBLEM | CAUSE | REMEDY |
| --- | --- | --- |
| Excessive cracked grain in grain tank. | Cylinder or rotor speed too fast for crop. | Decrease cylinder or rotor speed just enough to stop grain cracking but still do a good threshing job, and/or open concave slightly. |
| | Concave spacing too close. | Open concave spacing just enough to eliminate cracking. Decrease cylinder or rotor speed. |
| | Uneven feeding or slugs entering cylinder or rotor. | Check feeder conveyor chain tension and float in feeder house. |
| | Excessive clean grain in tailings, causing grain to crack when rethreshed. | Open sieve slightly to reduce tailings. Lower front of sieve. |
| | Not enough straw entering combine. | Increase ground speed to increase amount of material being taken into combine. |
| | Dented auger housings or bent auger shafts cracking grain between flights and housings. | Remove dents from auger housings and/or straighten bent auger shafts to eliminate cracking. |

## SEPARATING PROBLEMS TROUBLESHOOTING CHART

| PROBLEM | CAUSE | REMEDY |
| --- | --- | --- |
| Material lodging on straw walker and not being evenly discharged from rear of combine. | Walker slip clutch slipping. | Determine cause of slipping and correct. |
| | Material catching on straw walker curtain and building up on front of walkers. | Remove front curtain. |
| | Material not being delivered evenly to walkers. | See section on improper cylinder action. |
| Separator grain loss. | Straw walkers running at incorrect speed. | Check primary countershaft speed with engine at full throttle, no load. |
| | Damaged straw walker curtains. | Install new curtains. |
| | Straw walkers or rotor overloaded due to incomplete threshing or late threshing at the concave. | Reduce concave spacing and/or increase cylinder or rotor speed to increase threshing action. |
| | Straw walker or rotor overloaded due to incomplete threshing or late threshing at the concave. | Reduce ground travel speed to reduce amount of material entering combine. Check to be sure curtains are in place. Raise cutting platform to cut less material. |
| | Material overthreshed and pulverized. Cylinder or rotor speed too fast. | Reduce cylinder or rotor speed. |
| | Mat of straw not being broken up. | Remove one riser from same side of each straw walker. |

## CLEANING PROBLEMS TROUBLESHOOTING CHART

| PROBLEM | CAUSE | REMEDY |
| --- | --- | --- |
| Foreign material in clean grain. | Insufficient air blast from cleaning fan. | Increase fan speed. |
| | Sieve lips open too far, allowing foreign material to fall through with clean grain. | Consider more air blast; then close sieve lips so foreign material will be carried back to the tailings auger. |
| | Clean shoe overloaded with fine, chopped straw. | Open concave spacing to reduce amount of broken straw. Increase air blast. Reduce cylinder or rotor speed. Position sieve in upper position. |
| | Air deflectors or wind boards improperly adjusted. | Adjust air deflectors or wind boards. |
| Loss of grain over cleaning shoe. | Chaffer overload, causing carryover of grain. | Increase air blast. Open lips of chaffer. |
| | Grain blowing over shoe. | Reduce fan speed. |
| | Too much broken straw on chaffer for proper cleaning. | Open concave spacing and/or reduce cylinder or rotor speed. Reduce ground travel speed. |
| Excessive clean grain in tailings being returned to cylinder. | Incorrect setting of fan blast for condition of crop. | Reduce fan speed to permit clean grain to fall through sieve before it is carried back to tailings auger. |
| | | If increasing sieve openings and/or reducing fan speed results in foreign material in the clean grain, then lower the front of the sieve to the bottom hole and readjust the sieve openings and the fan speed. |

Continued on next page

DK75838,000020F -19-31JUL13-7/8



050217
PN=188

| Excessive trash in tailings. | Insufficient air blast. | Increase fan speed. |
| | Chaffer lips too far open. | Close chaffer spacing and/or reduce cylinder or rotor speed. |

DK75838,000020F -19-31JUL13-8/8

# TEST YOURSELF

## *QUESTIONS*

1. What are five clear signs of poor combine harvesting?
2. What are four things a combine operator should do prior to the harvest season when planning and preparing for harvest?
3. (Fill in the blank.) Turning, refueling, breakdowns, etc., will account for about __________ percent of the harvesting time during combine operation.
4. Preliminary settings on a combine are very important. What four areas could be set during the off season as a starter for proper combine adjustment?
5. In small grains, what four losses would you check to see if the combine was set correctly?
6. Identify the results of improper threshing. Place an "O" beside the following items which indicate overthreshing and a "U" beside the items which indicate underthreshing.
   - Grain losses over shoe
   - Broken and chewed straw
   - Overloaded straw walkers
   - Cracked kernels
   - Unthreshed heads
   - Grain losses over walkers
7. What are six sources of grain loss?
8. (Fill in the blank.) A __________ spreading device should not be used when checking grain losses.
9. (True or False) Acceptable losses in small grains range from 5 to 8 percent.
10. (Fill in the blank.) The majority of combine operating problems can be traced to improper ____________.

DK75838,0000210 -19-26AUG13-1/1

## INTRODUCTION

Proper maintenance and service adjustments are necessary to ensure efficient, safe operation of the combine.

Costly repairs, premature wear, loss of field time, and accidents can be reduced if the combine is properly maintained and adjusted.

DXP02706 —UN—23FEB11

DXP04624—UN—04MAR13

OUO1082,000624F -19-26AUG13-1/1

## MAINTENANCE

The operator's manual for the machine should be used in reference to specific maintenance intervals, location of service points, and instructions for the performance of maintenance and service adjustments. Because of differences in makes and models of combines, this chapter deals only with general service and maintenance requirements. Always study the operator's manual carefully to determine what maintenance is needed.

Fig. 1 — The Combine Requires Regular Maintenance and Service Adjustments

BB87125,00001C0 -19-26AUG13-1/1

## GENERAL MAINTENANCE

There are several practices that a good machine operator always follows. The operator knows that by following these rules, the job of operating and maintaining the combine will be much easier and safer.

1. Always keep the machine clean. Before starting the combine, clean all field trash, mud, and excess grease and oil from the machine. Not only is this a good safety practice, it also helps the combine to run more efficiently, prevents moisture accumulation and rust on metal parts, and cuts down on time lost in the field for repairs.

2. Make sure that nuts, cap screws, shields, and sheet metal parts are tight. A loose shield can vibrate, produce irritating noise, and cause a machine failure if it falls in the way of moving parts. Loose attaching hardware can cause breakdowns that take time the machine should be using for work.

3. Inspect the combine before starting every day. A brief look at all areas of the combine can help you spot potential machine failures and safety hazards.

4. Keep maintenance records (Fig. 2). A simple chart showing when lubrication and service adjustments were made can help you make sure that all needed maintenance has been performed.

5. Don't abuse the machine. Proper lubrication and adjustment is of little help if the operator abuses the machine. A good combine operator follows the operator's manual and doesn't overload the machine, operate it at speeds too fast for field conditions, or operate it under conditions that could cause damage to the machine.

Fig. 2 — Keeping Maintenance Records

BB87125,00001C1 -19-14MAR13-1/1



050217
PN=191

## ENGINE AND POWER TRAIN MAINTENANCE

Before starting the combine each day, the following checks should be made:

1. Check the oil level of the engine crankcase (Fig. 3). If the level is low, be sure to add the proper amount and type of engine oil. Refer to the operator's manual instructions as to how often the oil and filter should be changed.

2. Clean off any accumulation of grease, excess oil, or dirt. This will help the engine to run cooler and more efficiently.

3. Check the level of coolant in the radiator daily. If the level is low, add the proper type of coolant. In cold weather, make sure the engine has enough antifreeze to prevent freeze-up. Remove any trash that has accumulated on the radiator or air intakes. Use compressed air to blow out dust and leaves so enough air can pass through the radiator to cool the engine efficiently.

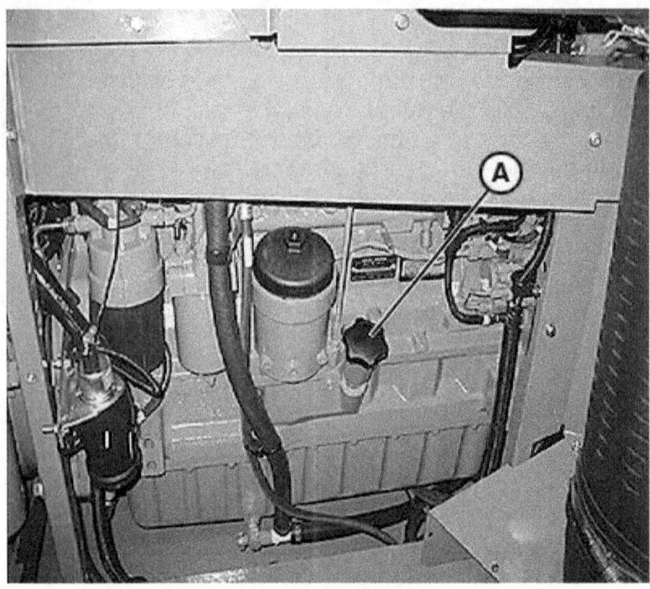

Fig. 3 — Checking Engine Oil Level

A— Engine Oil Dipstick

BB87125,00001C2 -19-26JUL13-1/3

4. Inspect engine belts and tensioner assemblies (Fig. 4).

A— Engine Belt          B— Belt Tensioner

Fig. 4 — Engine Belt and Tensioner

Continued on next page

BB87125,00001C2 -19-26JUL13-2/3

5. Inspect battery terminals and all other cable connections. Make sure they are tight and free of corrosion. Check the electrolyte level in the battery (Fig. 5). Running a battery with electrolyte too low will shorten its life and reduce its power output.

6. Examine the fuel system for leaks. Also drain and clean storage tanks periodically to eliminate the accumulation of dirt and water condensation.

   Never use fuel carried over from one season to another. Fuels with additives for improved hot weather performance will give reduced results if used in cold weather, and vice versa.

7. Check the hydraulic fluid levels and add fluid if needed. Be careful to wipe all dirt away from the inspection point before checking the level to prevent dirt from entering the system.

   Use a piece of cardboard or wood to check hydraulic lines for leaks. High-pressure leaks are sometimes invisible, but still can penetrate the skin. Be careful to avoid injury.

   If the combine has a separate power steering reservoir, check the fluid level.

8. Check the brakes for even brake pedal pressure as specified in the operator's manual.

9. Make sure the transmission and drive units have the proper amount of lubricant. Clean the inspection areas before checking lubricant levels.

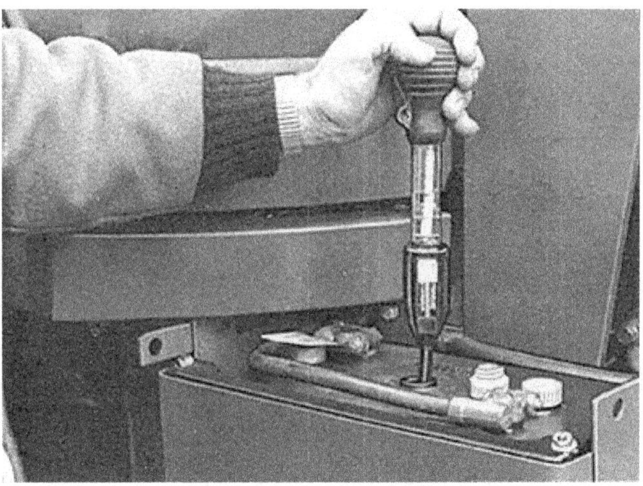

*Fig. 5 — Checking Battery Electrolyte Level*

10. Check variable-sheave and drive belts for proper alignment. Examine hydrostatic speed range lever to see that correct linkage and adjustment is maintained.

11. Periodically inspect wheel bearings. Clean and repack them as specified in the operator's manual.

12. Remember: Lubricate all points specified in the operator's manual.

BB87125,00001C2 -19-26JUL13-3/3

# HEADER MAINTENANCE

Before operating the header, perform the following procedures.

## *LUBRICATION*

Lubricate all header chains at frequent intervals. Operate chains for several minutes so they are warm when oil is applied. (Proper chain and belt maintenance is discussed later in this chapter.)

- Lubricate the reel shaft daily with a multi-purpose grease.
- Lubricate the cutterbar drive daily.
- Check gear cases on corn head for proper lubricant level (Fig. 6).
- Lubricate the cutterbar knife as needed.

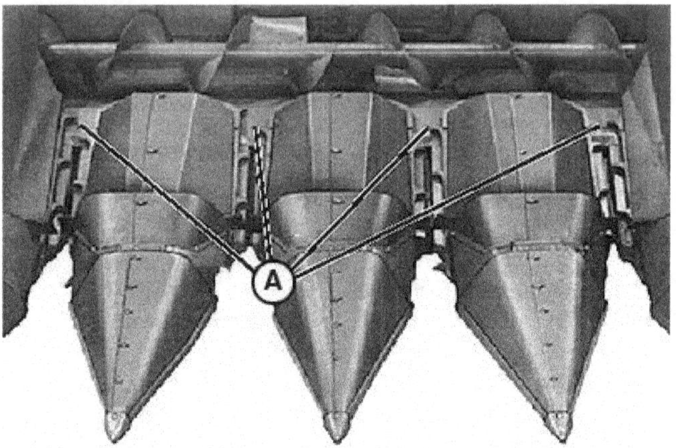

*Fig. 6 — Check Lubricant Level in Corn Head Gear Cases*

**A— Inspection Plugs**

Continued on next page

BB87125,00001C3 -19-26JUL13-1/15

## INSPECTION

1. Check the cutterbar frequently for:

   - Broken sections
   - Proper alignment of knife bar
   - Proper alignment of guards
   - Proper register

2. Also make sure cutterbar wear plates and knife clips are in good alignment (Fig. 7).

3. Check the knife blade for a sharp cutting edge.

4. Make sure the knife stroke is the proper length.

   **A— Wear Plate**

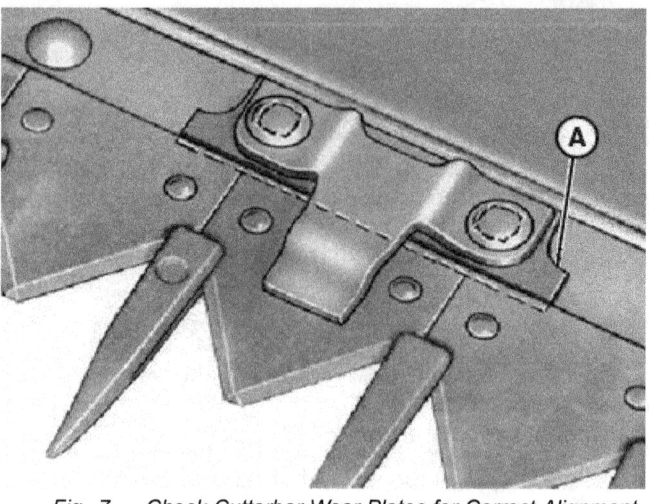


*Fig. 7 — Check Cutterbar Wear Plates for Correct Alignment*

BB87125,00001C3 -19-26JUL13-2/15

5. Check to see that all belts and chains are tightened properly and that the flights on the feeder conveyor are not loose, bent, or broken (Fig. 8).

6. Inspect all sealed bearings for excessive play and replace if necessary.

7. Inspect the header to see it is properly aligned and level.

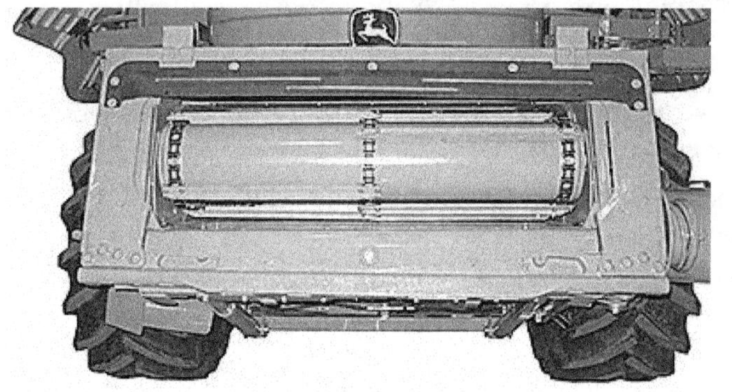

*Fig. 8 — Check for Loose or Broken Flights on Feeder Conveyor*

BB87125,00001C3 -19-26JUL13-3/15

## PLATFORM ADJUSTMENT

If the platform requires adjustment, perform the following steps.

### CUTTERBAR

To remove the knife for sharpening or repair, disconnect the cutterbar drive and pull the knife out of the cutterbar. Be very careful if it is necessary to drive the knife out of the cutterbar. Always use a wooden block or rubber mallet to strike the knife. Use care to avoid bending the knife.

Proper guard alignment is necessary for proper shearing of the crop. Guards that are bent or out of adjustment can also damage the knife. To adjust the guard, loosen the bolt connecting it to the cutterbar, then align the guard and tighten the bolt (Fig. 9).

**A— Adjusting Tool**       **B— Knife Guard**

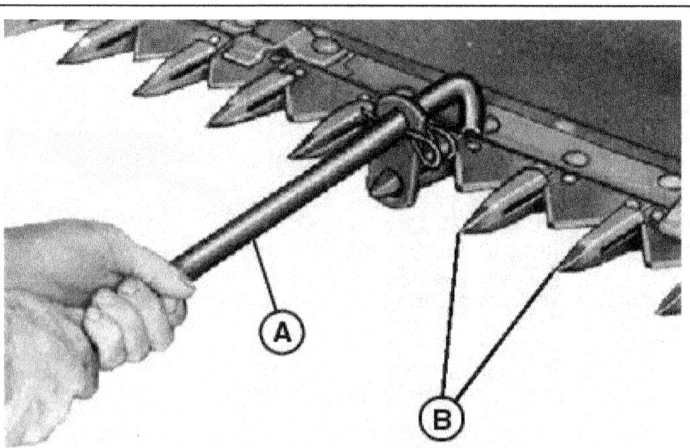


*Fig. 9 — Adjusting the Knife Guards*

**Continued on next page**

BB87125,00001C3 -19-26JUL13-4/15

Knife clips must keep sections from lifting off the guards and also permit the knife to slide without binding. Always set the knife clips after the guards are aligned. To set the clips, tap them up or down with a hammer until the knife will slide under them without binding (Fig. 10). Never bend the knife clip down when a knife section is directly under it.

Wear plates are located along the entire length of the knife back and are adjustable to compensate for wear on the knife back. The wear plates must line up with each other to give the knife back a straight bearing along its entire length.

To set the wear plates, loosen the bolts on the knife clip, hold the knife forward, and adjust the plates finger tight against the back of the knife. Then tighten the bolt securely.

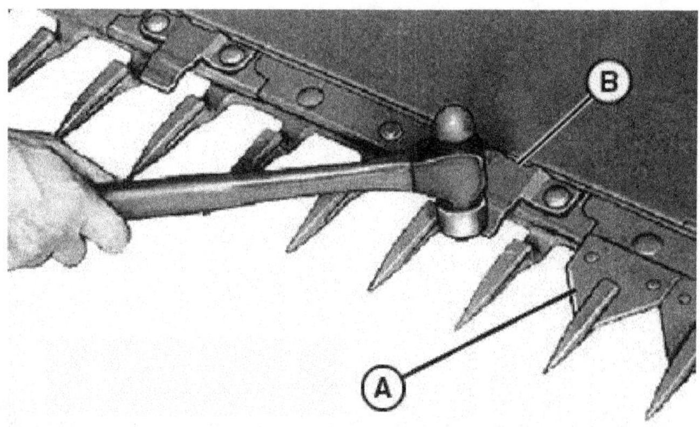

Fig. 10 — Adjusting Knife Clips

A— Knife Clip        B— Knife

BB87125,00001C3 -19-26JUL13-5/15

## PLATFORM

Because one side of the platform is usually heavier than the other, check the platform periodically to make sure it is level. Raise the platform to medium height and take a position about 15 feet in front of the combine. Sight from one end to the other, using the combine axle as a reference.

 **CAUTION: Make sure safety stops are in position before working around or under the header (Fig. 11).**

If the platform is not level, adjust with one of the following methods as follows.

Most combines are equipped with either slotted adjusting points or long threaded bolts with adjusting nuts.

If the combine is equipped with a slotted adjustment, loosen the bolts and level the platform. Then re-tighten the bolts.

If the combine is equipped with a bolt adjustment, loosen the jam nuts and then tighten or loosen the adjusting nuts

Fig. 11 — Always Make Sure Safety Stop Is in Position Before Working Under Header

A— Safety Stop

as needed. After adjustment, make sure jam nuts are tight against the adjusting nuts.

**Continued on next page**        BB87125,00001C3 -19-26JUL13-6/15

On the combine shown in Fig. 12, leveling the header is accomplished by adding or removing shims on each side of the feeder house.

## AUGER SLIP CLUTCH

The platform auger slip clutch is designed to prevent damage to the platform auger if it becomes clogged or if an obstruction enters the platform. The auger slip clutch on many platforms is not adjustable. Refer to the operator's manual to determine if adjustment is possible and for proper adjustment procedures. Improper adjustment may eliminate damage protection by not allowing the clutch to slip.

A— Platform Level Adjustment
    Points

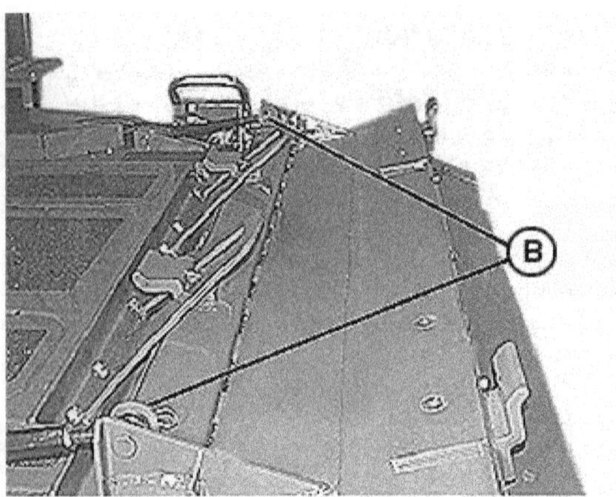

Fig. 12 — Adjusting Bolts for Leveling Platform

BB87125,00001C3 -19-26JUL13-7/15

## ROW-CROP HEAD ADJUSTMENT

Proper adjustment of the row-crop head reduces crop losses, saves operating time, and can help cut repair bills.

## DRIVE CHAIN TENSION

Keep all drive chains tight enough to keep chain links from climbing or jumping over sprocket teeth (Fig. 13). But, avoid overtightening, which increases wear on chains, sprockets, bearings, and shafts. See the operator's manual for proper tension on each chain.

Fig. 13 — Keep Chains Adequately Tensioned

BB87125,00001C3 -19-26JUL13-8/15

## TIMING THE GATHERING BELTS

Gathering belts must be correctly timed to properly grip crop stalks and feed them to the rotary knives. Belt timing can be disturbed if one of the belt drive slip clutches is activated. Always recheck gathering belt timing whenever an obstruction has caused the slip clutches to slip. To retime the belts, loosen belt drive chain tension and reposition the belt drive chain so that the belt loops are properly mated (Fig. 14).

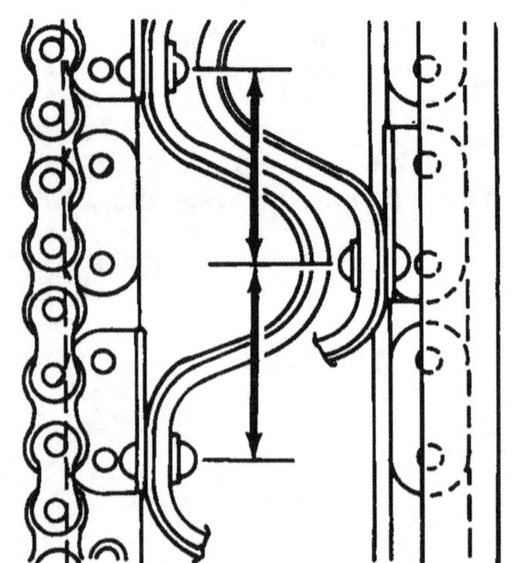

Fig. 14 — Keep Gathering Belts Properly Timed

Continued on next page

BB87125,00001C3 -19-26JUL13-9/15

## CUTTING KNIVES

Cutting knives (Fig. 15) must be sharp and clean for
smooth, efficient cutting, especially in weedy conditions,
and each section of the rotary knife must contact the
stationary knife evenly. When the stationary knife
becomes badly worn, reverse or replace it. As clearance
between stationary knife and rotary knife becomes
greater, the risk of uneven cutting and plugging increases.
To reduce clearance between knives, remove the
stationary knife and install special shims before replacing
the knife. Replace dull knife sections on the rotary knife
by removing the entire knife, taking off old sections, and
riveting on new ones.

A— Stationary Knife    C— Clearance
B— Rotary Knife

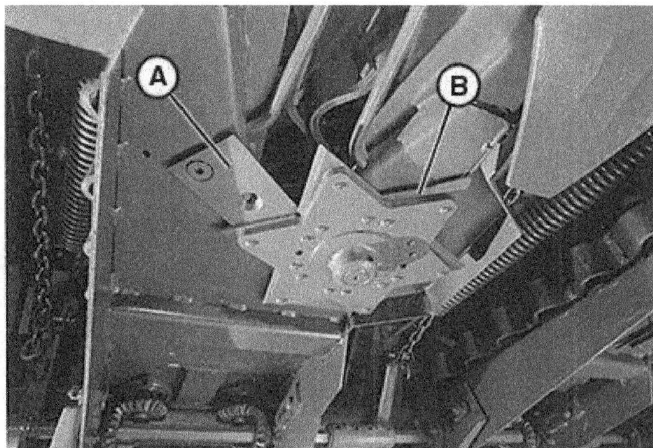

*Row-Crop Cutting Knives*

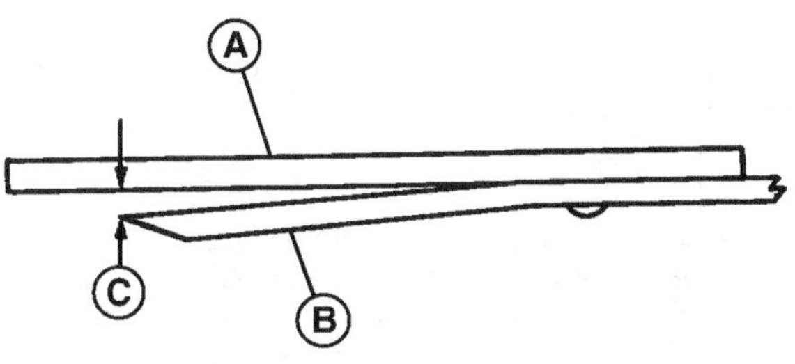

*Fig. 15 — Stationary and Rotary Knives Must Be Sharp*

BB87125,00001C3 -19-26JUL13-10/15

## AUGER SETTINGS

Set the clearance under the cross auger at approximately
1/4 to 1/2 inch (6 to 13 mm) for soybeans and crops with
smaller seed pods and stalks (Fig. 16), but increase
clearance to as much as 1 inch (25 mm) for sunflowers and
sorghum. Maintain proper clearance between the auger
and auger stripper across the back of the head to prevent
the crop from carrying over or wrapping around the auger.

A— Auger    C— Bottom of Auger Clearance
B— Stripper Clearance

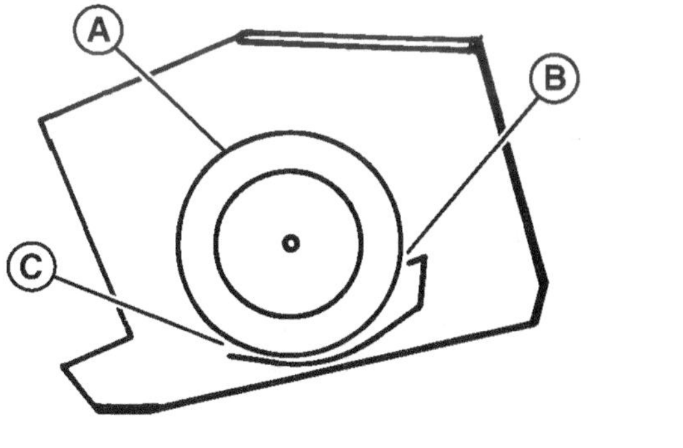

*Fig. 16 — Set Auger and Stripper Clearance for Crop Being Harvested*

**Continued on next page**    BB87125,00001C3 -19-26JUL13-11/15

## SLIP CLUTCHES

Slip clutches protect equipment from damage in case of plugging or sudden overload. However, if slip clutches slip too easily, time is wasted and plugging is increased. Many clutches are not adjustable. The slip clutch shown in Fig. 17 may be adjusted by adding or removing shims. Improper adjustment may result in damage to machine.

A— Slip Clutch

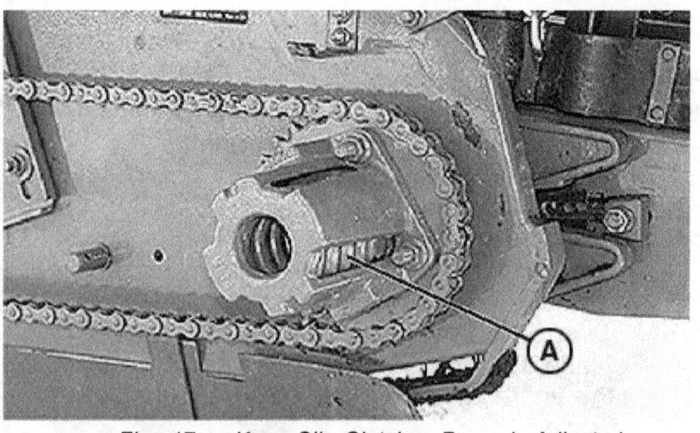

*Fig. 17 — Keep Slip Clutches Properly Adjusted*

BB87125,00001C3 -19-26JUL13-12/15

## *CORN HEAD ADJUSTMENT*

The following paragraphs describe adjustments that may be required on a typical corn head.

### GATHERER CHAINS

Proper tension on the gatherer chains should be maintained at all times. Most heads are equipped with adjustable tighteners. To adjust the tension, turn the adjusting nut (Fig. 18). To remove the chain, loosen the adjusting nut until the chain can be removed. See the operator's manual for specific procedures.

A— Adjusting Nut          C— Idler Sprocket
B— Gatherer Chain

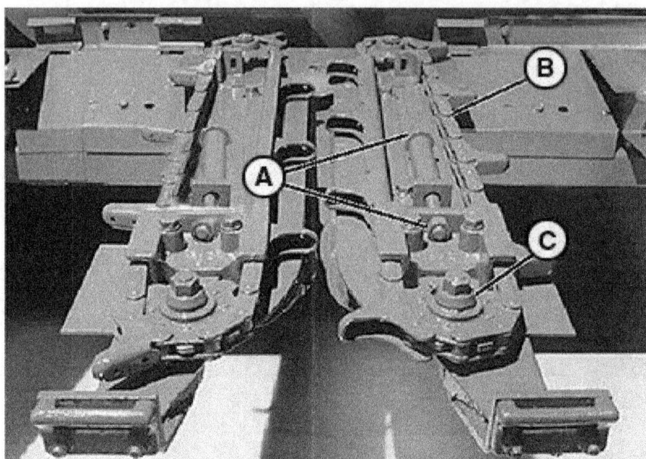

*Fig. 18 — Adjusting Gatherer Chain Tension*

BB87125,00001C3 -19-26JUL13-13/15

## TRASH KNIVES

Most trash knives have slotted holes that permit adjustment (Fig. 19). Loosen the bolts holding the knives in place and slide the knives until the proper cutting gap is reached. Then tighten the bolts.

A— Adjustment Bolts          B— Trash Knives

*Fig. 19 — Adjusting Bolts for Trash Knives*

**Continued on next page**                     BB87125,00001C3 -19-26JUL13-14/15

## FEEDER CONVEYOR ADJUSTMENTS

### CHAIN

The feeder conveyor should be kept in adjustment to ensure proper handling of the crop. Before adjusting the conveyor chain, remove the cover and clean away all trash. Most combines are equipped with eyebolt adjustments, located on the outside of the conveyor housing (Fig. 20). Tighten or loosen these bolts to obtain the correct tension. Make sure that both sides are adjusted equally.

A— Feeder Conveyor Chain
    Tension

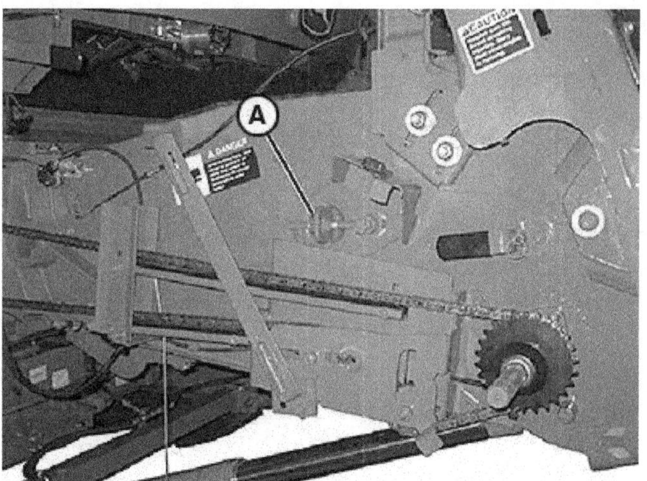

*Fig. 20 — Eyebolt Adjustment for Feeder Conveyor Chain Tension*

BB87125,00001C3 -19-26JUL13-15/15

## THRESHING UNIT MAINTENANCE

The following paragraphs describe maintenance of a typical threshing unit.

### LUBRICATION

Because makes and models vary greatly in lubrication requirements, always check the operator's manual to determine lubrication points and intervals and recommended lubricants.

### INSPECTION

1. Check all belts and chains for proper tension and for excessive wear. Refer to belt and chain maintenance procedures later in this chapter.

2. Examine the condition of the threshing cylinder or rotor bars periodically. Make sure cap screws and nuts are tight.

3. Check the alignment and clearance of spike-tooth cylinder teeth.

4. At each end of the cylinder or rotor, measure the distance from a rasp-bar to the concave. Measurements should be the same at both ends (Fig. 21).

5. Check for rasp-bar wear. After long use, the center of the rasp-bars may become worn so that equal clearance is not maintained along the concave. If rasp-bars become worn to this extent, replace them.

6. Make sure correct spacing proportions between the cylinder and concave are maintained. Usually the front spacing is twice the rear spacing. Refer to the operator's manual for specific instructions.

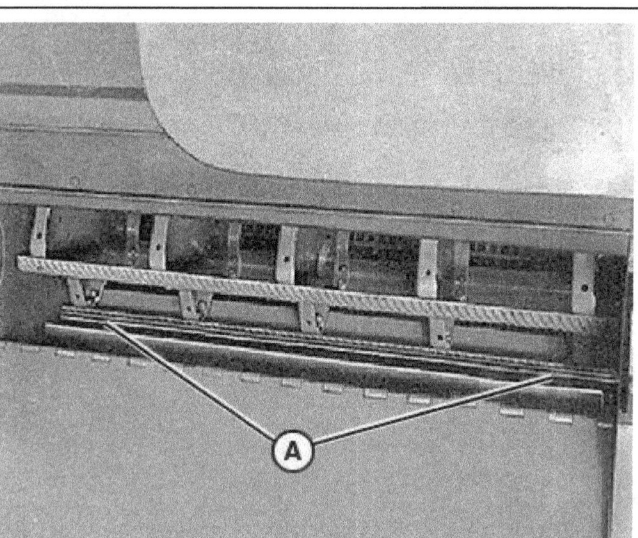

*Fig. 21 — The Distance from a Rasp-Bar to the concave Should Be the Same at Both Ends*

A— Equal Clearance at Both
    Ends of Cylinder

7. Rotary combines may have uniform spacing between all concave bars and rotor rasp-bars, or be set with wider space where the crop enters the concave, depending on crop and threshing conditions. Refer to the operator's manual for specific instructions.

8. Move concave adjustment throughout full adjustment range daily to ensure free movement of linkage when adjustment is necessary.

9. Inspect the stone trap and empty it if necessary.

Continued on next page

BB87125,00001C4 -19-26JUL13-1/4



050217
PN=199

## Adjustment

Typical adjustments of the grain handling units are described as follows.

## CONCAVE LEVELING

If the concave is not level in relation to the cylinder or rotor, it must be adjusted. A leveling bolt is provided on each side of the concave. Adjust it as needed. The concave must be kept in correct relation to the cylinder or rotor (Fig. 22).

A— 1-Inch Opening    C— 1/2-Inch Opening
B— Cylinder    D— Concave

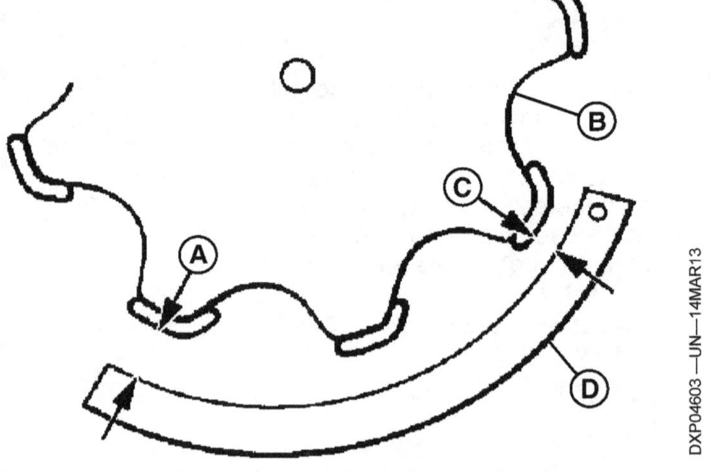


Fig. 22 — Typical Front and Rear Cylinder-Concave Spacing Proportions

BB87125,00001C4 -19-26JUL13-2/4

## RASP-BAR CYLINDER

If cylinder rasp-bars are bent, remove and straighten them. On most combines, this can be done without removing the cylinder or rotor. To gain access, open the cylinder or rotor housing. If rasp-bars must be replaced, the mating bar, which is located 180 degrees on the opposite side of the cylinder or rotor, must be replaced to maintain proper cylinder or rotor balance (Fig. 23).

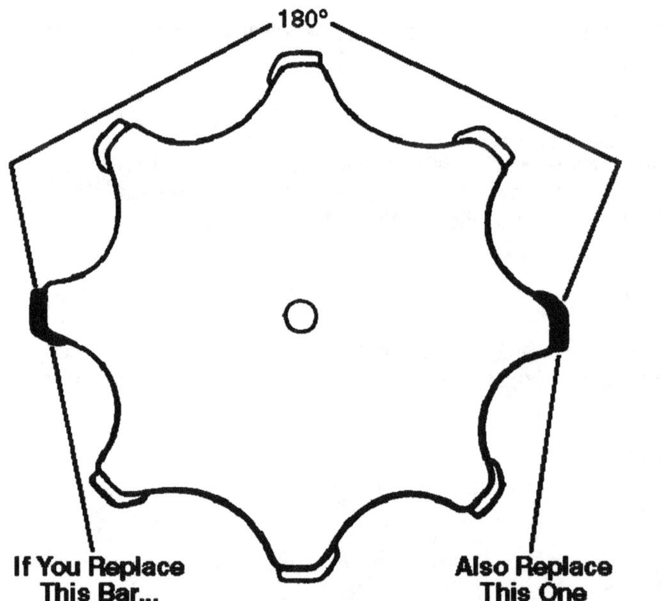


Fig. 23 — If One Rasp-Bar Needs Replacement, Also Replace the Bar 180 Degrees Opposite to Maintain Cylinder Balance

Continued on next page

BB87125,00001C4 -19-26JUL13-3/4

## SPIKE-TOOTH CYLINDER

With the concave in high position and the teeth straight and properly adjusted, there should be equal clearance on either side of the cylinder teeth as they pass through the concave teeth (Fig. 24). On each side of the cylinder housing, most combines have adjusting bolts that move the concave from side to side as shown. Adjust these bolts so the teeth have equal clearance on either side. Turn the cylinder at least one full revolution to make sure all teeth have proper clearance. It may be necessary to straighten or replace individual teeth if proper clearance cannot be obtained.

A— Cylinder
B— Concave Tooth (First Row)
C— Equal Clearances
D— Concave
E— Concave Tooth (Second Row)
F— Cylinder Tooth

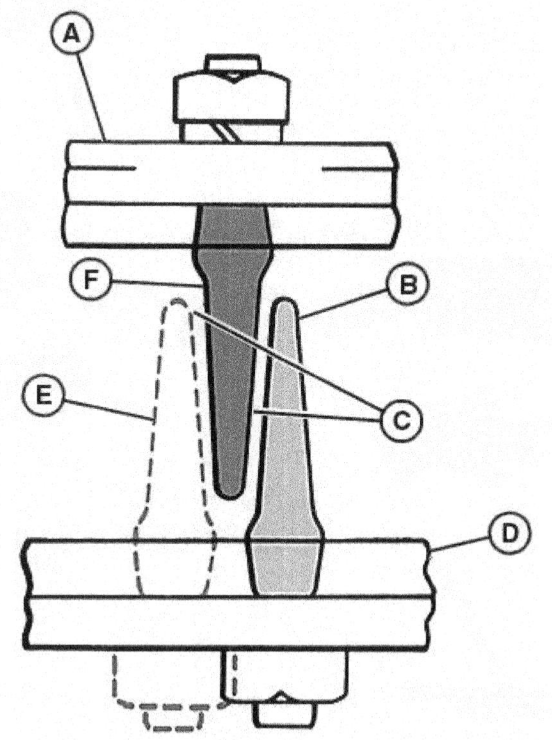

Fig. 24 — *Maintain Equal Clearance Between Cylinder and Concave Teeth*

BB87125,00001C4 -19-26JUL13-4/4

## SEPARATING UNIT MAINTENANCE

Typical maintenance of the separating unit is described as follows.

### LUBRICATION

Most separating units are designed with sealed bearings that do not require lubrication. For those parts that do require periodic lubrication, check the operator's manual.

### INSPECTION

1. Check all belts and chains for wear and proper adjustment. Refer to belt and chain maintenance procedures later in this chapter.

2. Inspect the slip clutches and the separator drive unit to see if they are adjusted properly. Many slip clutches are not adjustable. Refer to the operator's manual to determine if adjustment is necessary.

3. Check the straw walkers on conventional combines — rotary combines have no straw walkers. Make sure they are clean and are not bent or broken (Fig. 25).

4. Check the speed of straw walkers (if combine is so equipped and the operator's manual recommends it).

5. Open inspection doors and clean out field trash.

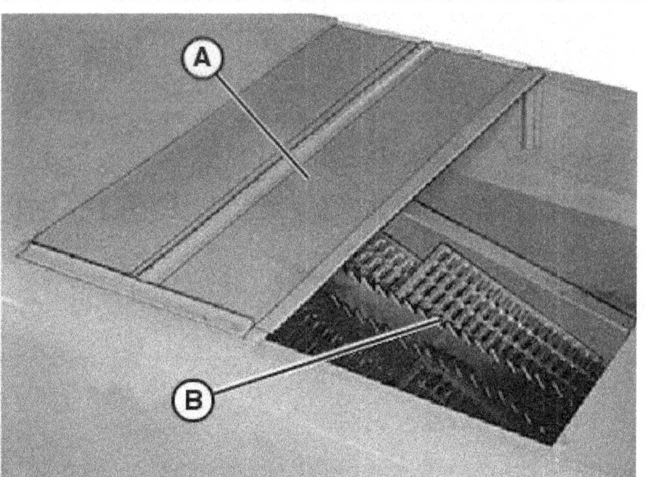

Fig. 25 — *Checking Straw Walkers Through Inspection Door*

A— Inspection Door
B— Straw Walkers

6. Make sure grain conveyor augers are in good condition.

7. Check grain conveyor bevel gears for proper alignment.

8. Check the grain conveyor chain adjustment.

BB87125,0000178 -19-26JUL13-1/1

## CLEANING UNIT MAINTENANCE

### *LUBRICATION*

Most cleaning units are designed with sealed bearings that do not require lubrication. For those parts that do require periodic lubrication, check the operator's manual.

### *INSPECTION*

1.  Check all belts and chains for proper tension and alignment. Refer to belt and chain maintenance procedures later in this chapter.

2.  Inspect the fan blades periodically to see that they are clean and straight.

3.  Inspect the fan screen and clean it if necessary.

4.  Check the chaffer for debris. Remove and clean it if necessary. Also check for bent or broken vanes.

5.  Examine rubber bushings on the cleaning shoe for excessive wear or cracking. Remove and replace those in bad condition.

6.  Inspect the sieve to see that screens are not plugged or broken.

7.  Examine all bearings for excessive play.

### *ADJUSTMENT*

When replacing worn or cracked rubber bushings, do not tighten fastening screws too tight. This may damage the new bushings. When installing a new bushing, it may be helpful to wet the shaft with water to act as a lubricant. Do not use oil, as it may damage the bushings.

BB87125,0000179 -19-23JAN13-1/1

## GRAIN HANDLING UNIT MAINTENANCE

The grain handling units usually require the following maintenance.

### *LUBRICATION*

Most grain handling units are designed with sealed bearings that do not require lubrication. For those parts that do require periodic lubrication, check the operator's manual.

### *INSPECTION*

1.  Check all belts and chains for proper tension and alignment. Refer to belt and chain maintenance procedures later in this chapter.

2.  Check elevator housings for breaks or weak points.

3.  Open inspection doors on elevators. Check for trash and dirt and also inspect for broken or missing paddles (Fig. 26).

Fig. 26 — Inspect Elevator Doors and Remove Trash

**Continued on next page**

BB87125,000017A -19-26JUL13-1/4

4. Inspect the grain tank (Fig. 27). Make sure it is kept dry and clean.

## *ADJUSTMENT*

Typical adjustments of the grain handling components are described as follows.

### ELEVATOR CHAINS

Elevator chains are usually equipped with adjustment bolts on each side. These are normally located at the top or the bottom of the elevator. Adjust the bolts to obtain proper tension. Always make sure bolts on both sides are adjusted equally.

### SLIP CLUTCHES

Many slip clutches are non-adjustable. Refer to the operator's manual to determine if adjustment is possible and for proper adjustment procedures. Improper adjustment may cause damage to machine.

Fig. 27 — Inspect the Grain Tank

BB87125,000017A -19-26JUL13-2/4

### UNLOADING AUGER

To remove the unloading auger, place the auger in the transport position (Fig. 28) and remove the cap screws or bolts holding it in position on the outer end of the auger. Then pull it from the tube. When replacing the auger in the tube, make certain it will properly align with the drive mechanism on the unloading auger.

**A— Unloading Auger**

Fig. 28 — Place Unloading Auger in Transport Position
Before Removing Auger from Tube

**Continued on next page**

BB87125,000017A -19-26JUL13-3/4

## GRAIN TANK

To clean or drain the grain tank, a door is usually provided at the bottom and to the outside of the bin (Fig. 29). Remove this door when cleaning or draining is necessary. When replacing the door, make sure it is fastened securely in the proper position to prevent grain leakage.

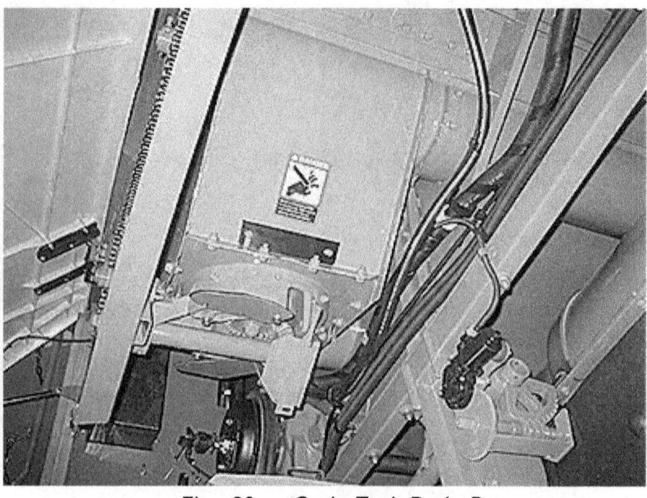

*Fig. 29 — Grain Tank Drain Door*

BB87125,000017A -19-26JUL13-4/4

# WHEEL AND TRACK MAINTENANCE

The following describes maintenance of typical wheels and tracks.

### LUBRICATION

Many steering mechanisms are equipped with sealed bearings at some points. However, almost all require some periodic lubrication. Refer to the operator's manual to make sure all points that need lubrication are serviced.

Most crawler tracks on rice combines are equipped with "button-head" grease fittings, which require a special gun for lubrication (Fig. 30).

When greasing this type of bearing, refer to the operator's manual to determine the correct type of lubricant. Be especially careful not to over-lubricate these bearings. Seal damage can result and cause bearing failure.

*Fig. 30 — Crawler Track for Rice Combine*

**A— Button-Head Grease Fittings**

BB87125,000017B -19-26JUL13-1/3

Most of the same maintenance procedures apply to rubber tracks as well (Fig. 31).

*Fig. 31 — Rubber Tracks*

**Continued on next page**

BB87125,000017B -19-26JUL13-2/3

## INSPECTION

1. Check the air pressure in both the rear and front tires.

2. Check the rear steering wheels for correct toe-in (Fig. 32). Correct toe-in is required for easy steering and stable operation at higher speeds. This can be done by marking the center of the tread of each tire at the rear of the tire (B), in line with the axle. Measure the distance between these marks.

3. Move the combine forward until the marks on the tires are at the front, in line with the axle. Measure the distance between the marks again (A).

4. Check the operator's manual to determine what the difference in the measurement should be. If necessary, adjust the toe-in.

5. If the combine is equipped with crawler tracks, check the tracks for correct tension.

## WHEEL TOE-IN ADJUSTMENT

The steering wheel toe-in can be adjusted by lengthening or shortening the tie rods. Loosen the jam nut on the tie rod and lengthen or shorten the tie rod as needed. Then tighten the jam nut securely.

 **CAUTION: To adjust the rear wheel tread, jack up the combine and block it securely. Never rely on jacks alone for support.**

Next, remove the pins in the telescoping axle and the tie rods and adjust to the width desired. Replace the pins in the proper holes.

## WHEEL SPACING

To widen front wheel tread on most combines, it is necessary to reverse the wheels. Always make sure the combine is blocked securely when reversing the wheels. Make sure proper clearance is maintained when wheels are reversed.

Some combines are equipped with wheel spacers. The width of the spacers varies but usually does not exceed 4 inches (102 mm). When using spacers to widen or

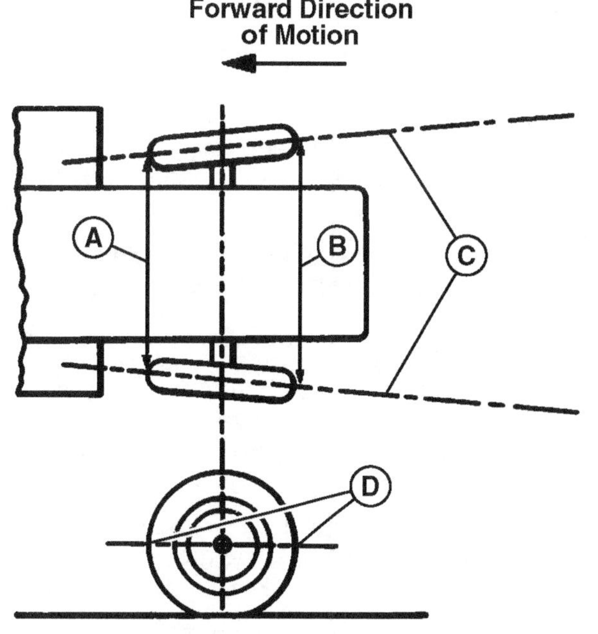


*Fig. 32 — Checking Steering Wheel Toe-In*

A— Front Wheel-to-Wheel
   Measurement
B— Rear Wheel-to-Wheel
   Measurement
C— Tire Centerline
D— Measuring Points

narrow the front tread, make sure the combine is blocked securely. It may be necessary to install dowel pins in these spacers to prevent shearing. Refer to the operator's manual for specific procedures.

## CRAWLER TRACKS

On combines equipped with crawler tracks, maintain proper track tension to prevent premature wear or machine failure. Some track mechanisms are equipped with large threaded bolts or use hydraulic pressure to adjust the track tension. Some units are also equipped with adjustable idlers that can be moved in or out to obtain the proper tension. Consult the operator's manual for proper adjustment.

BB87125,000017B -19-26JUL13-3/3

---

# CARE OF RUBBER TIRES

Always keep tires off oil-soaked floors and away from fuel or oil. Oil softens and deteriorates rubber, shortening the life of tires. Bright sunlight causes the surfaces of tires to crack and harden. During storage, keep the combine out of the sun if possible.

If the combine will sit for more than 60 days without being moved, jack up the axles to take the load off the tires.

Always refer to the operator's manual to determine the correct tire inflation pressures for the combine.

BB87125,00001C7 -19-26JUL13-1/1

## LUBRICATING WITH GREASE

Always wipe the grease fitting clean before placing the nozzle of the grease gun on the fitting (Fig. 33). This keeps foreign material from entering the lubrication point and damaging the bearing. After removing the grease gun, always wipe the fitting clean. Any grease left on the fitting will collect dirt and chaff.

If a fitting is missing, replace it immediately. Failure to do so will allow foreign matter to enter the bearing and could cause serious damage.

Never over-lubricate. Many bearings are equipped with seals that keep out dirt and abrasive material. Pumping too much grease into the fitting could cause this seal to rupture. Always use the recommended number of strokes or pump grease until increased pressure is felt.

Keep all lubricants in containers that protect them from dirt and moisture. Use care in filling the grease gun so that dirt does not enter the gun.

When checking the grease level in any gear housing, be sure to wipe away all dirt and grime so that none falls into the housing when the check or filler plug is removed. Always tighten filler plugs securely.

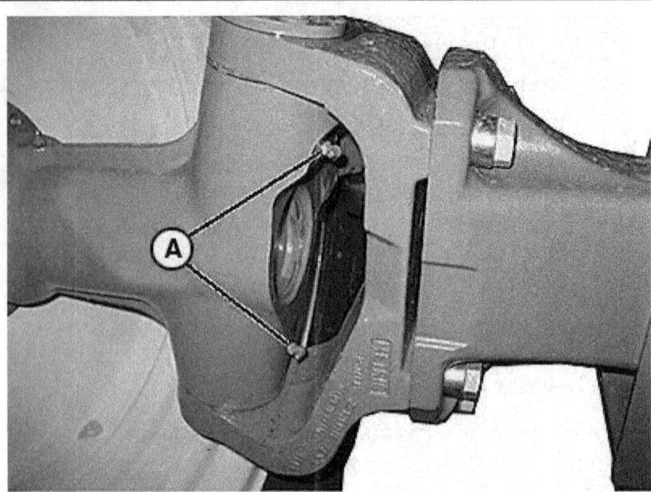

*Fig. 33 — Always Wipe the Grease Fitting Clean Before and After Greasing*

**A— Grease Fitting**

BB87125,00001C6 -19-26JUL13-1/1

## BELT AND CHAIN MAINTENANCE

Because many belts and chains are used on today's combines, proper maintenance is very important to keep them operating through the season.

### *BELTS*

The V-belts on a combine transmit power by friction and a wedging action against the sheaves. Belts are subject to increased wear through periodic heavy loads and should be checked often to be certain belt wear is normal. All belts and sheaves wear with use. Normal wear can be recognized as even wear — both on the belt and on the sides of the sheave.

A slight raveling or peeling of the belt at the lap does not indicate premature failure. Cut off the raveling if the covering peels at the lap.

Examine the sheaves for bent or chipped sidewalls (Fig. 34). Check for excessive sidewall wear. Damaged sheaves cause rapid belt wear. A bent sheave reduces the gripping power of the belt. Replace sheaves having any of the above defects.

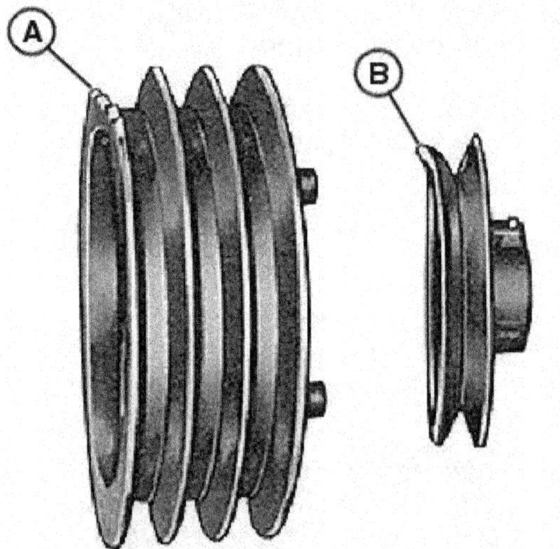

*Fig. 34 — Always Replace Belt Sheaves That Have Bent or Chipped Sidewalls*

**A— Chipped**　　　　　**B— Bent**

Continued on next page

BB87125,00001C5 -19-26JUL13-1/6

Sheaves must be properly aligned to ensure proper operation of all belts (Fig. 35). To check the alignment of sheaves, use a straightedge. Place a straightedge or wire so that it runs from one sheave to the next (Fig. 36). Then sight the alignment to see whether it is straight. If it is not, adjust or replace the sheaves.

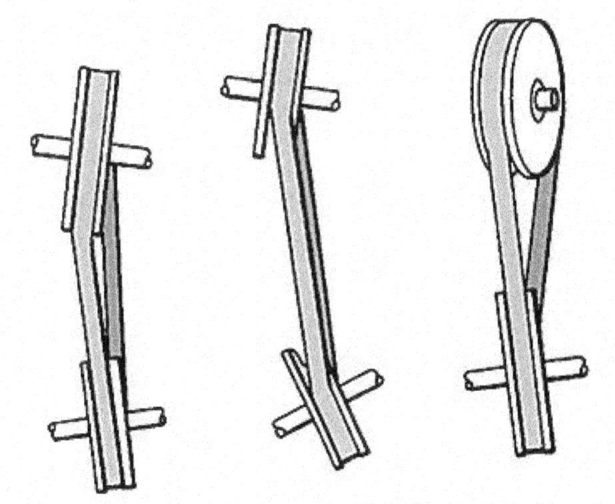

Fig. 35 — Belt Sheave Misalignment

Fig. 36 — Use a Straightedge to Determine Belt Sheave Alignment

BB87125,00001C5 -19-26JUL13-2/6

Make sure that dirt has not lodged and packed in sheave V-grooves. Excessive vibration may be caused by dirt collecting inside the sheaves. Loosen the dirt so it will fall out. (Fig. 37)

Vibration can also be caused by lumpy V-belts. Check the belts for swelling and lumps.

Always keep the belt clean. Oil or grease accumulating on the belts will cause them to deteriorate and slip.

Do not use belt dressings. Dressings often give only temporary gripping action while softening the belt and causing eventual deterioration, which shortens belt life.

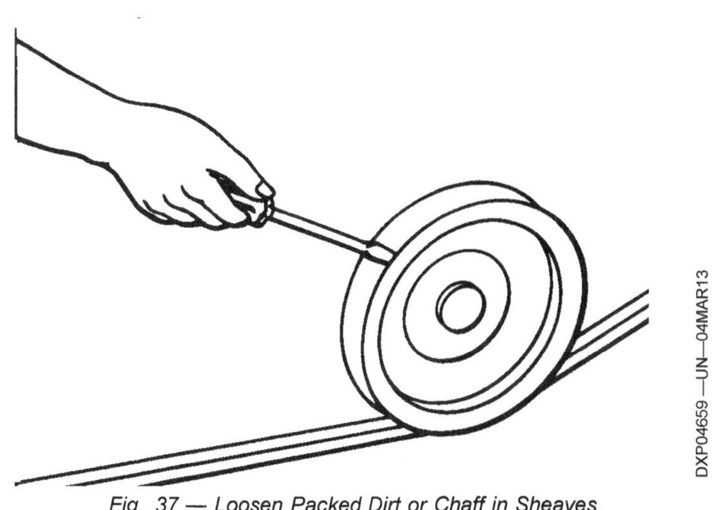

Fig. 37 — Loosen Packed Dirt or Chaff in Sheaves

Continued on next page

BB87125,00001C5 -19-26JUL13-3/6



050217
PN=207

Various types of tension devices are used for belt drives. For those with idler sheaves, the tension devices are found on belt drives. For belts equipped with idler sheaves, the tension may be adjusted by moving the idler sheave in or out (Fig. 38).

**A— Idler Sheave**

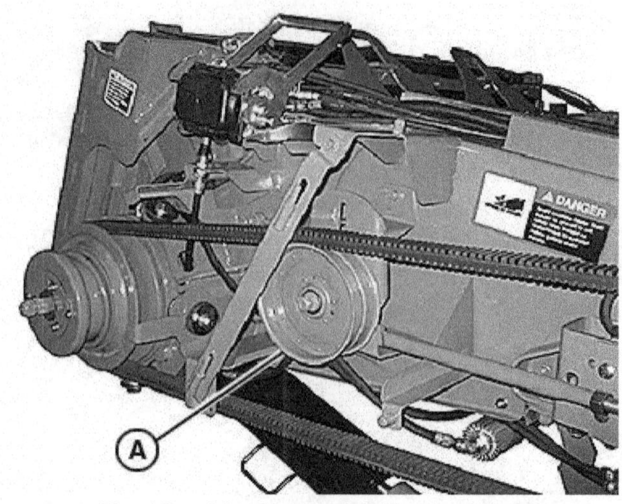

*Fig. 38 — Idler Sheave Belt Tension Adjuster*

BB87125,00001C5 -19-26JUL13-4/6

For belt drives equipped with spring and bolt adjusters, increase or decrease the spring tension by adjusting the bolt (Fig. 39). Proper adjustment tension varies from drive to drive. Always check the operator's manual to determine the correct tension.

**A— Adjust Tension Here**

*Fig. 39 — Spring-Type Belt Tension Adjuster*

**Continued on next page**

BB87125,00001C5 -19-26JUL13-5/6

When it becomes necessary to replace a belt, first loosen the idler or tension adjustment bolts and then remove the belt. Never pry a belt on or off, as this can stretch and damage the belts and bend the sheaves (Fig. 40).

## *CHAINS*

Chain drives normally have chain tighteners. Adjust chain tension so that chains are just tight enough to run without climbing or jumping sprockets and tight enough so that the chain doesn't whip or slap when running.

Be certain that all drive sprockets are properly aligned. Drive or driven sprockets may be moved in or out on shafts for proper alignment. Tightener sprockets may be aligned by increasing or decreasing the number of flat washers behind the sprocket hub.

Inspect the sprockets frequently to make sure teeth are not worn enough to cause damage to the chain.

Exposed chain drives should be cleaned regularly. Remove and clean them by soaking or dipping in diesel fuel. Dry and then oil the chain thoroughly after installing.

Before storing the combine, clean the chains and lubricate them with a heavy oil or grease. When removing the

Fig. 40 — Never Pry a Belt Onto or Off a Sheave

combine from storage, clean the chains again and lubricate them with motor oil.

BB87125,00001C5 -19-26JUL13-6/6

## PUTTING THE COMBINE INTO STORAGE

Proper storage of the combine will increase the life of the machine and save money on repairs. The following is a sample list of steps for storage. See the operator's manual for machine-specific recommendations. If the combine cannot be stored inside, cover it with a tarpaulin.

### PREPARING COMBINE FOR STORAGE

1. Swing out condenser and oil cooler. Clean radiator with air, water, or a vacuum cleaner. Clean condenser and oil cooler after cleaning radiator. Clean the charge air cooler.

2. Every 60 to 90 days start engine and turn air conditioning on. Run engine at slow idle for several minutes for compressor seal lubrication. Outside temperature should be above 40°F (5°C) for proper air conditioning operation.

3. Clean outside of engine with a safe solvent.

⚠ CAUTION: Do not use gasoline for cleaning.

4. Clean inside of air cleaner and install new elements.

5. With engine warm, drain crankcase. Replace filter and fill with correct oil. Add 22 oz. (0.66 L) of corrosion inhibitor to crankcase. Run engine to circulate.

6. Drain, flush, and refill cooling system with 50/50 mixture of antifreeze and water.

7. See lubrication and maintenance schedule.

8. Add 1 qt. (0.94 L) of RE23182 coolant conditioner. Run engine to circulate.

9. Charge batteries completely. Specific gravity will equal 1.260 volts. Remove negative lead to batteries to minimize load to the batteries.

10. Drain water separator.

11. Clean combine inside and out (Fig. 41). Leave elevator doors and drain covers open.

12. Cycle concave up and down several times to prevent material buildup in the concave area.

IMPORTANT: Do not use high pressure washer spray directly on bearings. High-pressure water can get past most seals and cause damage. Dry these areas, then lube and run combine.

13. Repaint areas where needed.

NOTE: It is not necessary to wax your combine. However, if desired, use a good clear wax that

Fig. 41 — Clean the Combine Thoroughly, Inside and Out

contains no abrasives. These type of waxes usually do not contain cleaners.

14. Lubricate combine and grease adjusting bolt threads.

15. Perform all 400 hour (yearly) service.

16. Fill fuel tank to prevent condensation.

17. Shelter the combine in a dry place.

IMPORTANT: Do not store hydraulic cylinders in the extended position. Heat may cause the oil to expand and burst the cylinders.

18. Make a list of the repairs that will be needed before the next season and order the parts immediately. These parts can be replaced in your spare time so the combine will be ready for the next harvesting season.

BB87125,00001C8 -19-26JUL13-1/1

## REMOVING THE COMBINE FROM STORAGE

Taking the combine from storage to ready it for harvesting involves several steps. Besides the steps mentioned here, make a thorough inspection of the combine to make sure all parts are in good operating condition. The following is a sample list of steps for removing from storage. See the operator's manual for machine-specific recommendations.

1. Check the engine, transmission, final drives, and the hydraulic system for leaks and add fluids as needed.

2. Reconnect and/or charge batteries.

3. Charge batteries completely. Specific gravity will equal 1.260 volts. Remove negative lead to batteries to minimize load to the batteries.

4. Check oil and coolant levels. Inspect for leaks and add oil and coolant if needed.

5. Close elevator doors and drain hole.

6. Check drive belt tensions. Adjust spring-loaded idlers to the gauge.

7. Check tire inflation.

8. Review the operator's manual.

BB87125,00001C9 -19-26JUL13-1/1

## TEST YOURSELF

### *QUESTIONS*

1. (Fill in the blank.) All operators should set up a ________________ chart to see that the combine receives adequate care.

2. (Fill in the blank.) Chains should be ________________ when they are lubricated.

3. (True or False) Some bearings do not need periodic greasing.

4. (Fill in the blank.) Whenever a person works under a combine header, the ________________ stop should be used.

5. (True or False) When lubricating grease fittings, pump grease gun until grease appears at seals.

6. (Fill in the blank.) Bent or chipped ________________ will cause excessive sidewall wear on V-belts.

7. (True or False) If rasp-bars must be replaced, the mating bar, which is located 90 degrees on the opposite side of the cylinder or rotor, must be replaced to maintain proper cylinder or rotor balance.

BB87125,00001CA -19-06SEP13-1/1

## INTRODUCTION

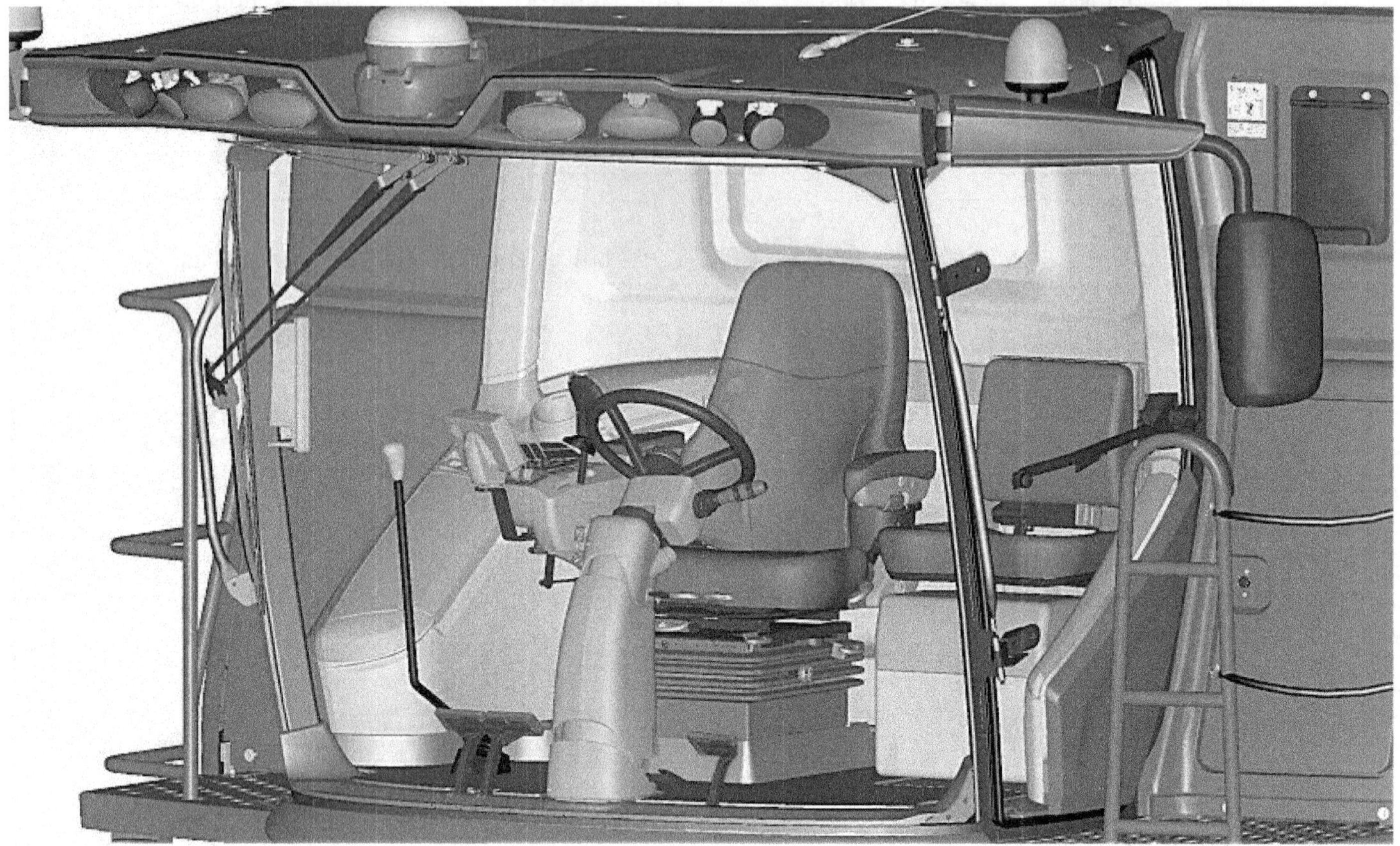

Many people are unaware that agriculture ranks as one of the most hazardous industries in the world. According to the Center for Disease Control, in 2010, there were 476 farmers and farm workers that died from agriculture-related injuries in the United States alone. Most farm-related injuries and fatalities can be avoided by using some basic attention to detail and the working environment.

OUO1082,0006250 -19-26AUG13-1/1

# SAFETY

Good safety habits are a must for anyone who operates or services one of today's combines. Every year engineers and technicians develop safety devices that make combines safer to operate, but the responsibility for safety still remains with the operator. To prepare workers for field safety, they need training on first aid, sun protection, heat/cold stress, and training geared toward the machines they operate.

Ergonomic safety and procedures should be emphasized to eliminate strains and sprains. Workers should get as much sleep as possible and avoid alcohol and stimulants that could impair judgment and reaction times. The operator must be aware of hazards and remain alert to situations that are potentially dangerous.

Operating equipment safely requires the full attention of the operator. Do not wear radio or music headphones while operating machine.

Belts and pulleys are needed for operating many components on a combine. They can pull something into them at speeds of up to 66 feet per second, which is beyond human ability to react.

When operating a combine, follow the safety precautions below:

- Never attempt to adjust, lubricate, unplug, or clean a combine when it is running. Always shut the machine off and remove the key.

- Never attempt to work underneath a raised header without first engaging the safety support lock and shutting off the engine.
- Do not allow anyone to climb onto machine while it is in motion and/or in operation.
- When working around machinery, wear clothing that fits snug to prevent entanglement.
- Stay alert. Take occasional breaks; get out of the cab and walk around every few hours.
- Always take note where power lines are during operation and unloading.
- Be aware of your surroundings; know where your coworkers are at all times.
- Use proper warning devices and placards when moving equipment on highways. Obey all of the rules of the road.
- Be sure that all lighting devices and warning devices are operational and are visible during operation and transport of combine after sunset.
- Always keep a fire extinguisher on all harvesting equipment, and check it daily.
- Develop a set of safety rules that everyone should follow and enforce them.
- Consider developing an emergency plan so everyone knows what to do if an emergency arises.

Continued on next page

DK75838,0000223 -19-26AUG13-1/18



050217
PN=213

## HAND SIGNALS

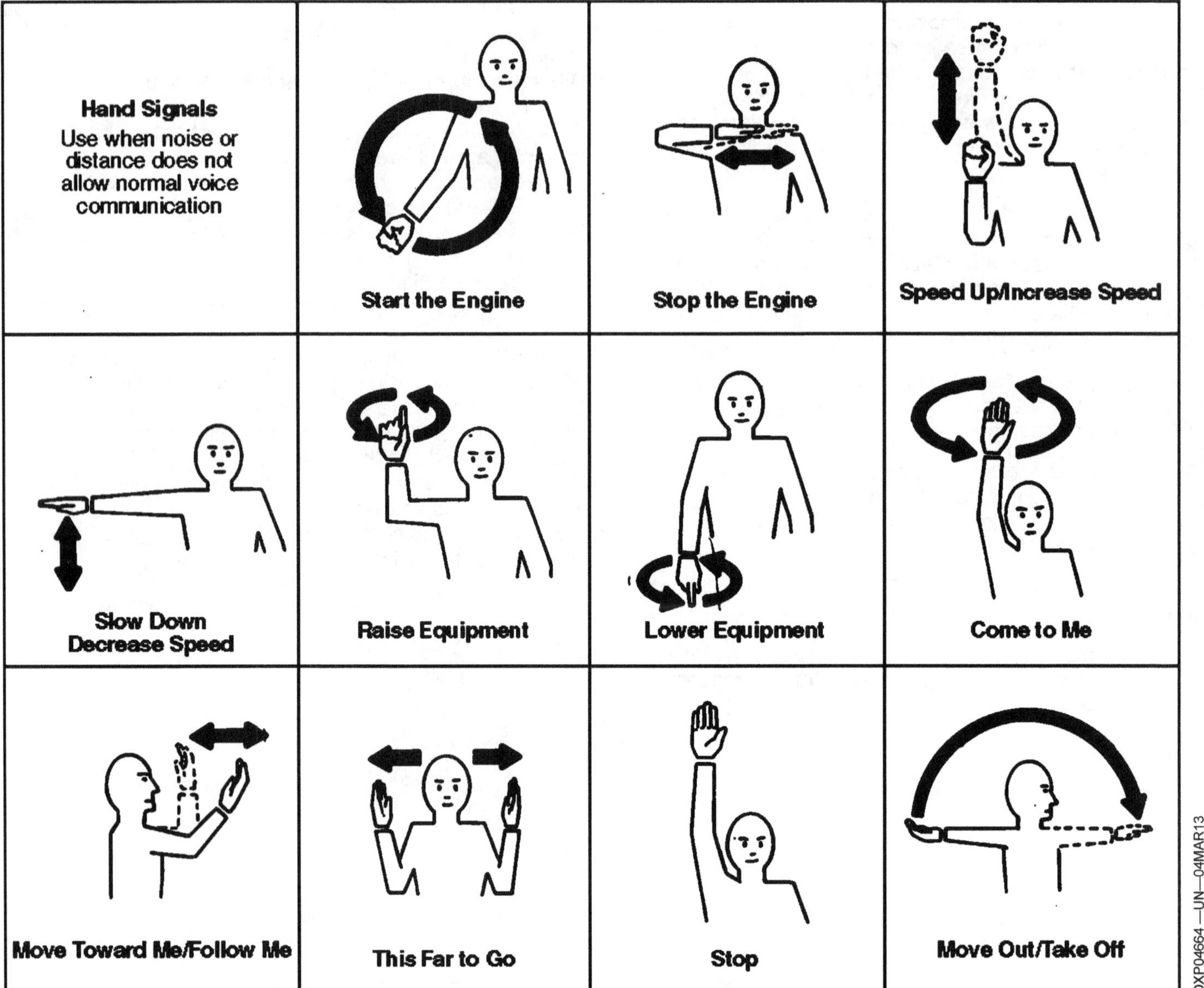



*Fig. 1 — Accepted Hand Signals*

Spoken signals are very difficult to hear over the sounds of a modern combine. Hand signals from a person on the ground can be helpful to the operator when maneuvering a combine in tight spaces. Several safety institutions have endorsed a set of universal signals, as shown in Fig. 1. Study these symbols and use them to prevent accidents.

**Continued on next page**

DK75838,0000223 -19-26AUG13-2/18

## SAFETY BEFORE STARTING

Before attempting to operate a combine, study the operator's manual. It has information on general safety rules, plus specific safety recommendations for the particular machine. The more you know about the combine, the better prepared you will be to safely operate it.

The exhaust fumes from a gasoline or diesel engine are very poisonous. If the combine is run inside, be sure to open the doors to provide good ventilation.

Always clean the combine before starting. Trash around the exhaust system can cause fires. Oil, grease, or mud on ladders or the platform can cause serious falls. Clean the cab glass to provide safe visibility.

Check the tire pressure each day (Fig. 2). Under-inflation can cause buckling of the sidewall, which can cause dangerous tire failure. Over-inflated tires have a great deal of "bounce" and cause upsets more readily than tires with correct pressure.

Check the brakes once a week. With hydraulic brakes, make sure that the master cylinder is full of fluid and that no air is present in the lines. Adjust the pedal free travel, if necessary, so that the brakes are engaged with the pedals an equal distance from the floor of the platform. Check the operator's manual for specific instructions.

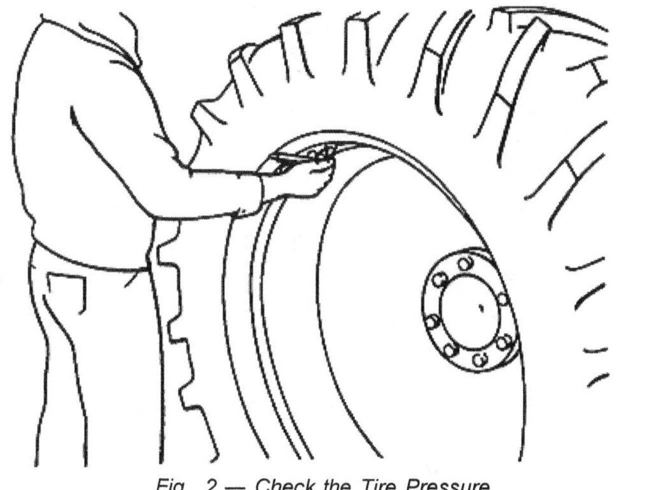

*Fig. 2 — Check the Tire Pressure*

Check the threshing cylinder rocking bar to see that it is clear of the cylinder.

Make sure that all shields and covers are in place and fastened securely.

Remove or stow all service equipment.

Use the seat belt whenever you operate the combine or ride as an observer.

DK75838,0000223 -19-26AUG13-3/18

Always use the handrails and ladders provided on the combine for safe mounting and dismounting (Fig. 3).

### *STARTING THE COMBINE*

1. Before mounting the combine, make sure that everyone is clear of the machine. Do not allow anyone to ride with you, unless the combine is equipped with a passenger seat.

2. Before starting the combine:

   - Disengage header drive
   - Disengage separator drive
   - Place control lever in neutral position
   - Place engine speed control in slow idle position
   - Sound horn to alert bystanders that machine is being started

3. Be careful when using diesel starting fluid. It is extremely flammable.

4. If it is necessary to use jumper cables to start the combine, be careful to avoid sparks around the

*Fig. 3 — Use Handrails and Ladders for Safe Mounting and Dismounting*

battery. Hydrogen gas escaping from the battery can explode. Follow the operator's manual instructions for using jumper cables.

**Continued on next page**　　　DK75838,0000223 -19-26AUG13-4/18

## TRANSPORTING THE COMBINE

Always keep your mind on the dangers of driving the combine on public roads. Besides maintaining control of the machine, you must watch for obstacles on the road, pedestrians, and traffic.

High speed is the leading cause of accidents. Never drive faster than road conditions allow for safe operation. Anticipate dangers and slow down to avoid accidents.

Make sure you are familiar with local traffic laws. Check the safety flashers and SMV emblems to be sure they are clean and visible. Always lock the brake pedals together (Fig. 4). If the combine is not equipped with a locking mechanism, be sure to depress both pedals at the same time evenly. Applying only one brake, or applying one harder than the other, can cause the combine to swerve and perhaps roll over.

Be careful when applying brakes when a header is attached to the combine. The added weight up front can cause the combine to tip forward if the brakes are applied abruptly. Always drive slow enough to allow controlled application of brakes at all times.

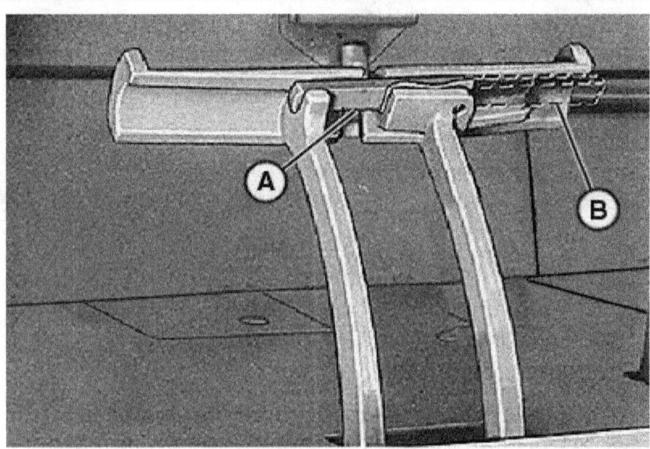

*Fig. 4 — Lock Brake Pedals Together When Transporting Combine*

**A— Brake Lock Position for Both Brakes**

**B— Brake Lock Position for Single Brake**

Always check headlights and safety flashers to make sure they are properly adjusted and in working order.

DK75838,0000223 -19-26AUG13-5/18

Put the unloading auger in the transport position (Fig. 5). Be certain it is not blocking a safety flasher or an SMV emblem.

On self-propelled combines, never use the header safety support when transporting the machine. Raise the header enough for safe ground clearance, but not high enough to reduce visibility.

**A— Unloading Auger in Transport Position**

*Fig. 5 — Unloading Auger in Transport Position*

**Continued on next page**

DK75838,0000223 -19-26AUG13-6/18

Watch for low power or telephone lines, bridges, buildings, and any other obstacles, to make sure you can pass under them safely (Fig. 6). Always keep as far to the right of the roadway as possible. Keep a careful watch to see that you have safe clearance on both sides.

Always sit down when traveling at high speeds or going over rough terrain.

Be careful when making turns. Make sure that the rear of the combine will clear obstacles when it swings around. Avoid sharp turns. Turning too sharply at high speeds can cause the machine to turn over.

Because the wheels for steering are in the back, self-propelled machines often fishtail when turned too quickly at transport speeds. Steering to the right will whip the rear to the left, and vice versa. Steering suddenly to the right when meeting oncoming traffic causes the back of the combine to swing out into the path of oncoming traffic.

Slowing or braking too rapidly could cause loss of some steering control (weight on rear wheels). This is most noticeable when driving with a corn head or some other heavy header raised high. In this case, most of the weight will be on the drive wheels. Install rear wheel weights. Keep header as low as possible. Use the variable-speed drive or engine throttle to slow the machine. Reduce speed before you need to apply brakes and always lock brake pedals together.

Never depress the clutch pedal or take the combine out of gear to coast downhill. When the combine is moving, it is impossible to shift the transmission back in gear. Always maintain complete control of the combine.

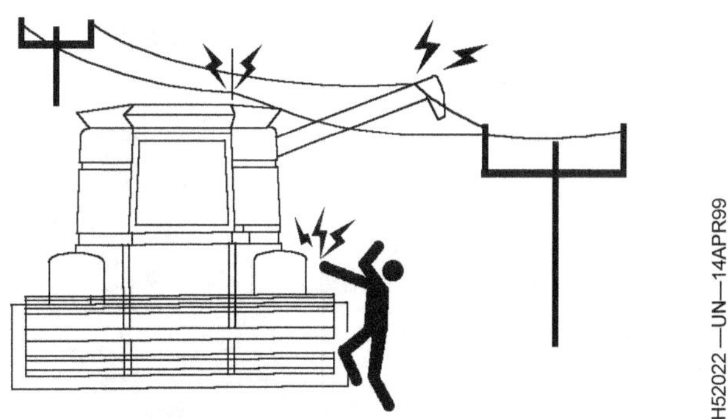

Fig. 6— Avoid Electrical Power Lines at All Times

## TRANSPORT THE COMBINE WITH HEADER SAFELY

Whenever possible, avoid transporting on public roadways with the header attached.

If the combine must be transported with the header attached, make sure that the flashing warning lights on the header are operating and the reflective material is clean and visible.

The use of a spotter or pilot vehicle is recommended on busy, narrow, or hilly roads and when crossing bridges. Drive at a speed that is safe for conditions.

Secure radio aerial in its transport position before driving on public roads; it may come into contact with low-hanging electrical cables. This would result in the operator suffering a severe electrical shock or damage to the machine.

DK75838,0000223 -19-26AUG13-7/18

---

## TOWING THE COMBINE SAFELY

If the combine must be transported over long distances, it is safer to haul it on a large truck or a special low trailer (Fig. 7).

The combine can be towed in an emergency situation, and then only at a slow speed over a short distance. Refer to the machine operator's manual for specific safety instructions and recommended towing procedures for each machine.

Fig. 7 — When Transporting over Long Distances, It Is Safer to Haul the Combine

Continued on next page

DK75838,0000223 -19-26AUG13-8/18

## OPERATING SAFETY

Operate machine only when all guards are correctly installed.

Never operate the combine if you are ill or sleepy.

Operating safety depends on alert, efficient handling of the combine.

Wear safety glasses at all times.

Wear clothing that fits snugly to avoid catching clothing in moving parts.

Never let anyone ride on the combine unless it is equipped with a passenger seat. A rider's clothing may become entangled in moving parts, or the rider may be thrown off the machine.

Before starting to harvest a field, check it carefully for ditches, fences, or other obstacles. Be aware of weather conditions that present safety hazards.

Be especially careful when operating on hillsides (Fig. 8). Avoid sharp turns that could cause the combine to tip

Fig. 8 — Be Very Careful When Operating on Hillsides

over. Beware of ditches or obstacles — they are doubly dangerous on slopes.

If grain tank extensions are used, remember that the added weight may make the combine top-heavy and more subject to upsets.

DK75838,0000223 -19-26AUG13-9/18

Use extra caution when operating during times of low light (Fig. 9) or after dark. A good operator is aware of what is around the combine at all times. During dark or extremely dusty conditions, slow down to ensure the operator is not outrunning the lighting system.

Lights and warning flashers must be turned ON when transporting combine.

Never travel over 10 mph (16 km/h) with a full grain tank. The added weight makes the combine more difficult to maneuver and easier to upset.

Always sit down when traveling over rough terrain. A sharp jolt can throw you from the platform or away from the controls.

Hillside combines are equipped with automatic or manual leveling devices. Hydraulic cylinders act to level these combines on steep slopes. These machines are equipped with a warning signal that indicates when the leveling system has reached its limit. Be especially careful after the device activates.

When using the steering brakes, always turn the steering wheel before applying the steering brakes. Failure to do so can cause the combine to swerve and turn dangerously.

Use a laser thermometer occasionally to check the temperature of shafts and bearings. High temperatures may be caused by excessive bearing wear. Check and

Fig. 9 — Use Lighting Devices When Operating During Low Light Conditions or While Transporting.

replace worn bearings before they become a fire hazard or cause damage to the machine.

Attach a metal chain to the underside of the combine to reduce static buildup and to prevent fires.

Before moving away, always check immediate vicinity of machine (e.g., for children).

Always adapt ground speed to road or field conditions.

Use the horn as a warning device immediately before moving away.

When making turns, always take into consideration the width of the attachment and the fact that the rear end of the machine swings out.

Before descending a steep hill, shift to a lower gear.

**Continued on next page**

DK75838,0000223 -19-26AUG13-10/18



PN=218

Operating, steering, and braking performance of the combine can be considerably affected by heavy front end attachments, which alter the center of gravity of the combine.

To maintain the necessary ground contact, ballast the combine at the rear end as necessary (Fig. 9). Observe the maximum permissible axle loads and total weights.

### FIELD REPAIR AND MAINTENANCE SAFETY

Always keep the machine clean. Field trash around the exhaust system can cause fires. Mud, grease, or oil on the operator platform or ladders can cause falls.

Before lubricating or adjusting the combine, disengage all drives and stop the engine. Never leave the operator platform with the engine running.

Make sure that the header drive and separator drive are disengaged before attempting to clean the combine.

Always stop the machine before opening inspection doors.

Keep all shields in place. After working on the combine, make sure the shields are fastened securely.

When operating in very dusty or noisy locations, wear goggles and ear plugs to ensure safe visibility and prevent hearing loss. Never wear loose clothing that can become entangled in moving parts.

Stay clear of moving parts at all times.

Keep belts and chains properly adjusted and aligned.

Don't rely on the hydraulic system for support when working under the machine header. Always use the stops

Fig. 10 — Use Ballast Weights for Safe Ground Contact

or supports provided on the machine. If no safety device is provided, block the header securely.

When adjusting the wheel spacing, make certain the machine is blocked. Never rely on jacks alone for support.

Always support the reel arm securely when adjustments are being made.

Be careful when removing heavy parts. Make certain they are held firmly to avoid dropping them. Have someone help you with heavy jobs.

When operating in dry fields, install a spark arresting muffler to prevent fires.

Avoid sparks or open flames when working with the battery. Hydrogen gas escaping from the battery may explode.

DK75838,0000223 -19-26AUG13-11/18

---

When possible, always refuel the combine outside the field. Let the engine cool before attempting to refuel, and never smoke around fuels (Fig. 11).

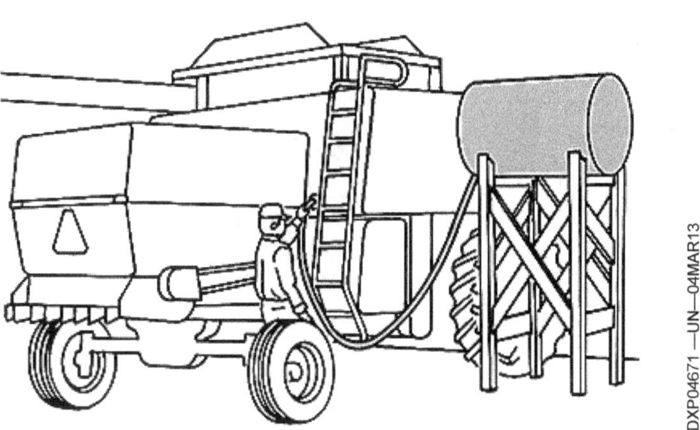

Fig. 11 — Be Careful When Refueling Combine

Continued on next page

DK75838,0000223 -19-26AUG13-12/18

Allow the system to cool and remove the radiator cap slowly, turning it until pressure escapes through the overflow pipe (Fig. 12). Make sure all pressure is relieved before removing the cap.

Stay clear of the exhaust system until it cools.

Fig. 12 — Always Remove the Radiator Cap Slowly

DK75838,0000223 -19-26AUG13-13/18

Avoid serious injury or death from entanglement in the grain tank cross augers. For functional purposes the cross augers cannot be completely covered.

Do not enter the grain tank when the engine is running (Fig. 13). Before entering the tank to clean out residual grain, always shut off the engine, set the parking brake, and remove the key.

If grain bridges and fails to flow into the cross augers, shut off the engine, remove the key and, from a position on the engine compartment door, use a rod, broom, or shovel to break the bridge and restore grain flow.

Fig. 13 — Be Aware and Keep Clear of Augers

DK75838,0000223 -19-26AUG13-14/18

Cutterbar, auger, reel, and feed rolls cannot be completely shielded due to their function (Fig. 14). Stay clear of these moving elements during operation. Never try to unclog the machine with a stick or pole while the machine is running. The stalk rolls on a corn head can pull a 12 foot (3.6 m) stick through in 1 second, shorter sticks or stalks even faster, long before you can let go of it. Always disengage the main clutch, shut off the engine, set the parking brake, and remove the key before servicing or unclogging the machine.

Fig. 14 — Stay Clear of Harvesting Units When in Operation

Continued on next page

DK75838,0000223 -19-26AUG13-15/18

Never attempt to clear obstructions in front of or on the header unless the separator is disengaged, the parking brake is set, the engine is shut off, and the key is removed. Everyone must be clear of the machine before starting the engine.

*Fig. 15 — Keep Hands Away from Knives at All Times During Operation*

DK75838,0000223 -19-26AUG13-16/18

## AVOID HIGH-PRESSURE FLUIDS

The hydraulic system and the fuel injection system contain fluids under very high pressure. Small leaks in these systems will allow oil or fuel to escape with high velocity.

Escaping fluid under pressure can penetrate the skin, causing serious injury (Fig. 16).

Avoid the hazard by relieving pressure before disconnecting hydraulic or other lines. Tighten all connections before applying pressure.

Search for leaks with a piece of cardboard. Protect hands and body from high-pressure fluids.

If an accident occurs, see a doctor immediately. Any fluid injected into the skin must be surgically removed within a few hours, or gangrene may result. Doctors unfamiliar with this type of injury should contact other knowledgeable medical sources.

Always carry a first aid kit and fire extinguisher kept in working condition on the combine at all times.

## STOPPING THE COMBINE SAFELY

Many machines are equipped with safety features that will prevent the combine from starting if the header or separator engage switches are engaged or if the control lever is not in the neutral position.

To make sure drive units do not cause injury when the machine is started again, do the following when stopping the combine.

1. Disengage header drive.

2. Disengage separator drive.

3. Place control lever in neutral.

*Fig. 16 — Never Check for Leaks with Hands*

4. Lower header.

5. Apply parking brake.

6. Remove ignition key to prevent tampering or accidental starting.

**IMPORTANT: Remember, the hydrostatic drive unit is not an effective parking brake.**

## PERSONAL PROTECTIVE EQUIPMENT

Combine operators also should keep in mind the protective clothing and accessories related to the task at hand. Such devices may include work boots, gloves, coveralls, hard hats, eye protection, hearing protection, and respiratory protection; and operators should be trained in their proper use.

Continued on next page

DK75838,0000223 -19-26AUG13-17/18

## SAFETY FEATURES OF COMBINES

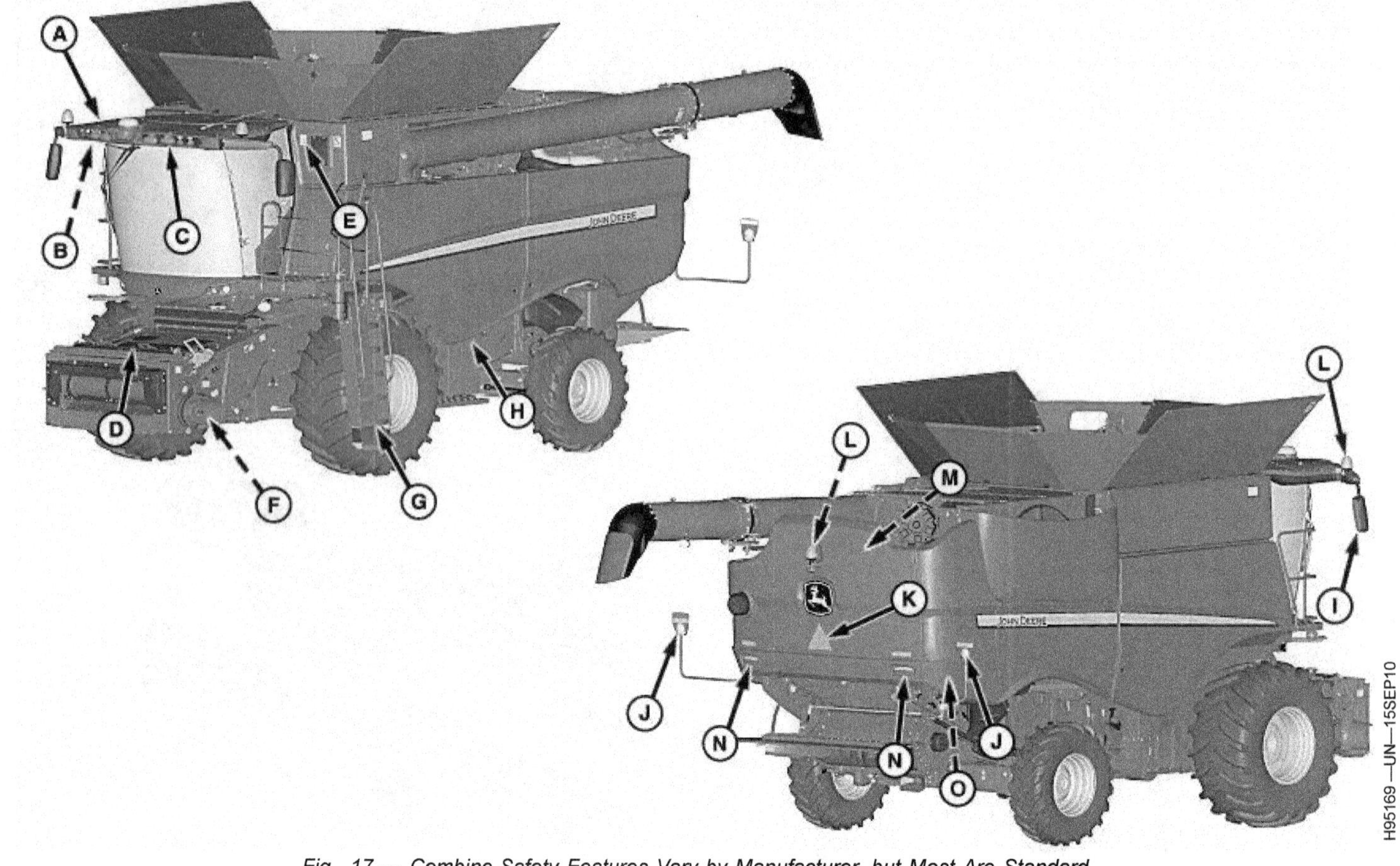

*Fig. 17 — Combine Safety Features Vary by Manufacturer, but Most Are Standard*

A—Cab Safety Features:
   Operator's Presence System,
   Electronic Engine Start
   Lockout, Seat Belts, Horn,
   Emergency Exit Window,
   Parking Brake, Turn Signals
B—Handholds
C—Headlights
D—Slip Resistant Skid Mats
E—Safety Signs
F—Mechanical Safety Stop
   (Feeder House)
G—Slip Resistant Steps and
   Platform with Handrails
H—Shields
I— Rear View Mirrors
J— Warning Lights and Reflective
   Tape
K—Slow Moving Vehicle Emblem
L—Beacon Lights
M—Backup Alarm
N—Tail Lights
O—Slip Resistant Service
   Platform with Handrails

In addition to the safety features shown in Fig. 17, other components and systems, safety lights on the machine, and safety messages and instructions in the operator's manual contribute to the safe operation of a combine when combined with the care and concern of a capable operator.

DK75838,0000223 -19-26AUG13-18/18

# FIRE PREVENTION

Fires caused by harvesting equipment cost farmers more than an estimated $50 million in machinery losses each year. This number excludes crop losses from resulting field fires, downtime related to combine damage, personal injuries, and other related costs.

In order for a fire to occur, three things must be present: air, a fuel source (chaff, fuel, oil, or anything else that will readily burn), and heat. Remove any one of these three components and the risk of fire is drastically reduced. It is impossible to remove air, and a little more difficult to control sources of heat, but the fuel source is something that can be controlled.

Even through your best maintenance and fire prevention practices, a fire may still occur at any given time and the best defense is being prepared for such occurrences. Before first start-up for the day, after checking the engine oil and other fluids, scan the engine compartment for anything unusual or check areas around the exhaust manifold, turbocharger, and anywhere else that might become a problem area. Check the wiring harness for spots where the protective jacket has been worn away by vibrating against something or by rodent problems. Any and all leaks found should be repaired before operating the combine to reduce the risk of fire.

Being able to quickly extinguish a fire is anything but an easy task on a combine, but taking the necessary steps to prevent one is an easier task. Such tasks include keeping engine compartment clean of dried chaff, trash, and dust. This general area, due to the high amount of heat generated by the engine, is where the majority of fires start. See the combine operator's manual for specific instructions for the prevention of fires.

The combine must be inspected periodically throughout the harvest day. Buildup of chaff and other debris must be removed to ensure proper machine function and to reduce the risk of fire and costly downtime. Some operators use hand-held leaf blowers to rapidly clean large areas of the combine of dust and dried chaff. Leaf blowers can be used almost anywhere, fairly effort-free, and they will run in almost any position.

Regular and thorough cleaning of machine combined with other routine maintenance procedures listed in the operator's manual will greatly reduce the risk of fire and the chance of costly downtime, and will improve machine performance.

Always follow all safety procedures posted on the machine and in the operator's manual. Before carrying out any inspection or cleaning, always shut off engine, set parking brake, and remove key.

The machine should be equipped with a general-purpose fire extinguisher and a pressurized water fire extinguisher. Extinguishers must be checked daily when entering or exiting the cab and when working around machine to ensure that they are in working condition. Dry-chemical fire extinguishers must be replaced or professionally serviced after any usage.

FIRE SUPPRESSION SYSTEMS

Several companies offer automatic or manual fire suppression systems that can be fitted into a combine. These systems are usually composed of a dry chemical pressurized tank, release valve (either manually operated, automatic, or both), supply line tubing, and discharge nozzles.

Automatic systems use different discharge methods. In some systems, the tubing is always charged, and when a fire starts, it melts the tubing and releases the extinguishing agent in the direct area of the fire. In other systems, tubing is routed into different fire prone areas of the combine, and a centrally located dry chemical tank with a release valve will expel the extinguishing agent throughout the system when it is opened.

These systems work well, according to tests and demonstrations, but are not a replacement of hand-held extinguishers and keeping fire prone areas clean of chaff and other combustible materials.

Continued on next page

DK75838,0000247 -19-31JUL13-1/8

## FIRE EXTINGUISHERS

Fire extinguishers are made in three general class types:

| Fire Extinguisher Class | Combustible Material Type | Fire Extinguisher Recommended to Use |
|---|---|---|
| A | **Ordinary combustibles such as paper, wood, and cloth** | **Use water or class ABC dry-chemical fire extinguisher** |
| B | Flammable combustibles such as gasoline, diesel fuel, and grease | Use only class ABC or BC fire extinguisher |
| C | Energized electrical fires | Use only class ABC or BC fire extinguisher. |

A 10 lb. general-purpose fire extinguisher and a pressurized water fire extinguisher (Fig. 18) with mounting brackets should be installed on the machine.

Read labels on extinguishers and become familiar with instructions on how to use and maintain them.

**IMPORTANT: Pressurized water fire extinguisher must not be exposed to freezing temperatures unless protected with antifreeze. See instructions decal on extinguisher for further information.**

Extinguishers must be checked daily when entering or exiting the cab and when working around machine to ensure that they are in working condition. Dry-chemical fire extinguishers must be replaced or professionally serviced after any usage. After a dry-chemical fire extinguisher has been discharged, a small amount of extinguishing agent remains within the valve assembly and allows the propellant to slowly leak out, rendering the fire extinguisher useless.

Fire Extinguisher Recommendations

A 10 lb. general-purpose dry-chemical fire extinguisher with a minimum of an ABC rating.

- Use this extinguisher for grease, oil, electrical, and chemical fires.

A pressurized water fire extinguisher.

- Use pressurized water extinguisher on crop material buildup or crop debris fires that do not involve oils or electrical components.

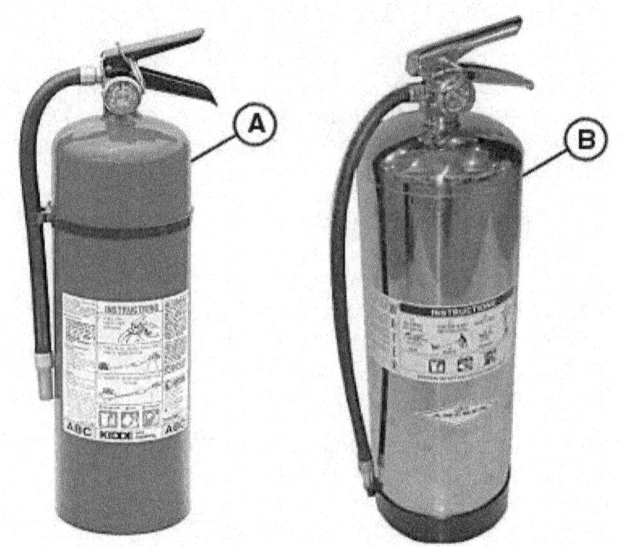


*Fig. 18 — General-Purpose and Pressurized Water Fire Extinguishers*

The following are basic steps for the basic inspection of a fire extinguisher:

INSPECTION CHECKLIST

At least once per month, inspect fire extinguishers and ensure the following:

1. Are fire extinguishers positioned in designated location on cab ladder landing and at rear of machine?

2. Are there any obstructions to proper access or visibility?

3. Are operating instructions on nameplate legible and facing outward?

4. Are safety seals broken or missing?

5. Is the extinguisher full? (Determined by pressure gauge (if present) and weighing or "hefting")

6. Is there any physical damage, corrosion, leakage, or a clogged nozzle?

When inspection of fire extinguisher reveals a deficiency, extinguisher must be serviced or replaced.

Continued on next page

DK75838,0000247 -19-31JUL13-2/8

## PRESSURIZED WATER FIRE EXTINGUISHER CHARGING

⚠ **CAUTION: Before attempting to recharge, ensure that extinguisher is completely depressurized.**

1. Discharge all remaining pressure and water (or antifreeze solution), making sure that there is no remaining air pressure.

2. Loosen nut (A) and remove valve assembly (B) from cylinder (C) as shown in Fig. 19.

**IMPORTANT: Fire extinguisher must not be exposed to freezing temperatures unless protected with antifreeze. See instructions decal on extinguisher for further information.**

3. Fill cylinder with 2.5 gal. (9.5 L) of clean water or antifreeze solution.

*NOTE: Fluid level will be approximately 6 in. (15.2 cm) below the top of the cylinder.*

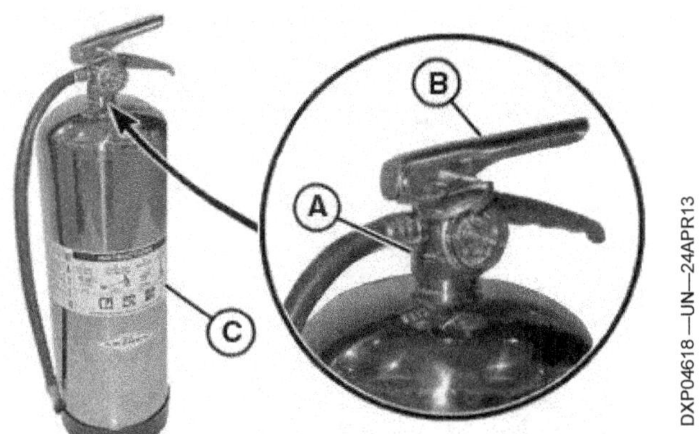

*Fig. 19 — Removing Valve Assembly from Cylinder*

A—Nut  
B—Valve Assembly  
C—Cylinder



---

4. Install valve assembly (A) and tighten nut (B) hand tight.

5. Remove cap from pressurizing valve (C) as shown in Fig. 20.

*NOTE: Set pressure regulator on air compressor to no more than 25 psi (175 kPa)(1.75 bar ) higher than gauge operating pressure.*

⚠ **CAUTION: Never leave fire extinguisher connected to a regulator of a high-pressure source for an extended period of time. Do not over-pressurize fire extinguisher. Fire extinguisher may rupture if over-pressurized.**

6. Pressurize fire extinguisher to specification on cylinder using air or nitrogen.

*NOTE: Check nut, gauge, pressurizing valve, cylinder welds, and valve orifice for leaks using leak detection fluid or a solution of soapy water.*

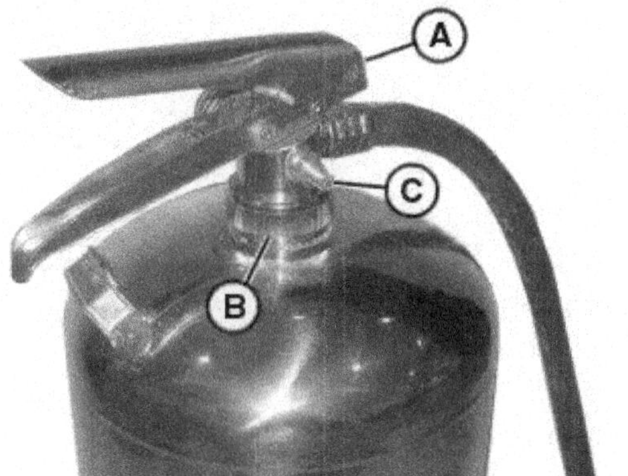

*Fig. 20 — Install Valve Assembly onto Cylinder*

A—Valve Assembly  
B—Nut  
C—Pressurizing Valve

7. Install previously removed cap on pressurizing valve.

**Continued on next page**   

8. Install pin (A) with ring facing toward front of extinguisher and install tamper seal.

9. Install hose and nozzle assembly (B) in holder (C) as shown in Fig. 21.

10. Install fire extinguisher on machine.

A—Pin  
B—Hose and Nozzle Assembly  
C—Holder

Fig. 21 — Install Safety Pin and Nozzle in Holder

DK75838,0000247 -19-31JUL13-5/8

IN CASE OF FIRE

Fig. 22 — Recommended Stance While Using Fire Extinguisher

Stop harvesting immediately at the first sign of fire. This may be the smell of smoke or the sight of smoke or flames.

Always have a cellular phone or a two-way radio handy to summon help quickly.

Do not risk personal injury. If a fire is too far advanced or in a difficult area to get to, do not try to extinguish it.

The diagram (Fig. 22) shows the recommended method to extinguish a fire. Always aim toward base of fire and maintain a safe distance from the fire.

If fire can be safely extinguished, proceed carefully and follow these guidelines:

1. Remove fire extinguisher from bracket and carry to area of fire.

2. Approach area of fire with wind to your back.

3. Pull safety pin from top of extinguisher.

4. Hold extinguisher upright by handles and aim hose at base of flames.

5. Squeeze handles to discharge fire extinguisher.

6. Move nozzle back and forth, covering flames with a cloud of powder or water.

Another easy way to remember how to use a fire extinguisher is to use the PASS system:

**P** – Pull the pin.

**A** – Aim the nozzle at the base of the fire.

**S** – Squeeze the handle.

**S** – Sweep side to side.

**Continued on next page**

DK75838,0000247 -19-31JUL13-6/8



PN=226

# TEST YOURSELF

## *QUESTIONS*

1. (Fill in the blank.)  Consider developing an
   ________________ plan so everyone knows what
   to do if an emergency arises

2. (Fill in the blank.) Wear clothing that fits _________ to
   avoid catching clothing in moving parts.

3. (True or False) Always adapt ground speed to road or
   field conditions.

4. (Fill in the blank.) Whenever a person works under a
   combine header, the ________________ stop should
   be used.

5. (True or False) Fire extinguishers must be replaced or
   professionally serviced after any usage.

6. (Fill in the blanks.) ________________ and
   ________________ fire extinguishers are recommended
   for use on a combine.

7. (True or False) A safe machine is more important than
   a safe operator.

8. Match the hand signal commands with the proper
   description below.

   ____1. Stop the engine

   ____2. Start the engine

   ____3. This far to go

   ____4. Lower equipment

   a. Hands held apart in front of face

   b. Arm and finger pointing down
   with circular motion

   c. Hand motion back and forth
   across throat.

   d. Circular cranking motion of
   hand.

DK75838,0000224 -19-30JUL13-1/1

# APPENDIX

## MEASUREMENT CONVERSION CHART

| METRIC TO ENGLISH | ENGLISH TO METRIC |
|---|---|
| **Length** | **Length** |
| 1 millimeter = 0.03937 inches (in.) | 1 inch = 25.4 millimeters (mm) |
| 1 meter = 3.281 feet (ft.) | 1 foot = 0.3048 meters (m) |
| 1 kilometer = 0.621 miles (mi) | 1 mile = 1.608 kilometers (km) |
| **Area** | **Area** |
| 1 meter$^2$ = 10.76 square feet (sq. ft.) | 1 square foot = 0.0929 meter$^2$ (m$^2$) |
| 1 hectare = 2.471 acres (acre)<br>(1 hectare = 10 000 m$^2$) | 1 acre = 0.4047 hectare (ha)<br>(1 hectare = 10 000 m$^2$) |
| **Mass (Weight)** | **Mass (Weight)** |
| 1 kilogram = 2.205 pounds (lb) | 1 pound = 0.4535 kilograms (kg) |
| 1 tonne (1000 kg) = 1.102 ton (tn) | 1 ton (2000 lb) = 0.9071 tonnes (t) |
| **Volume** | **Volume** |
| 1 meter$^3$ = 35.31 cubic feet (cu. ft.) | 1 cubic foot = 0.02832 meter$^3$ (m$^3$) |
| 1 meter$^3$ = 1.308 cubic yards (cu. yd.) | 1 cubic yard = 0.7646 meter$^3$ (m$^3$) |
| 1 meter$^3$ = 28.38 bushel (bu.) | 1 bushel = 0.03524 meter$^3$ (m$^3$) |
| 1 liter = 0.02838 bushel (bu.) | 1 bushel = 35.24 liter (L) |
| 1 liter = 1.057 quart (qt.) | 1 quart = 0.9464 liter (L) |
|  | 1 gallon = 3.785 liters (L) |
| **Pressure** | **Pressure** |
| 1 kilopascal = 0.145 pounds per square inch (psi)<br>(1 bar = 101.325 kilopascals) | 1 psi = 6.895 kilopascals (kPa)<br>1 psi = 0.06895 bars (bar) |
| **Stress** | **Stress** |
| 1 megapascal = 145 pounds per square inch (psi)<br>(1 megapascal = 1 newton/millimeter$^2$) | 1 psi = 0.006895 megapascal (MPa)<br>1 psi = 0.006895 newton/millimeter$^2$ (N/mm$^2$) |
| **Power** | **Power** |
| 1 kilowatt = 1.341 horsepower (hp)<br>(1 watt = 1 N•m/s) | 1 horsepower (550 ft-lb/sec) = 0.7457 kilowatt (kW)<br>(1 watt = 1 N•m/sec) |
| **Energy (Work)** | **Energy (Work)** |
| 1 joule + 0.0009478 British thermal units (Btu) | 1 British thermal unit = 1055 joules (J) |
| **Force** | **Force** |
| 1 newton = 0.2248 pounds force (lb.-force) | 1 pound = 4.448 newtons (N) |
| **Torque or Bending Moment** | **Torque or Bending Moment** |
| 1 newton-meter = 0.7376 pound-foot (lb.-ft.) | 1 pound-foot = 1.356 newton-meters (N•m) |
| **Temperature** | **Temperature** |
| °F = °C x 1.8 + 32 | °C = (°F − 32) / 1.8 |

OUO1082,000003F -19-27AUG13-1/1

## METRIC BOLT AND SCREW TORQUE VALUES

TS1670 —UN—01MAY03

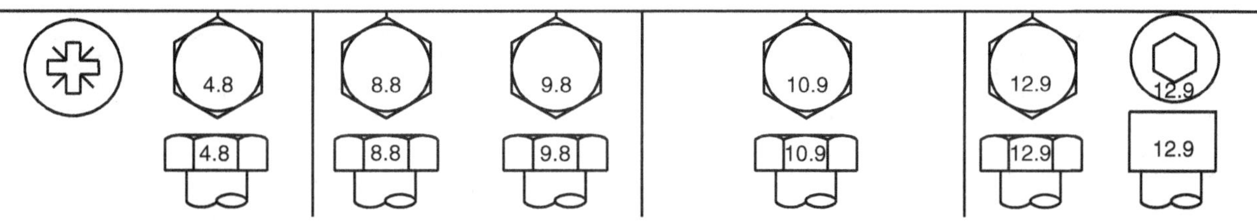


| Bolt or Screw Size | Class 4.8 | | | | Class 8.8 or 9.8 | | | | Class 10.9 | | | | Class 12.9 | | | |
|---|---|---|---|---|---|---|---|---|---|---|---|---|---|---|---|---|
| | Lubricated[a] | | Dry[b] | | Lubricated[a] | | Dry[b] | | Lubricated[a] | | Dry[b] | | Lubricated[a] | | Dry[b] | |
| | N·m | lb.-in. | N·m | lb.-in. | N·m | lb.-in. | N·m | lb.-in. | N·m | lb.-in. | N·m | lb.-in. | N·m | lb.-in. | N·m | lb.-in. |
| M6 | 4.7 | 42 | 6 | 53 | 8.9 | 79 | 11.3 | 100 | 13 | 115 | 16.5 | 146 | 15.5 | 137 | 19.5 | 172 |
| | | | | | | | | | N·m | lb.-ft. | N·m | lb.-ft. | N·m | lb.-ft. | N·m | lb.-ft. |
| M8 | 11.5 | 102 | 14.5 | 128 | 22 | 194 | 27.5 | 243 | 32 | 23.5 | 40 | 29.5 | 37 | 27.5 | 47 | 35 |
| | | | N·m | lb.-ft. | N·m | lb.-ft. | N·m | lb.-ft. | | | | | | | | |
| M10 | 23 | 204 | 29 | 21 | 43 | 32 | 55 | 40 | 63 | 46 | 80 | 59 | 75 | 55 | 95 | 70 |
| | N·m | lb.-ft. | | | | | | | | | | | | | | |
| M12 | 40 | 29.5 | 50 | 37 | 75 | 55 | 95 | 70 | 110 | 80 | 140 | 105 | 130 | 95 | 165 | 120 |
| M14 | 63 | 46 | 80 | 59 | 120 | 88 | 150 | 110 | 175 | 130 | 220 | 165 | 205 | 150 | 260 | 190 |
| M16 | 100 | 74 | 125 | 92 | 190 | 140 | 240 | 175 | 275 | 200 | 350 | 255 | 320 | 235 | 400 | 300 |
| M18 | 135 | 100 | 170 | 125 | 265 | 195 | 330 | 245 | 375 | 275 | 475 | 350 | 440 | 325 | 560 | 410 |
| M20 | 190 | 140 | 245 | 180 | 375 | 275 | 475 | 350 | 530 | 390 | 675 | 500 | 625 | 460 | 790 | 580 |
| M22 | 265 | 195 | 330 | 245 | 510 | 375 | 650 | 480 | 725 | 535 | 920 | 680 | 850 | 625 | 1080 | 800 |
| M24 | 330 | 245 | 425 | 315 | 650 | 480 | 820 | 600 | 920 | 680 | 1150 | 850 | 1080 | 800 | 1350 | 1000 |
| M27 | 490 | 360 | 625 | 460 | 950 | 700 | 1200 | 885 | 1350 | 1000 | 1700 | 1250 | 1580 | 1160 | 2000 | 1475 |
| M30 | 660 | 490 | 850 | 625 | 1290 | 950 | 1630 | 1200 | 1850 | 1350 | 2300 | 1700 | 2140 | 1580 | 2700 | 2000 |
| M33 | 900 | 665 | 1150 | 850 | 1750 | 1300 | 2200 | 1625 | 2500 | 1850 | 3150 | 2325 | 2900 | 2150 | 3700 | 2730 |
| M36 | 1150 | 850 | 1450 | 1075 | 2250 | 1650 | 2850 | 2100 | 3200 | 2350 | 4050 | 3000 | 3750 | 2770 | 4750 | 3500 |

Torque values listed are for general use only, based on the strength of the bolt or screw. DO NOT use these values if a different torque value or tightening procedure is given for a specific application. For stainless steel fasteners or for nuts on U-bolts, see the tightening instructions for the specific application. Tighten plastic insert or crimped steel type lock nuts by turning the nut to the dry torque shown in the chart, unless different instructions are given for the specific application.

Shear bolts are designed to fail under predetermined loads. Always replace shear bolts with identical property class. Replace fasteners with the same or higher property class. If higher property class fasteners are used, tighten these to the strength of the original. Make sure fastener threads are clean and that you properly start thread engagement. When possible, lubricate plain or zinc plated fasteners other than lock nuts, wheel bolts or wheel nuts, unless different instructions are given for the specific application.

[a]*"Lubricated" means coated with a lubricant such as engine oil, fasteners with phosphate and oil coatings, or M20 and larger fasteners with JDM F13C, F13F or F13J zinc flake coating.*
[b]*"Dry" means plain or zinc plated without any lubrication, or M6 to M18 fasteners with JDM F13B, F13E or F13H zinc flake coating.*

OUO1082,0000040 -19-27AUG13-1/1

## METRIC CAP SCREW TORQUE VALUES—GRADE 7

*NOTE: When bolting aluminum parts, tighten to 80% of torque specified in table.*

| Size | N·m | (lb.-ft.) |
|---|---|---|
| M6 | 9.5—12.2 | (7—9) |
| M8 | 20.3—27.1 | (15—20) |
| M10 | 47.5—54.2 | (35—40) |
| M12 | 81.4—94.9 | (60—70) |
| M14 | 128.8—146.4 | (95—108) |
| M16 | 210.2—240 | (155—177) |

OUO1082,0000041 -19-27AUG13-1/1

## UNIFIED INCH BOLT AND SCREW TORQUE VALUES

TS1671 —UN—01MAY03

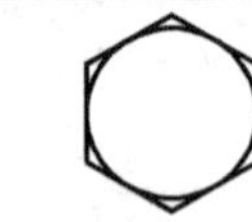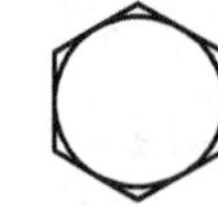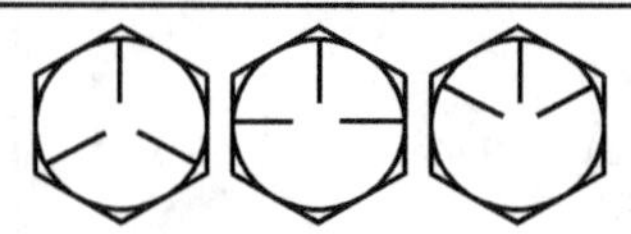

| Bolt or Screw Size | SAE Grade 1 | | | | SAE Grade 2[a] | | | | SAE Grade 5, 5.1 or 5.2 | | | | SAE Grade 8 or 8.2 | | | |
|---|---|---|---|---|---|---|---|---|---|---|---|---|---|---|---|---|
| | Lubricated[b] | | Dry[c] | | Lubricated[b] | | Dry[c] | | Lubricated[b] | | Dry[c] | | Lubricated[b] | | Dry[c] | |
| | N·m | lb.-in. | N·m | lb.-in. | N·m | lb.-in. | N·m | lb.-in. | N·m | lb.-in. | N·m | lb.-in. | N·m | lb.-in. | N·m | lb.-in. |
| 1/4 | 3.7 | 33 | 4.7 | 42 | 6 | 53 | 7.5 | 66 | 9.5 | 84 | 12 | 106 | 13.5 | 120 | 17 | 150 |
| | | | | | | | | | | | | | N·m | lb.-ft. | N·m | lb.-ft. |
| 5/16 | 7.7 | 68 | 9.8 | 86 | 12 | 106 | 15.5 | 137 | 19.5 | 172 | 25 | 221 | 28 | 20.5 | 35 | 26 |
| | | | | | | | | | N·m | lb.-ft. | N·m | lb.-ft. | | | | |
| 3/8 | 13.5 | 120 | 17.5 | 155 | 22 | 194 | 27 | 240 | 35 | 26 | 44 | 32.5 | 49 | 36 | 63 | 46 |
| | | | N·m | lb.-ft. | N·m | lb.-ft. | N·m | lb.-ft. | | | | | | | | |
| 7/16 | 22 | 194 | 28 | 20.5 | 35 | 26 | 44 | 32.5 | 56 | 41 | 70 | 52 | 80 | 59 | 100 | 74 |
| | N·m | lb.-ft. | | | | | | | | | | | | | | |
| 1/2 | 34 | 25 | 42 | 31 | 53 | 39 | 67 | 49 | 85 | 63 | 110 | 80 | 120 | 88 | 155 | 115 |
| 9/16 | 48 | 35.5 | 60 | 45 | 76 | 56 | 95 | 70 | 125 | 92 | 155 | 115 | 175 | 130 | 220 | 165 |
| 5/8 | 67 | 49 | 85 | 63 | 105 | 77 | 135 | 100 | 170 | 125 | 215 | 160 | 240 | 175 | 305 | 225 |
| 3/4 | 120 | 88 | 150 | 110 | 190 | 140 | 240 | 175 | 300 | 220 | 380 | 280 | 425 | 315 | 540 | 400 |
| 7/8 | 190 | 140 | 240 | 175 | 190 | 140 | 240 | 175 | 490 | 360 | 615 | 455 | 690 | 510 | 870 | 640 |
| 1 | 285 | 210 | 360 | 265 | 285 | 210 | 360 | 265 | 730 | 540 | 920 | 680 | 1030 | 760 | 1300 | 960 |
| 1-1/8 | 400 | 300 | 510 | 375 | 400 | 300 | 510 | 375 | 910 | 670 | 1150 | 850 | 1450 | 1075 | 1850 | 1350 |
| 1-1/4 | 570 | 420 | 725 | 535 | 570 | 420 | 725 | 535 | 1280 | 945 | 1630 | 1200 | 2050 | 1500 | 2600 | 1920 |
| 1-3/8 | 750 | 550 | 950 | 700 | 750 | 550 | 950 | 700 | 1700 | 1250 | 2140 | 1580 | 2700 | 2000 | 3400 | 2500 |
| 1-1/2 | 990 | 730 | 1250 | 930 | 990 | 730 | 1250 | 930 | 2250 | 1650 | 2850 | 2100 | 3600 | 2650 | 4550 | 3350 |

Torque values listed are for general use only, based on the strength of the bolt or screw. DO NOT use these values if a different torque value or tightening procedure is given for a specific application. For plastic insert or crimped steel type lock nuts, for stainless steel fasteners, or for nuts on U-bolts, see the tightening instructions for the specific application. Shear bolts are designed to fail under predetermined loads. Always replace shear bolts with identical grade.

Replace fasteners with the same or higher grade. If higher grade fasteners are used, tighten these to the strength of the original. Make sure fastener threads are clean and that you properly start thread engagement. When possible, lubricate plain or zinc plated fasteners other than lock nuts, wheel bolts or wheel nuts, unless different instructions are given for the specific application.

[a]*Grade 2 applies for hex cap screws (not hex bolts) up to 6 in. (152 mm) long. Grade 1 applies for hex cap screws over 6 in. (152 mm) long, and for all other types of bolts and screws of any length.*
[b]*"Lubricated" means coated with a lubricant such as engine oil, fasteners with phosphate and oil coatings, or 7/8 in. and larger fasteners with JDM F13C, F13F or F13J zinc flake coating.*
[c]*"Dry" means plain or zinc plated without any lubrication, or 1/4 to 3/4 in. fasteners with JDM F13B, F13E or F13H zinc flake coating.*

OUO1082,0000042 -19-27AUG13-1/1

## SERVICE RECOMMENDATIONS FOR FLAT FACE O-RING SEAL FITTINGS

1. Inspect the fitting sealing surfaces. They must be free of dirt or defects.

2. Inspect O-ring. It must be free of damage or defects.

3. Lubricate O-rings and install into groove using petroleum jelly to hold in place.

4. Push O-ring into the groove with plenty of petroleum jelly so O-ring is not displaced during assembly.

5. Index angle fittings and tighten by hand, by pressing joint together to ensure O-ring remains in place.

6. Tighten fitting or nut to torque value shown on the chart per dash size stamped on the fitting. Do not allow hoses to twist when tightening fittings.

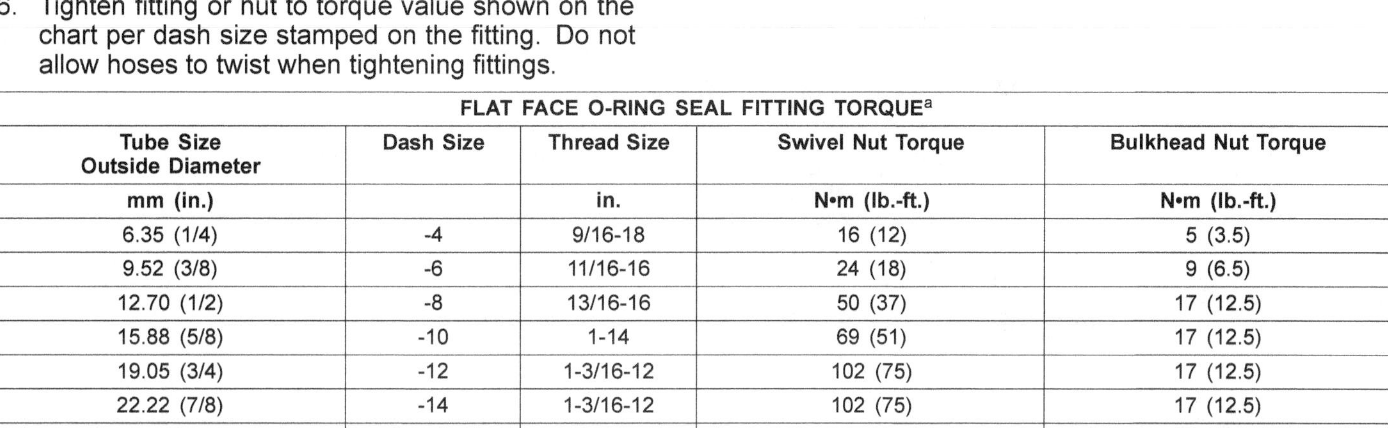

| FLAT FACE O-RING SEAL FITTING TORQUE[a] | | | | |
|---|---|---|---|---|
| Tube Size Outside Diameter | Dash Size | Thread Size | Swivel Nut Torque | Bulkhead Nut Torque |
| mm (in.) | | in. | N•m (lb.-ft.) | N•m (lb.-ft.) |
| 6.35 (1/4) | -4 | 9/16-18 | 16 (12) | 5 (3.5) |
| 9.52 (3/8) | -6 | 11/16-16 | 24 (18) | 9 (6.5) |
| 12.70 (1/2) | -8 | 13/16-16 | 50 (37) | 17 (12.5) |
| 15.88 (5/8) | -10 | 1-14 | 69 (51) | 17 (12.5) |
| 19.05 (3/4) | -12 | 1-3/16-12 | 102 (75) | 17 (12.5) |
| 22.22 (7/8) | -14 | 1-3/16-12 | 102 (75) | 17 (12.5) |
| 25.40 (1) | -16 | 1-7/16-12 | 142 (105) | 17 (12.5) |
| 31.75 (1-1/4) | -20 | 1-11/16-12 | 190 (140) | 17 (12.5) |
| 38.10 (1-1/2) | -24 | 2-12 | 217 (160) | 17 (12.5) |

[a]*The torque values shown are based on lubricated connections as in reassembly.*

OUO1082,0000035 -19-27AUG13-1/1

## SERVICE RECOMMENDATIONS FOR O-RING BOSS FITTINGS

### STRAIGHT FITTING

1. Inspect O-ring boss seat for dirt or defects.

2. Lubricate O-ring with petroleum jelly. Place electrical tape over threads to protect O-ring. Slide O-ring over tape and into O-ring groove of fitting. Remove tape.

3. Tighten fitting to torque value shown on chart.

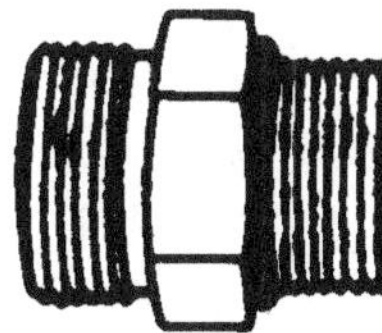

**Continued on next page**

OUO1082,0000043 -19-27AUG13-1/2

## ANGLE FITTING

1. Back off lock nut (A) and back-up washer (B) completely to head end (C) of fitting.

2. Turn fitting into threaded boss until back-up washer contacts face of boss.

3. Turn fitting head end counterclockwise to proper index (maximum of one turn).

4. Hold fitting head end with a wrench and tighten lock nut and back-up washer to proper torque value.

*NOTE: Do not allow hoses to twist when tightening fittings.*

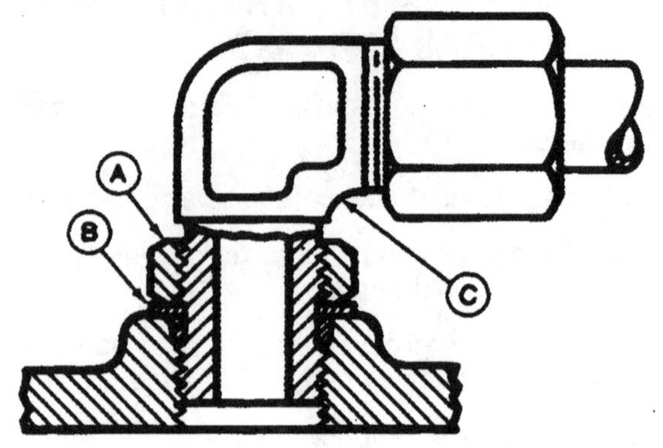

| TORQUE VALUE | | | |
|---|---|---|---|
| Thread Size | | N·m | lb.-ft. |
| 3/8-24 | UNF | 8 | 6 |
| 7/16-20 | UNF | 12 | 9 |
| 1/2-20 | UNF | 16 | 12 |
| 9/16-18 | UNF | 24 | 18 |
| 3/4-16 | UNF | 46 | 34 |
| 7/8-14 | UNF | 62 | 46 |
| 1-1/16-12 | UN | 102 | 75 |
| 1-3/16-12 | UN | 122 | 90 |
| 1-5/16-12 | UN | 142 | 105 |
| 1-5/8-12 | UN | 190 | 140 |
| 1-7/8-12 | UN | 217 | 160 |

*NOTE: Torque tolerance is ± 10%.*

OUO1082,0000043 -19-27AUG13-2/2

# ANSWERS TO CHAPTER QUESTIONS

## ANSWERS TO CHAPTER 1 QUESTIONS

1. Combine:

   a. Harvesting

   b. Threshing

2. Self-propelled combines:

   a. Hillside

   b. Level-land

3. Two classifications of combines:

   a. Conventional

   b. Rotary

4. Today's easier combine harvesting (any three):

   a. Convenient controls

   b. Monitoring devices

   c. Hydraulic power

   d. Operator cab

5. Capacity and size of combines (any four):

   a. Engine horsepower

   b. Separator width and length

   c. Type of threshing cylinder

   d. Header size

   e. Grain tank capacity

6. Attachments (any three):

   a. Straw chopper

   b. Straw spreader

   c. Shaft speed sensing units

   d. Overload signal devices

   e. Operator cab

7. Windrowed crops:

   a. Weedy conditions

   b. Excessive moisture

   c. Uneven ripening

## ANSWERS TO CHAPTER 2 QUESTIONS

1. Combine functions:

   a. Cutting and feeding

   b. Threshing

   c. Separating

   d. Cleaning

   e. Handling

   f. Residue disposal

2. Header

3. Pickup

4. Small

5. Concave

6. Adjustments:

   a. Cylinder or rotor speed

   b. Concave-cylinder spacing or concave-rotor spacing

7. False. Reducing the ground speed also reduces the speed of the platform, straw walkers, shoe, and elevators.

8. False. The straw walkers move straw to the rear of the machine as the grain falls through openings in the bottom of the straw walker.

9. Cleaning unit:

   a. Fan

   b. Chaffer

   c. Sieve

10. Tailings

11. Attachments for residue disposal:

    a. Chopper

    b. Spreader

## ANSWERS TO CHAPTER 3 QUESTIONS

1. Power systems:

   a. Power train

   b. Engine

   c. Hydraulic system

2. A - 2; B - 4; C - 1; D - 3

3. Basic engine systems (any five):

   a. Fuel system

   b. Intake and exhaust systems

   c. Lubricating system

   d. Cooling system

   e. Electrical system

   f. Governing system

4. False. The engine lubricating system has five major parts: crankcase oil reservoir, oil pump, oil filter, oil passages, and the pressure regulating valve.

5. Cooling

**Continued on next page**

OUO1082,0000044 -19-28AUG13-1/3

6. Drive systems:

   a. Propulsion

   b. Header

   c. Separator

7. Hydraulic systems (any four):

   a. Control ground speed

   b. Raise and lower the reel

   c. Control speed of reel

   d. Drive pickup

   e. Vary speed of conveyor

   f. Swing unloading auger

   g. Steady the machine

   h. Others as described in chapter 3

## ANSWERS TO CHAPTER 4 QUESTIONS

1. Indicator lights (any of following):

   a. Alternator

   b. Oil pressure

   c. Parking brake

   d. Water temperature

   e. Others as described in operator's manual

2. Throttle

3. Allow engine to idle for two to three minutes to allow turbocharger to cool.

4. Fluid levels:

   a. Oil

   b. Coolant

   c. Hydraulic

   d. Fuel

5. False. The combine should be operated at the full-open position in the field so all units of the combine are operating at the same speed, as regulated by the governor.

6. Combine transporting:

   a. Driving

   b. Hauling

   c. Towing

7. False. Most combine operator's manuals recommend that the cylinder speed be adjusted only when the cylinder is running.

8. Cleaning units (any two):

   a. Fan

   b. Sieve

   c. Chaffer

9. True

10. To keep the separator level and prevent material from building up on the downside.

## ANSWERS TO CHAPTER 5 QUESTIONS

1. Poor combine harvesting (any five):

   a. Grain losses on ground

   b. Unthreshed grain

   c. Straw chewed up

   d. Grain lost from walkers or shoe

   e. Excessive tailings

   f. Cracked grain in tank

   g. Chaff and trash in tank

   h. Marketing discounts

2. Prior to harvesting:

   a. Plan and prepare for harvesting

   b. Maintain condition of machine

   c. Determine when to combine

   d. Learn consequences of early and late harvest

3. 25

4. Preliminary settings:

   a. Cylinder speed

   b. Concave clearance

   c. Sieve spacing

   d. Chaffer spacing

5. Small grain losses:

   a. Header losses

   b. Grain tank losses

   c. Separator losses

   d. Tailings returns

6. A - O; B - O; C - U; D - O; E - U; F - U

7. Sources of grain loss:

   a. Preharvest losses

   b. Header losses

   c. Threshing losses

   d. Separating losses

Continued on next page

OUO1082,0000044 -19-28AUG13-2/3

e. Shoe losses

f. Leakage losses

8. Straw

9. False. Typical header losses for small grain crops, when the combine is adjusted and operated correctly, may vary from 1/2 to 2 percent of the average yield.

10. Adjustment

## ANSWERS TO CHAPTER 6 QUESTIONS

1. Maintenance

2. Warm

3. True. Some bearings are sealed and do not require greasing.

4. Safety

5. False. Use recommended number of strokes or pump grease until increased pressure is felt.

6. Sheaves

7. False. If rasp-bars must be replaced, the mating bar, which is located 180 degrees on the opposite side of the cylinder or rotor, must be replaced to maintain proper cylinder or rotor balance.

## ANSWERS TO CHAPTER 7 QUESTIONS

1. Emergency

2. Snug

3. True

4. Safety

5. True

6. Fire Extinguishers:

a. General-purpose

b. Pressurized water

7. True

8. Match the hand signal commands with the proper description below.

| | | |
|---|---|---|
| C | 1. Stop the engine | a. Hands held apart in front of face |
| D | 2. Start the engine | b. Arm and finger pointing down with circular motion |
| A | 3. This far to go | c. Hand motion back and forth across throat |
| B | 4. Lower equipment | d. Circular cranking motion of hand |

OUO1082,0000044 -19-28AUG13-3/3

# Index



Continued on next page

Index-3